SPECTRAL ANALYSIS OF MUSICAL SOUNDS WITH EMPHASIS ON THE PIANO

Spectral Analysis of Musical Sounds with Emphasis on the Piano

David M. Koenig

Engineering Associate (Retired), Corning, Inc.

With comments by

Delwin D. Fandrich

Piano Research, Design, and Manufacturing Consultant,
Fandrich Piano Company

OXFORD
UNIVERSITY PRESS

OXFORD
UNIVERSITY PRESS

Great Clarendon Street, Oxford, OX2 6DP,
United Kingdom

Oxford University Press is a department of the University of Oxford.
It furthers the University's objective of excellence in research, scholarship,
and education by publishing worldwide. Oxford is a registered trade mark of
Oxford University Press in the UK and in certain other countries

First Edition published in 2015

Impression: 1

Published in the United States of America by Oxford University Press
198 Madison Avenue, New York, NY 10016, United States of America

British Library Cataloguing in Publication Data
Data available

Library of Congress Control Number: 2014942361

ISBN 978–0–19–872290–8

Printed in Great Britain by
Clays Ltd, St Ives plc

Preface

Intended Audience

It is hoped that music professionals, performers, teachers, and retailers will think about the common sounds in their professional life in a different way after reading this book. On the other hand, engineers and scientists will perhaps be intrigued with the way the technical tools of their profession can be used to study musical phenomena that they usually enjoy after the work day is over. As you probably noticed, I have just specified two groups of target readers who have little in common other than a love of music. For future reference I will refer to those two groups as "the musician-reader" and "the engineer-reader." To be able to write a book that can be useful to both groups is quite a challenge, but it is my first goal.

I am a member of the engineer-reader group. My interest in musical sounds followed a renewed interest in pianos and the acquisition of a grand piano which took place after I retired from a career in engineering. During the search for my piano I read and posted on PianoWorld (<http://www.pianoworld.com>). I also took a course in piano technology at Ithaca College. From these two sources I obtained some insight into interesting problems that appeared to be begging for analysis from a spectral viewpoint. In my previous life as an engineer, I analyzed signals collected from manufacturing processes. To apply the same tools to musical sounds, especially those from the piano, is a natural progression.

Although the calculus and algebra behind the graphs in this book are important, I emphasize the *visualization* of musical sounds rather than the mathematics or the programming. If you are in the musician-reader group then you have experience (and talent) in visualizing the graph in Figure P.1, which to me is a plot of frequency vs. time. The note symbols are positioned vertically on the page in proportion to their frequency. This vertical axis, labeled "frequency," is often called the y-axis. The horizontal axis, or x-axis, represents time and is a little more subtle. The notes are positioned horizontally in the order of their occurrence but their duration depends on the type of the symbol (quarter note, eighth note, etc.). The "musician-reader" is assumed to be able to look at this graph and, perhaps with a little thought, translate it into music—in this case, part of Louis Armstrong's famous introduction to "West End Blues."

Figure P.1 *Musical stave for the beginning of "West End Blues"*

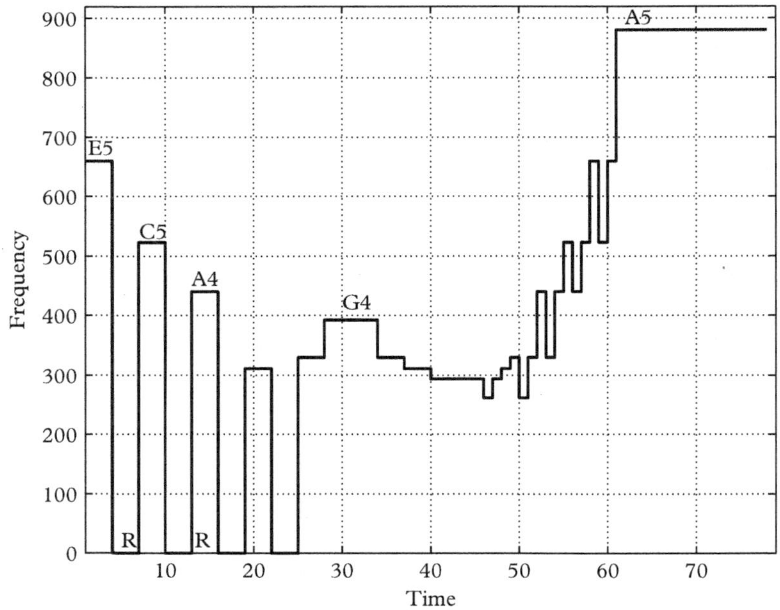

Figure P.2 *Frequency versus time graph for "West End Blues"*

Figure P.2 shows an alternative presentation of the same music. The y-axis (the vertical axis) in Figure P.2 is the fundamental frequency in Hz of the note, and the x-axis (horizontal axis) is the duration of the note. The first note is E5 at 659 Hz and it is an eighth note. It is worth three units of time. This is followed by an eighth note rest (look for the R in the lower left of the graph). The eighth note rests are represented by zero frequency and three units of time. The next note is C5 at 523 Hz and it is also an eighth note. A couple of notes later there is a G4 which is a quarter note and is worth six units of time. Hopefully, whether you are a musician-reader or an engineer-reader, you can follow the sequence of triplets (worth one unit of time), eighth notes, rests, and quarter notes through to the final dotted half note which is an A5 at 880 Hz and is worth 18 time units. If the musician-reader can see how this latter presentation relates to the former then he/she should be able to understand the graphs in this book.

Appendices are used to cover the mathematics involved. If the musician-reader does not understand algebra or calculus and is not interested (and I am assuming this to be the case), it will be easy to make a short detour around the appendices with no loss in understanding. In those cases I will offer verbiage as an alternative to the mathematics. Please note that I am not trying to patronize the musician-reader, but I am going to expend extra effort to make him/her comfortable. Notwithstanding, there are also a few portions of the book where the musician-reader will have a significant advantage over the engineer-reader.

In between the musician-reader and the engineer-reader is a vast array of readers with differing abilities and interests. One middle-ground group would contain young people who are interested in both music and physics but have not become accomplished in either field. They might find this book enjoyable as an introduction to *both* an application of mathematics and an analysis of musical sounds. If they have access to Matlab® their enjoyment can be further enhanced (most of the Matlab scripts used to generate the figures are available on the website that accompanies this book, which is mentioned at the end of this Preface). They can also download a free software package called Octave, which can run many of the Matlab scripts used in this book.

Another group in this middle ground is the musical instrument manufacturer, and especially the piano builder/designer and the piano technician. It is my hope that some of the visualization tools presented in this book could be used for problem diagnosis and/or quality control.

Meanwhile, the musician, music teacher, vocalist, piano technician, or musical instrument salesman who is curious about a different way of visualizing musical sounds could read this book with comprehension and enjoyment even if they never construct a single line of Matlab code.

This book could be used for a course taught in a college or university that has a piano technology program or a course in the physics of musical sounds. Not only could the book be a basis for lectures, but the Matlab scripts could be used in a laboratory environment where the students would collect data such as recorded sounds from a piano, and carry out analyses. At the end of some of the chapters there are exercises that require the reader to either write a simple script or modify one that has been used to generate a book figure or is available on this book's website. More exercises will be posted on the book's website.

As mentioned above, readers without access to Matlab can download Octave, which is free. There is a suite of Octave scripts available from the book's website that the reader can use, along with Audacity (which is also free) to gather sound wave data and construct some of the two- and three-dimensional graphs that appear in the book. To have a piano technician or piano technology student run this suite of scripts on his/her netbook or tablet to collect data from a piano and then to construct their own cumulative line spectrum map is my second goal.

Book Tools

Most of the tools in this book involve the frequency domain, hence the use of the word "spectral" in the title. At first blush, the following will perhaps be an off-putting list of tools, especially to the musician-reader. Please do not let unfamiliarity deter you. I will go to great pains to explain and demonstrate these tools early in the book and in the appendices. *You will not need to know any of the mathematical or programmatic details supporting these tools, graphs, videos, or sound tracks anymore than you might need to understand the chemistry of digestion in order to eat.* Treat these tools as black boxes that accept input and generate insight.

The tools of visualization are:

1. The time domain plot: plots of sound waves versus time.
2. The spatial domain plot where the amplitude of a standing wave is plotted against position. Some of the spatial domain plots will be snapshots at a particular time.
3. The line spectrum: a plot of sound strength vs. the frequencies in the signal.
4. The cumulative line spectrum: an alternative and often insightful way of presenting the information in the line spectrum.
5. The spectrogram: a 3D plot that graphs the sound strength against frequency *and* time, that is, a kind of transient line spectrum. I will also present snapshots of line spectra which are an alternative to the spectrogram.
6. The spectral centroid: a weighted average of the frequencies that appear in a line spectrum. Do not be put off by the fancy name.

7. The time–spatial domain 3D plot, where a quantity like the amplitude of a wave is plotted against time and position.
8. The 3D piano map, where the cumulative line spectrum is plotted against frequency *and* key number. There will also be plots of the amplitude vs. key number and time as well as line spectra vs. frequency and key number.
9. Sound files in mp3 format for most of the sounds studied (for "aural visualization").
10. The autocorrelation. Strictly speaking, this is not a "spectral" tool but it is related to the line spectrum and it can be useful as an indicator of pitch.
11. Although several time-sequential snapshots might give the reader a hint of the dynamics involved, a movie is significantly better. To address that problem, these snapshots will be supplemented by YouTube videos referenced from the website. Alternatively, the reader can run selected Matlab scripts to generate the movies.

Thumbnail Sketch of the Book

There are three parts to this book. The reader is introduced to a variety of waves and a variety of ways of presenting and analyzing them in the first part. The first chapter surveys several simple but common musical sound generators: a sinusoidal wave, an artificial wave made up of a fundamental and its harmonics, and finally a square wave.

The second chapter presents a tutorial on the tools used throughout the book. This chapter may be a stumbling block for some readers. If you are having trouble with Chapter 2 I suggest "strategic reading": when you get hung up on something, skip it and go to the next section or even the next chapter. You can always come back. Do not give up on the book because of Chapter 2.

The third chapter presents different visualizations of several musical instruments: a tuning fork, a bell chime, a chromatic pitch pipe (followed by my mimicking hum), a cornet, a flute, a trombone, a Didjeribone®, and several others. Chapter 4 closes Part 1 with a discussion of the harmonics in many musical sounds and some simulation experiments on a tube closed at one end.

Part 2 is a graphical and pictorial survey of the classical areas of acoustics. Chapter 5 is devoted to the vibrating string, both flexible and stiff (a component of stringed instruments such as the piano, guitar, and string bass). This chapter is probably the most important of Part 2 if you are partial to the piano, as I am. Chapter 6 presents an analysis of vibrating bars (a component of vibraphones and marimbas). Chapter 7 deals with vibrating membranes (drums). Chapter 8 looks at vibrating plates (piano soundboards). There are links to YouTube videos of vibrating string, bar, membrane, and plate models responding to different initial conditions and disturbances on the book's website.

Part 3 is devoted almost exclusively to the piano. Chapter 9 begins with some note naming conventions. Chapter 10 contains an analysis of several individual piano notes and interactions between them. The strange subject of the missing (or weak) fundamental frequency is the subject of Chapter 11. Chapter 12 delves into the rather perplexing subject of inharmonicity or octave stretching. Chapter 13 uses spectral analysis to follow some tuning exercises. Chapter 14 applies a two-dimensional metric, the spectral centroid, to the comparison of pianos. This segues to a novel three-dimensional visualization of all 88 notes of a piano in the time and frequency domains by means of a CLSM (cumulative line spectrum map) in Chapter 15. Chapter 16 takes a look at hammer knock. Chapter 17 applies several of the tools to the evaluation of a Wapin conversion. Chapter 18 studies the variation between two Steinway pianos and the variation between two data runs on a Knabe grand piano. Chapter 19 applies the CLSM and the spectral centroid to the instruments introduced in Part 1. Chapter 20 summarizes the book and discusses ways that the tools could be used for quality control in piano manufacturing.

Chapter 21 is written by Delwin D. Fandrich, Piano Researcher, Design and Manufacturing Consultant of Fandrich Piano Co., Olympia, WA, and contains his comments on the methods presented in the book.

There are appendices to cover the mathematics used in Parts 1, 2, and 3, an appendix on the experimental setup and an appendix giving a brief introduction to Matlab and Octave.

The book's website, <https://sites.google.com/site/analysisofsoundsandvibrations/>, contains (1) Matlab and Octave scripts used to produce the figures in the book and used in the exercises, (2) mp3 files of sounds, (3) references to YouTube videos, (4) more exercises, (5) results of studies carried out after the book goes to print and (6) improvements on data collection procedures and scripts. If the website should change its address after the book goes to press please use my email, spectraldmk@gmail.com, or the Oxford University Press homepage to get a link to the new website.

The object of this book is not to produce pretty graphs, although I think some of them are quite beautiful, but to pose questions and gain insight. Some of the presentations are developed from experimental data gathered using a variety of transducers, mostly the Zoom H2. None was collected under pristine laboratory conditions—remember, the goal is to gain insight, not to present numbers.

I would enjoy hearing from you.

Acknowledgments

Robert Murphy, pianoforte curator, Oberlin University, brought my Knabe grand piano to life and willingly shared many insights. Don McKechnie, head piano technician, Ithaca College, taught a wonderful course in piano technology that I audited. Don gave me access to several Ithaca College pianos. Rich Galassini, co-owner of Cunningham Pianos, allowed me to record several of his pianos. Larry Buck, E. J. Buck & Sons, piano rebuilders, Lowell, MA, hosted me on a couple of visits, gave me access to the Steingräber Phoenix grand piano, and shared his piano knowledge. Ron Hollenbeck, Distinguished Associate, Corning, Inc., provided suggestions for the structure of the book. Dan Janisch, Engineering Director, Corning, Inc., a former boss of mine, drilled the 88 holes in the board mentioned in Appendix 4. Delwin D. Fandrich supported many of my early posts on the web forum PianoWorld and supplied Chapter 21. Barbara Fandrich edited the manuscript and found many opportunities for improvement.

Contents

Part 1

Some Basic Sound Waves and Some Simple Tools

1

An Introduction to Some Elementary Waves

Several different kinds of sound waves are presented graphically in the time and frequency domains. While doing so we will qualitatively introduce the reader to the line spectrum, the cumulative line spectrum, and the autocorrelation, all of which may be new.

1.1 A Simple Sinusoidal Sound Wave

Transverse and Longitudinal Realizations

A localized pressure disturbance in air is generated when something makes a sound. This pressure disturbance is propagated outward from the source of the sound as a *longitudinal* wave. When viewed from a fixed location, say at the ear, this sound wave consists of oscillatory variations in pressure with time in the direction of propagation. The human eardrum is a stretched membrane, like the skin of a drum. When these pressure variations arrive at the eardrum, it vibrates and the brain interprets these vibrations as sound. The eardrum vibrations are *transverse* in the sense that there is a displacement in a direction perpendicular or transverse to the surface of eardrum.

Figure 1.1.1 suggests that the sound wave consists of packets of air molecules colliding with one another thereby transmitting disturbances toward the ear at the speed of sound, which at room temperature and atmospheric pressure is approximately 343 m s^{-1}. The concept of a packet of air molecules is somewhat elusive, but keep in mind that a cubic centimeter of air contains approximately 2.5×10^{19} molecules. Also, note that these molecular collisions only transmit the pressure disturbance. The medium, air, does not move on the macroscopic scale.

Therefore, the origin of a sound is manifested in a localized pressure peak, which in turn means there is a locally higher concentration of molecules than at neighboring locations. The collisions of the molecules tend to dissipate the high concentration. But, the excess number of molecules must go someplace so they move a differential amount in both directions along the path of propagation and generate other localized pressure maxima, which then dissipate and generate others, and so on. The dissipation moves molecules both ways along the propagation path. Consequently, there is a localized oscillatory displacement of the molecules back and forth as pressure maxima pass through an area along the propagation path. However, there is no net movement of molecules along this path.

The simplest mathematical model for a vibrating wave is the sine wave. Figure 1.1.2(a) shows a sound wave that vibrates at a rate of (or has a frequency of) 440 cycles per second or 440 Hz

Spectral Analysis of Musical Sounds with Emphasis on the Piano. First Edition.
David M. Koenig, with comments by Delwin D. Fandrich.
Published in 2015 by Oxford University Press.

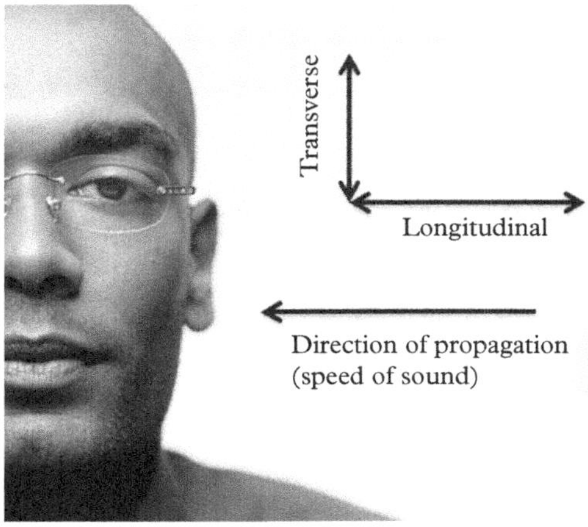

Figure 1.1.1 *Longitudinal and transverse sound waves and the ear*

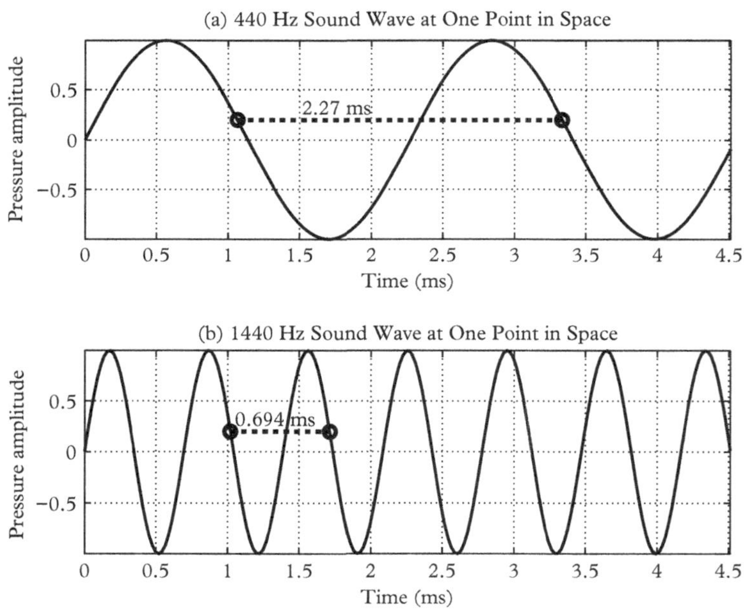

Figure 1.1.2 *Two sine waves*

or 440 s^{-1} (where "s" refers to seconds), and it has an amplitude of 1.0. The graph shows the variation in time of the pressure *at one point* along the wave. The A4 note on the piano (the 49th key up from the lowest key) has a fundamental frequency of 440 Hz—many tuning forks are designed to resonate at 440 Hz. Figure 1.1.2(b) shows a sound wave that vibrates at a rate of 1440 Hz, which is about 40 Hz sharp of the F6 note. (The note notation convention is as follows: F6 represents the F note in the sixth octave on a piano.)

The time axis is plotted in milliseconds (ms) and the 440 Hz wave repeats itself every 2.27 ms, which is equivalent to 0.00227 s, which is also equivalent to $1/(440\ s^{-1})$. The wave amplitude, although plotted perpendicularly to the time axis, represents the *longitudinal* variation of the pressure in the air directed at the eardrum. Figure 1.1.1 illustrated the difference between longitudinal motion and transverse. (In Chapter 5 the motion of the vibrating string will be transverse.)

To construct Figure 1.1.2(a), a Matlab® script was written to generate 44,100 samples of a 440 Hz sine wave over a time span of 1.0 s. The sampling frequency was 44,100 Hz and the sampling interval was 1/44100=0.0000227 s or 0.0227 ms. That is, every 0.0227 ms the sound wave is sampled and the collected value is stored. The period of the wave is 2.27 ms, which contains 100 samples. The figure plots only 200 of the samples (or two cycles) because of the high frequency of the wave. If many more samples were plotted, the graph would resemble dense hash and would not be particularly insightful. The sound of the sine wave is available in the file PureSineWave.mp3.

If one were to take a snapshot in time of the 440 Hz wave's pressure from a vantage point at the ear, the wave would look like Figure 1.1.3.

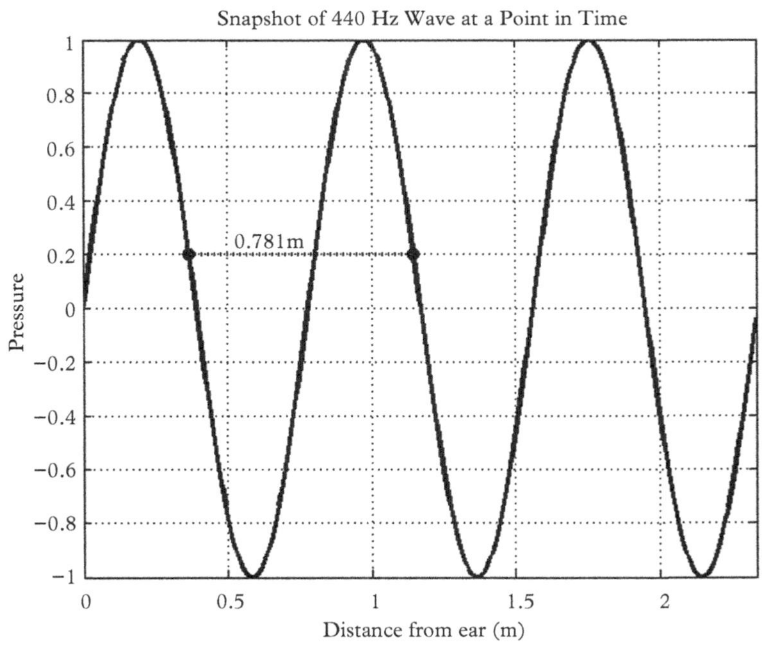

Figure 1.1.3 *A snapshot of a sine wave*

Now the amplitude is plotted perpendicularly to the distance axis but in fact the variations are along the direction of the propagation. The distance axis is plotted in meters and the wavelength of this particular 440 Hz sound wave is 0.78 m (or about 2½ feet). This means that the points of maximum pressure are 0.78 m apart. In general, the wavelength λ is given by

$$\lambda = \frac{c}{f},$$

where f is the frequency (Hz) and is the speed of sound in air (m s^{-1}). At the risk of beating the concept to death, the amplitude, although plotted perpendicularly to the distance axis, represents the longitudinal variation in air pressure in the direction of the wave's propagation.

It is important to note that the speed of propagation c is a constant, independent of frequency. Therefore, all sound waves, no matter what their frequency, propagate at the same speed. If this were not the case then higher frequency waves would arrive at the sensor or ear at different times, causing cacophony.

Figure 1.1.4 shows a picture of a longitudinal sound wave exhibiting the alternating packets of high pressure and low pressure. It also shows the wavelength λ. As noted above, a sound wave having a frequency of 440 Hz, i.e., that vibrates at 440 cycles per second (a period of 2.27 ms), has a wavelength of 2½ feet. The amplitude of the longitudinal wave is a measure of the difference between the values of the maximum pressure and the minimum pressure. It is graphed perpendicularly to the axis of propagation, but in fact the pressure variations lie along the axis of propagation.

An alternative view of the wave takes into account both time and distance and is shown in Figure 1.1.5, where there is a plane containing time and distance. Perpendicular to the plane is an axis denoting the longitudinal amplitude or the longitudinal pressure variation that takes place in

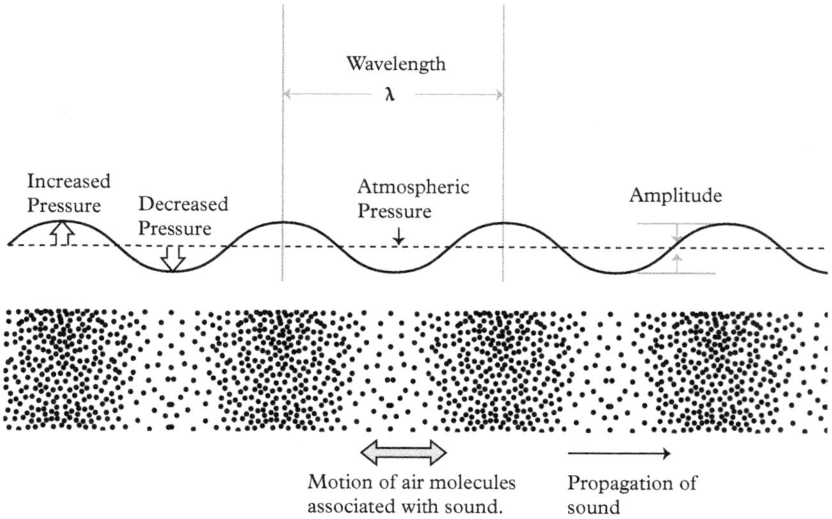

Figure 1.1.4 *A longitudinal wave with alternating compression and rarefaction*

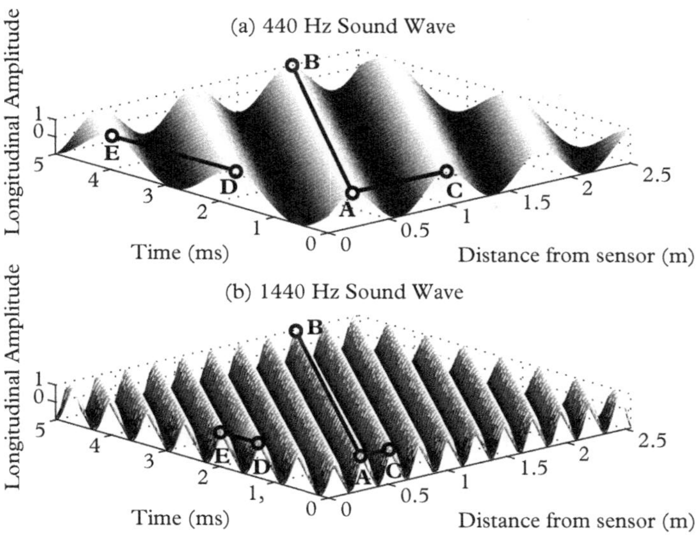

Figure 1.1.5 *Two sine waves in time and space*

the time–distance plane. Figure 1.1.5(a) shows a wave with a frequency of 440 Hz while Figure 1.1.5(b) shows a wave with a frequency of 1440 Hz.

In Figure 1.1.5(a), the point (0, 0) in the time–distance plane might represent the entrance to the ear at zero distance and at zero time. If, as time progresses, one stays at zero distance then one would move along the time axis right to left as time varies between 0.0 ms and 5.0 ms.

If, at zero time, one could hypothetically move away from the entrance to the ear then one would move along the distance axis from 0.0 m to 2.5 m. The wavelength of the 440 Hz wave has a length equal to 0.78 m, represented by the line from A to C. The period of the 440 Hz wave has a time equal to the length of the line from D to E, in this case 2.27 ms.

If you positioned yourself at 0.2 m from the ear at time zero on the peak of the wave (point A) and then started to move away from the ear at a speed of 343 m s^{-1} (the speed of sound in air), you would ride along the ridge on a diagonal path to point B in the time – distance plane. As you can see from Figure 1.1.5(a) it would take 5.0 ms to make the journey.

A similar argument applies to Figure 1.1.5(b), except that the period from point D to E is 0.694 ms. The wavelength from point A to C is now 0.238 m. However, it would still take 5.0 ms to make the journey from A to B.

Therefore, consistent with the equations $\tau = 1/f$ and $\lambda = c/f$, we see in Figure 1.1.5 that the higher frequency wave has a shorter wavelength λ and a shorter period τ but it propagates at the same speed as the lower frequency wave.

Note that the time unit is milliseconds while the distance unit is meters. Also, keep in mind that the back and forth displacement of the waves is represented by the vertical axis in the graph. In fact, the actual pressure variation lies in the time–distance plane.

There are several insightful images of transverse and longitudinal waves on <http://www.bing.com/images/search?q=longitudinal+waves&qpvt=longitudinal+waves&FORM=IGRE#a>. One of them, comparing longitudinal and transverse waves, is shown in Figure 1.1.6.

The reader is invited to look at a YouTube video showing a movie of the longitudinal pressure variation at no. 17 on the book's webpage.

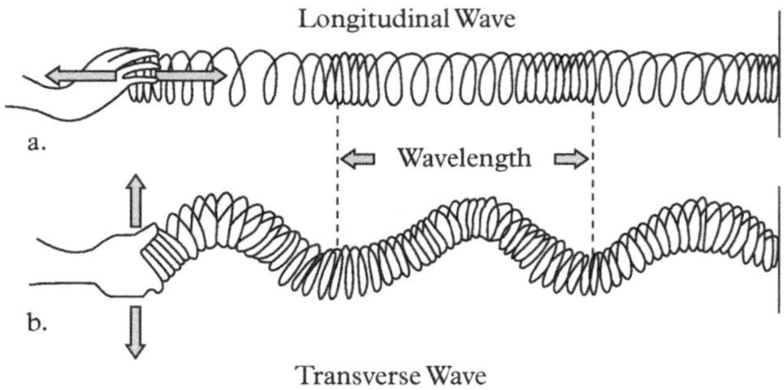

Figure 1.1.6 *Longitudinal and transverse waves*

The Line Spectrum and the Cumulative Line Spectrum of the Sine Wave

A line spectrum can be constructed from the 44,100 time domain samples of the 440 Hz wave, as shown in Figure 1.1.7, which is a plot of the sine wave's intensity versus frequency.

Since this sound wave consists of a single 440 Hz sine wave, all of its signal power is represented by a solitary line or spike at 440 Hz. The magnitude of the spike is 1.0, consistent with the amplitude of the wave in Figure 1.1.1. There are several alternative ways to plot a line spectrum and we will deal with them in Chapter 2. This is perhaps the first encounter of a line spectrum for the musician-reader. Therefore, it would be beneficial for this reader to make sure she/he is

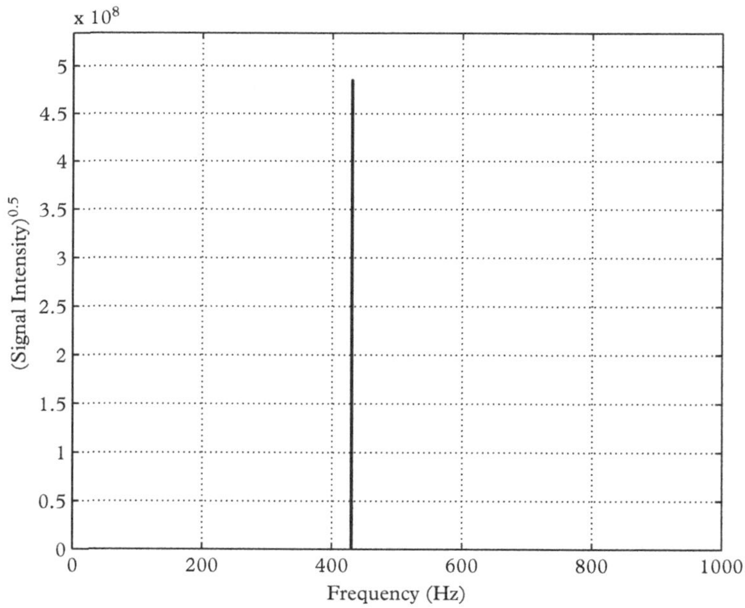

Figure 1.1.7 *Line spectrum of a 440 Hz sine wave*

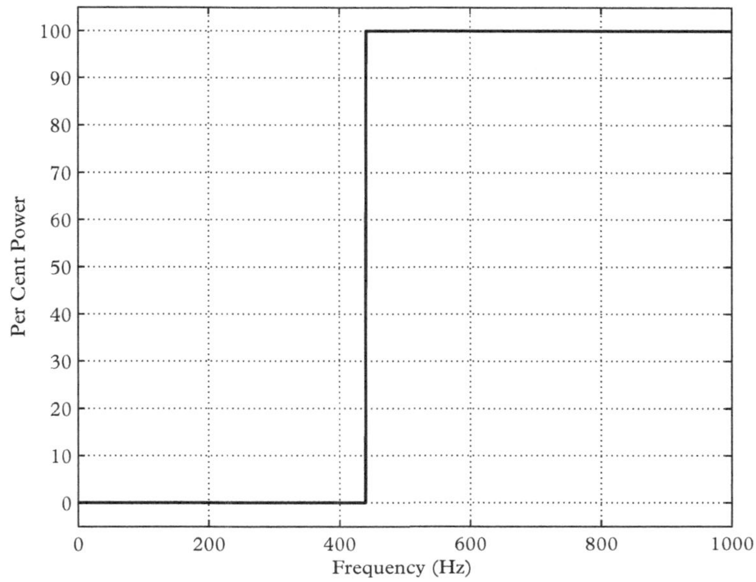

Figure 1.1.8 *Cumulative line spectrum of 440 Hz sine wave*

comfortable with how it is laid out. Note that (signal intensity)$^{0.5}$, which is plotted on the y-axis, is equivalent to $\sqrt{\text{intensity}}$; that is, the square root of the intensity.

The *cumulative* line spectrum of the 440 Hz sine wave is shown in Figure 1.1.8, where the accumulated signal power is plotted versus the frequency. Moving from left to right, the graph shows that no power is accumulated until 440 Hz, at which point the trace jumps to 100%, indicating that all of the signal power resides at that frequency. There is no more accumulation after 440 Hz, as would be expected since there is only one frequency present in the signal.

The cumulative line spectrum will be discussed in more detail in Chapter 2. Again, I will use the cumulative line spectrum repeatedly throughout the book, so the musician-reader needs to be comfortable with Figures 1.1.7 and 1.1.8. Other than in this book, the cumulative line spectrum is not widely used so it may be new to the engineer-reader also.

The Autocorrelation of the Sine Wave and Some Monthly Temperatures

Autocorrelation is a measure of a wave's tendency to repeat its shape in time. The time interval over which the repetition occurs has significance for the estimation of that wave's pitch. The autocorrelation is plotted versus the time that it takes for peaks in pressure to reach your ear. If the wave has a high pitch then the time between when the air near your ear experiences compressions will be small. For example, if the pitch is 440 Hz then every 2.27 ms the air near your ear will experience a compression followed by a rarefaction. The autocorrelation plot contains peaks at the time lags between the arriving peaks. For the 440 Hz wave there will be a peak at 2.27 ms, another one at 4.54 ms, and so on. The first and strongest peak is indicative of the lag between compression

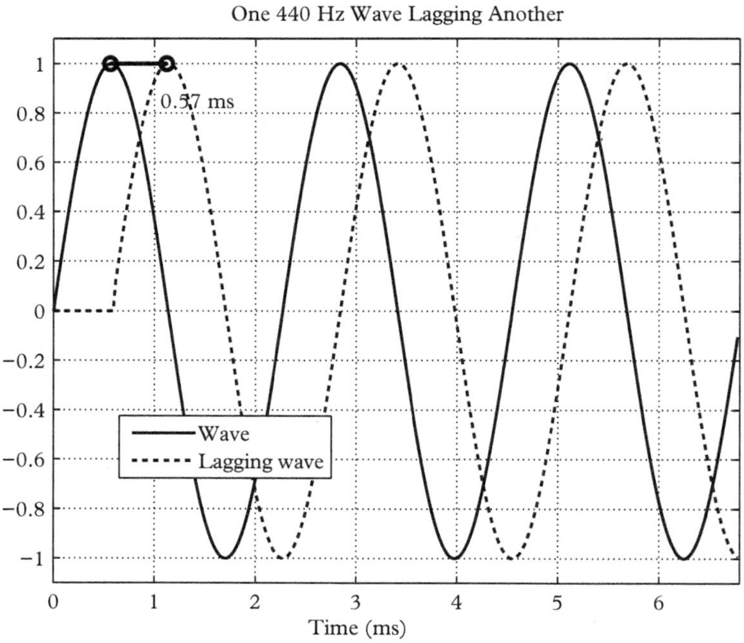

Figure 1.1.9 *Time domain plot of two sine waves, one displaced by 0.57 ms*

experiences. The reciprocal of that time lag, having units of frequency, is often interpreted as the pitch of the wave.

Given that introduction, it may be best just to wade through the steps of an autocorrelation calculation for a sine wave. Consider the two sine waves shown in Figure 1.1.9. One of the waves, the solid line, starts at time zero. The second wave, represented by the dotted line, starts a little later at time 0.57 ms. The second wave is displaced in time by 0.57 ms relative to the first wave. How do these two waves "correlate"?

To find out we sample each wave, say, every 0.0227 ms (this is equivalent to sampling at 44,100 Hz and it means that there will be about 309 samples taken over the 7.0 ms shown in Figure 1.1.9). We multiply each pair of samples and then add up the products and divide by the number of samples. Save the result and call it *A*. Now, change the displacement and repeat the calculation. Save the result. Do this for every displacement from 0.0 ms to, say, 4.5 ms. Since the sampling interval is 0.0227 ms, and since 4.5 ms/0.0227 ms = 198.45, we will have calculated about 200 values of *A*, one for each displacement from 0.0 ms all the way out to about 4.5 ms in steps of 0.0227 ms. Before doing anything else, divide each of the 200 *A* values by the value of *A* for zero displacement. So, now the *A*'s are normalized or scaled. Plot these *A*'s against the displacements and you will get a curve that looks like Figure 1.1.10 — an autocorrelation plot for a 440 Hz sine wave. The plot has peaks at the times at which pressure compressions hit the ear. The peaks are separated by 2.27 ms.

Notice that the autocorrelation in Figure 1.1.10 is unity at zero displacement. That makes sense because two sine waves with the same displacement (or phase) correlate with each other exactly. What about a displacement of 2.27 ms? In this case both curves are still in phase with each other

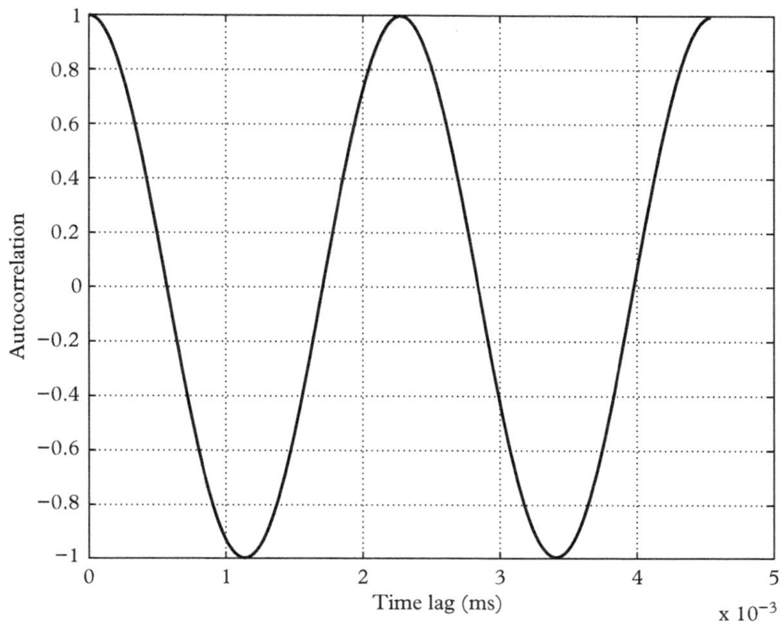

Figure 1.1.10 *Autocorrelation of a 440 Hz sine wave*

because the period of the sine wave is 2.27 ms. Therefore the autocorrelation will be also unity. This will happen for all displacements that are integer multiples of the wave's period.

At the other extreme, consider the case where the displacement is 1.1364 ms, as shown in Figure 1.1.11. Now the two waves are 180° out of phase, so every product will be negative and the value of A for this displacement will be −1.0. One would say that these two sine waves negatively correlate with each other. The reader might spend a moment and consider how two sine waves displaced by 0.57 ms would correlate (look at Figure 1.1.10 to get the answer—then reason why you got that number).

After all the dust has settled, the autocorrelation is a measure of how a wave correlates with itself at different displacements or "time lags." For our perfect sine wave the first peak occurs at 2.27 ms, which says that every 2.27 ms our ears will experience a compression followed by a rarefaction. Subsequent peaks are just as strong because the wave is a pure sine wave and there is no noise or attenuation.

Several researchers in acoustics claim that the first and tallest peak in the autocorrelation plot is an indicator of the pitch of the wave. (An interesting book on this subject is *Why You Hear What You Hear* by Eric J. Heller.) Figure 1.1.10 shows the first and tallest peak to occur at a displacement of 2.27 ms, which is the period of 440 Hz. Therefore, the pitch of this wave, according to the autocorrelation, is 440 Hz. In the next section we will deal with a more complicated wave and again use the autocorrelation to determine its pitch. Some of the details of calculating the autocorrelation will be discussed in Chapter 2 and in Appendix 1.

To perhaps add to your understanding, consider 53 years of data for an average monthly temperature. Figure 1.1.12 shows the time domain plot, the line spectrum, and the autocorrelation plot.

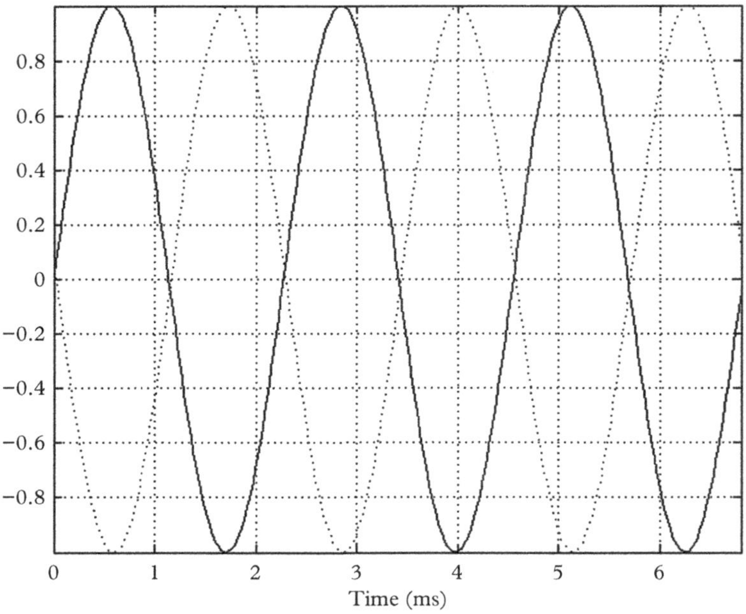

Figure 1.1.11 *Time domain plot of two sine waves, one displaced by 1.13 ms*

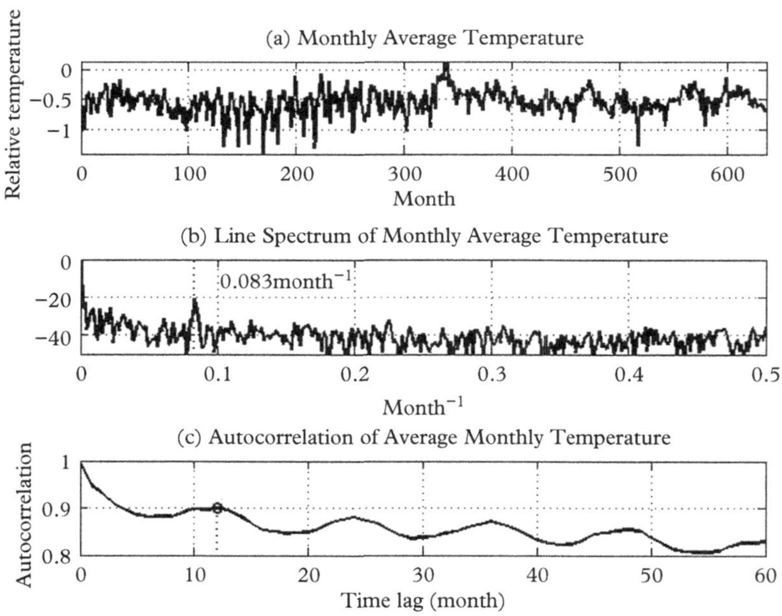

Figure 1.1.12 *Monthly average temperatures: the time domain plot, the line spectrum, and the autocorrelation*

Figure 1.1.12(b), the line spectrum, shows a peak at $1/12$ month^{-1}. Figure 1.1.12(c), the autocorrelation, shows the first and strongest peak at a lag of 12 months. This peak suggests that the temperature data correlates with itself on a 12-month interval. The value of the autocorrelation is 0.9. There is a second peak at 24 months where the autocorrelation has a value of about 0.88, which suggests that the temperatures replicate themselves to a lesser degree on a two-year interval and even less so on a three-year interval, as would be expected.

How useful the autocorrelation will be to us when we analyze sound waves coming from musical instruments remains to be seen. Keep in mind that the jury is still out on this matter.

1.2 A Built-Up Signal Consisting of the Fundamental and Harmonics

The 440 Hz sine wave of Section 1.1 will make up the *fundamental* component of a built-up wave that we will construct in this section. *Harmonics*, consisting of sine waves with frequencies that are an integer multiple of 440 Hz, are added to the fundamental. The amplitudes of the harmonics are scaled inversely proportionally to the harmonic number, that is, the higher the harmonic frequency, the less the contribution. Furthermore, the amplitudes will decay with time at a rate proportional to the harmonic number. That is, the higher harmonics decay more quickly.

The next couple of paragraphs discuss the mathematics of the built-up wave. You can skip around them with no loss in understanding if they do not make sense. There will be a test at the end of this section to see if you grasped the concepts by viewing the graphs even though you did not understand the equations.

The built-up signal is the sum of *nine* sine waves, the first being the fundamental or first harmonic,

$$A_1 \sin(2\pi f_1 t)$$

where $f_1 = 440$ Hz represents the fundamental frequency and A_1 represents the amplitude. The symbol

$$\sin(\ )$$

stands for the trigonometric sine function. The argument of the sine function inside the parentheses, $2\pi f_1 t$, contains the frequency f_1 and the time t. This function was graphed in Figure 1.1.2 for two frequencies.

The reader unfamiliar with mathematical notation might look upon sin( ) as a two-stage machine that accepts an input and generates an output, as shown in Figure 1.2.1. Note that the input, the time, varies from 0.0 to 4.5 ms and the output of the sin( ) machine ranges from −1.0 to 1.0. In our example we are multiplying the sine function by the coefficient A_1 which amplifies or attenuates the wave depending on whether A_1 is less than or greater than one.

The second term in the sum is the second harmonic, as in

$$A_2 \sin(2\pi f_2 t) \quad \text{or} \quad A_2 \sin[2\pi (2f_1)\, t],$$

where $f_2 = 2f_1 = 800 = 2 \times 440$ Hz. Figure 1.2.2 shows the first two harmonics and their sum.

Note how the second harmonic oscillates twice as fast as the first harmonic and how the sum of the two harmonics produces a wave that appears to repeat every 2.27 ms, which is the period

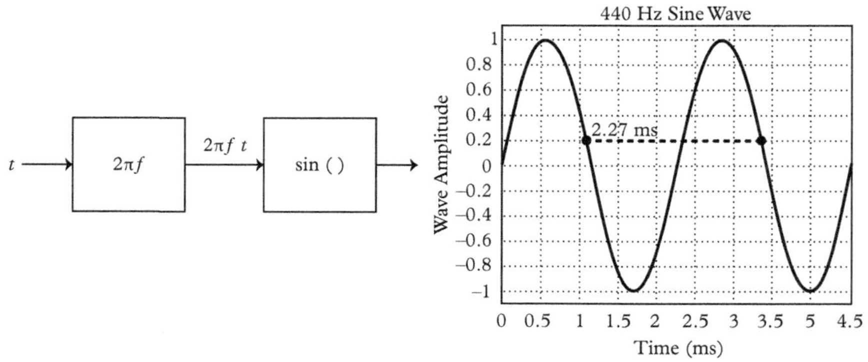

Figure 1.2.1 *The sine function as a machine that accepts an input and generates an output*

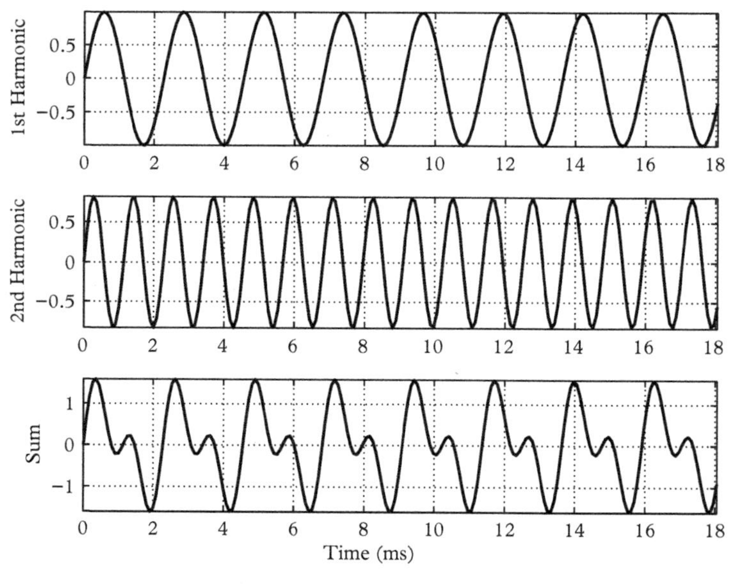

Figure 1.2.2 *The first two harmonics of a built-up wave and their sum*

of the first harmonic. This is an interesting feature that we will touch on again in Part 3 when the case of the missing fundamental is studied.

Each of the nine sine wave terms will have an amplitude represented by coefficients

$$A_k,\ k = 1, 2, 3, \ldots, 9,$$

and each term will have frequencies

$$f_1, f_2, \ldots, f_9 \quad \text{or} \quad f_1, 2f_1, 3f_1, \ldots, 9f_1.$$

The sum will look like:

$$y(t) = A_1 \sin(2\pi f_1 t) + A_2 \sin(2\pi f_2 t) + \cdots + A_9 \sin(2\pi f_9 t)$$
$$\text{built-up signal} = \text{fundamental} + \text{2nd harmonic} + \cdots + \text{9th harmonic.} \tag{1.2.1}$$

The amplitudes vary with the harmonic number and with time, as in

$$A_k(t) = \frac{1}{k^{0.25}} e^{-\frac{t}{\tau_k}} \qquad \tau_k = \frac{2}{k}. \tag{1.2.2}$$

To the musician-reader: Understanding Eq. 1.2.1 or Eq. 1.2.2 is not necessary to the comprehension of the idea of these built-up waves and how they behave. In words, Eq. 1.2.2 says that the amplitude of the kth harmonic, namely $A_k(t)$, decreases with increasing harmonic number k and with time t. The rate of decay is related to τ_k, which in turn decreases with increasing harmonic number. Equation 1.2.2 is an artificially concocted formula designed to produce a wave of a desired shape. It has no physical significance. Do not worry if Eq. 1.2.2 is not clear. Just remember that the amplitudes decrease with harmonic number and decay with time.

Figure 1.2.3 shows the time domain picture of the built-up wave. Figure 1.2.3(a) is a close-up and Figure 1.2.3(b) is a view from a larger (in time) perspective. Note the different time span on the two graphs: the first spans 18 ms while the second spans 4000 ms or 4 s. In Figure 1.2.3(a) there is a horizontal dotted line indicating a period of 2.27 ms corresponding to the fundamental frequency of 440 Hz. The second graph has the dense hash referred to above because of the larger time span.

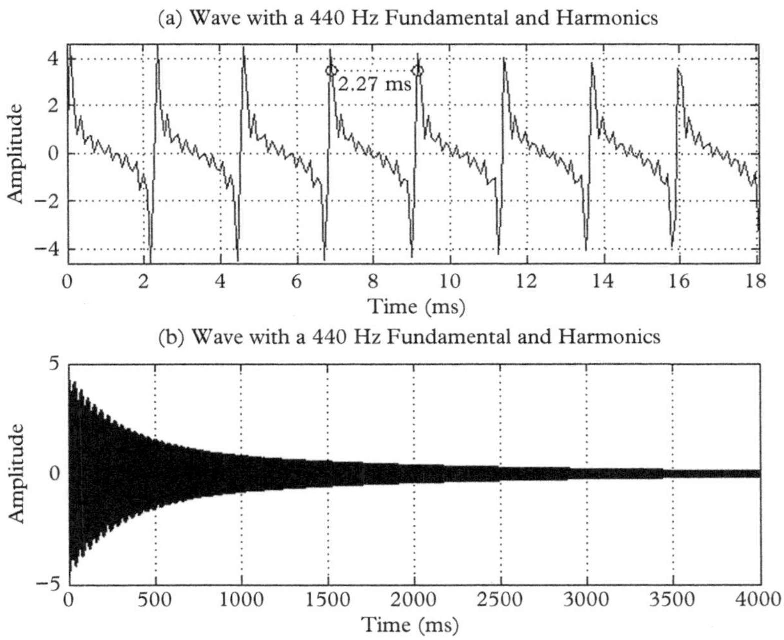

Figure 1.2.3 *Two views of the built-up wave: (a) a close up and (b) a longer-term view*

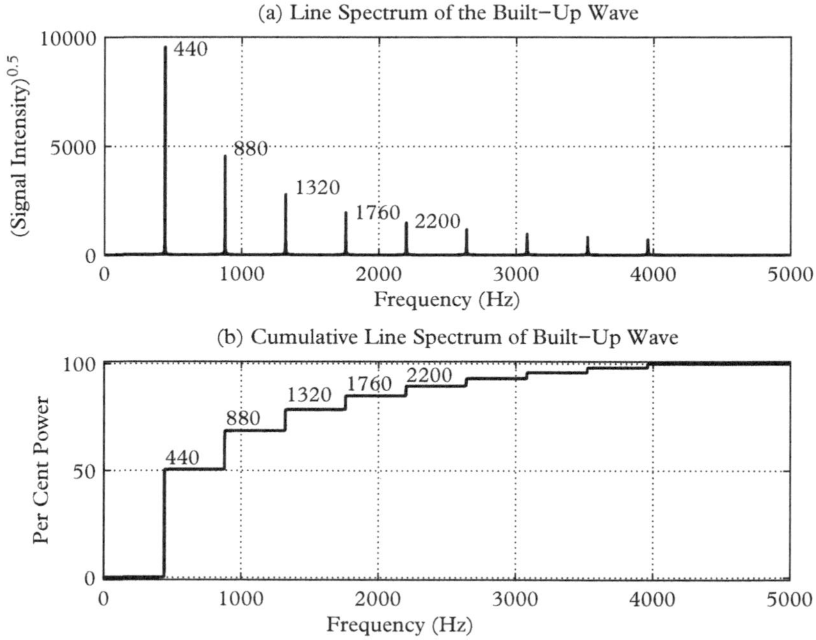

Figure 1.2.4 *Line spectrum and cumulative line spectrum of the built-up wave*

Unlike the wave in Section 1.1, this wave decays with time, as seen in Figure 1.2.3(b), and has a more complex-appearing structure. The sound of the built-up wave is available in the file BuiltUpWave.mp3. The built-up wave sounds similar to the single sine wave shown in Figure 1.1.2.

The line spectrum and the cumulative line spectrum of the built-up wave are shown in Figure 1.2.4, where the fundamental (also the first harmonic) at 440 Hz is the strongest (50% of the total power) and the other eight harmonics at 880 Hz, 1320 Hz, etc., have progressively less strength (as designed).

Figure 1.2.4 is somewhat similar to Figures 1.1.6 and 1.1.7. It is important that both the engineer-reader and the musician-reader understand how the two graphs in Figure 1.2.4 are related to each other and how they differ from Figures 1.1.6 and 1.1.7.

The spectrogram of the built-up signal is shown in Figure 1.2.5.

Figure 1.2.5 plots the signal power or strength versus frequency, as does the line spectrum, but it adds another dimension, time, and shows how the wave's behavior in the frequency domain changes with time. Each of the nine harmonics has a different initial amplitude and a different decay rate. Many real-life sounds that we will look at later in the book, especially those resulting from striking piano keys, have similar characteristics. This kind of graph is probably new to the reader so it bears some study. For example, if you remove the time dependence from Figure 1.2.5 the result will be just the "back wall" where frequency varies from 0 to 5000 Hz and the power density varies from 0.0 to about 0.03. This remaining "back wall" will resemble Figure 1.2.4(a).

An alternative approach to the spectrogram divides the time axis into n equal segments each of length T/n, where T is the total time span of the wave. Then m overlapping intervals of length

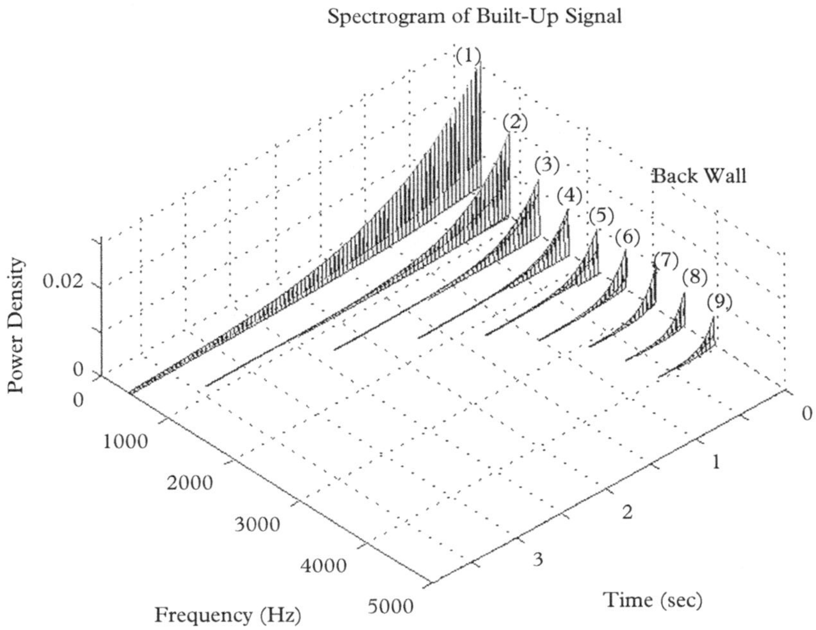

Figure 1.2.5 *Spectrogram of the built-up wave showing the nine harmonics*

T/m, where m is less than n, are constructed and the line spectrum is computed for each of the m overlapping intervals. Figure 1.2.6 shows such a sequence of line spectra for the built-up signal.

The line spectrum in Figure 1.2.4 uses the same data as the spectrogram in Figure 1.2.5 and the sequence of line spectrum snapshots in Figure 1.2.6. Therefore, if the nature of the signal changes with time (as it does here) then the line spectrum in Figure 1.2.4 *may* not present an accurate picture of the wave. For the case of the built-up wave, the decaying amplitude, strictly speaking, might cast some suspicion on the line spectrum in Figure 1.2.4. However, we will apply the line spectrum to decaying waves throughout the book with impunity. Thus, a line spectrum will use the overall wave including its transient behavior. If there are problems with this approach then the spectrogram can be used to aid in gathering insight.

To complete our analysis of the built-up wave, Figure 1.2.7 shows the autocorrelation. Even with the presence of the higher harmonics the autocorrelation has its first and strongest peak (barely) at 2.27 ms, which corresponds to 440 Hz. Therefore, according to the autocorrelation, the pitch of this wave is 440 Hz.

The built-up wave has many characteristics of the sound waves that result from striking a piano key. You should put yourself at ease and gaze on the five representations of the built-up wave: the time domain plot, the line spectrum, the cumulative line spectrum, the spectrogram, and the autocorrelation. At the risk of sounding spiritual, let those pictures talk to you and tell you about the built-up wave.

This section has shown that one can gain insight into the nature of a wave by using the time domain (the time domain plot and the autocorrelation) and the frequency domain (the line spectrum and the cumulative line spectrum), the latter domain perhaps being new to some readers. We have also accompanied the time and frequency domain pictures with an aural

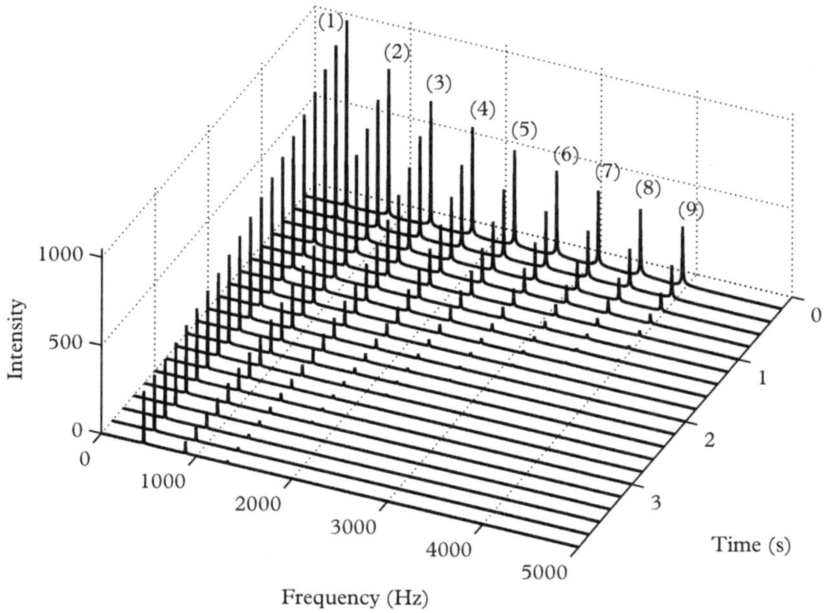

Figure 1.2.6 *Snapshots of the line spectrum of the built-up wave*

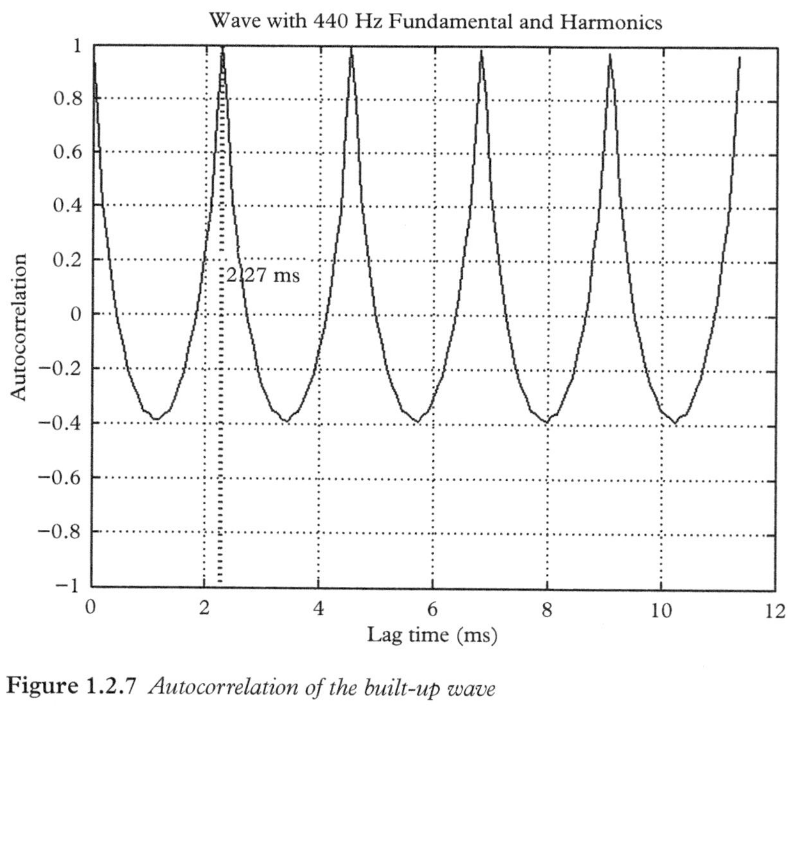

Figure 1.2.7 *Autocorrelation of the built-up wave*

"picture" using mp3 files. The reader can "see" the wave from at least two different perspectives as well as listen to it.

The "Test"

Did you grasp the concept of constructing a sound wave by adding several sine waves (or harmonics)? Does it make sense that, for this particular built-up wave, the amplitude or strength of the harmonics decreases as their harmonic number increases? Does it make sense that the amplitude of each harmonic decreases with time? Does the built-up wave sound different when compared to the simple 440 Hz sine wave? Did Eqs. 1.2.1 and 1.2.2 assist in your understanding?

Can you look at the line spectrum and cumulative line spectrum and gain insight into the built-up wave? Do these graphs make sense, given that you know how the wave was constructed? Finally, does the autocorrelation assist you in gaining insight? If you have a little trouble with the autocorrelation you probably have some company.

1.3 A Square Wave

Instead of a sine wave with a frequency of 440 Hz, we generate a square wave with the same fundamental frequency and plot it in Figure 1.3.1.

This wave repeats itself with the same period as the wave plotted in Figure 1.1.1 but its shape is different. The horizontal dotted line with circles at each end illustrates that the period of the square

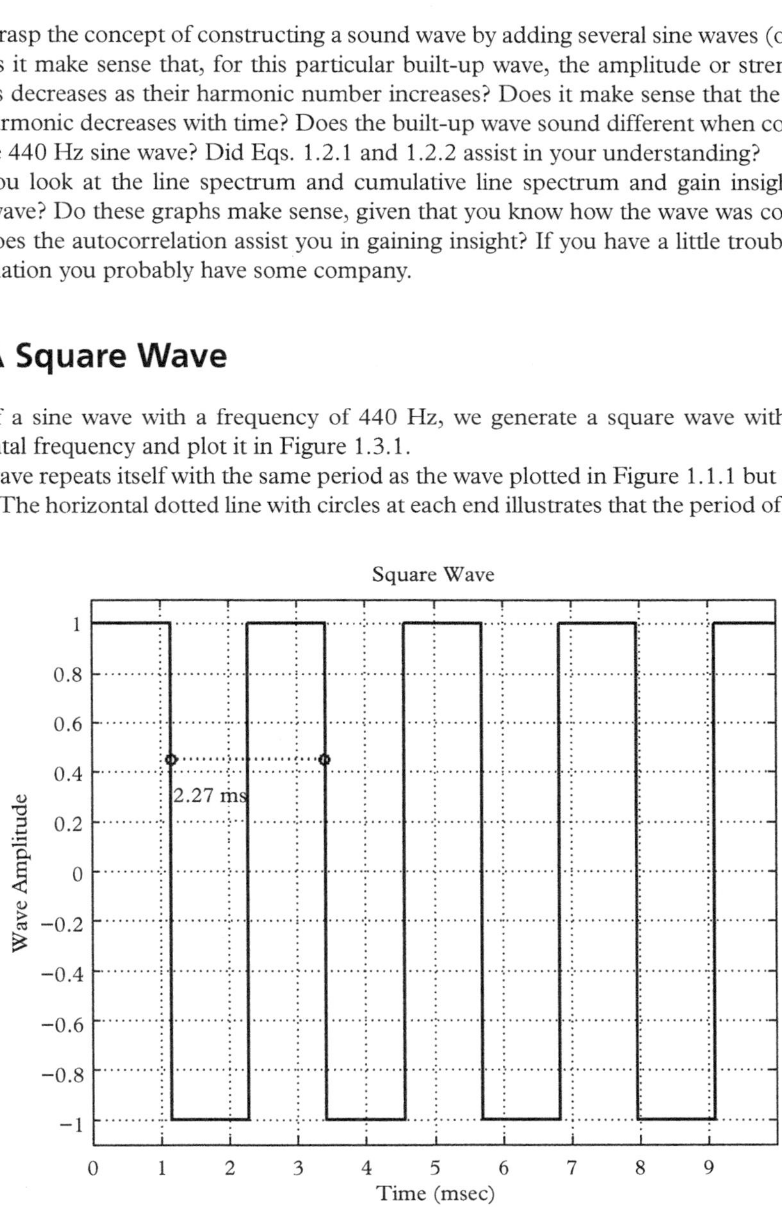

Figure 1.3.1 *Time domain plot of a square wave*

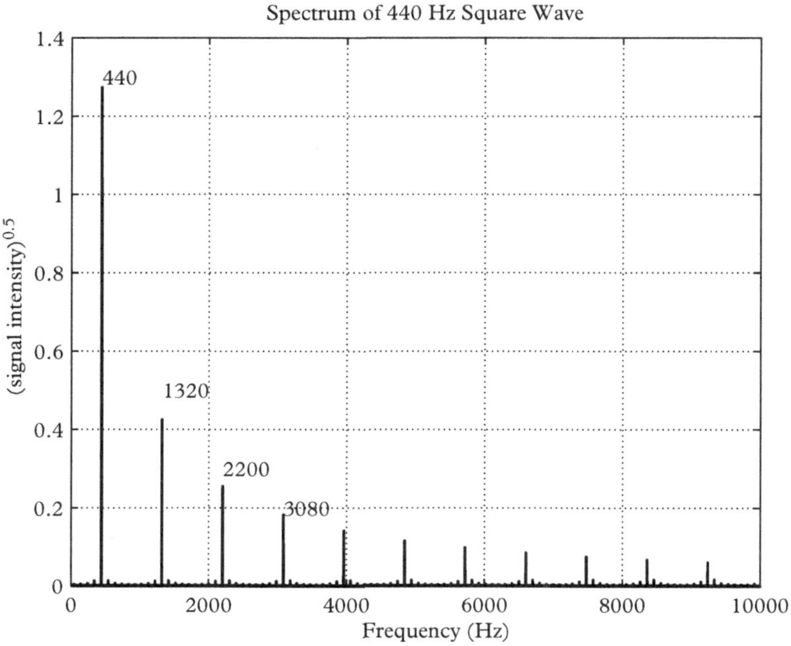

Figure 1.3.2 *Line spectrum of a square wave*

wave, τ, is 2.27 ms or 1⁄440 s. The sound of the square wave is available in the file SquareWave.mp3. Note that it does not sound that different from the built-up wave and the single sine wave.

The line spectrum of this wave, sampled at 44,100 Hz, is shown in Figure 1.3.2.

Like that for the 440 Hz sine wave and the built-up wave, the line spectrum for the square wave has a fundamental (and a strong) line at 440 Hz. There are also relatively strong lines at *odd* multiples of the fundamental but not at the *even* multiples. If the reader listens to the mp3 files for the sine wave, the built-up wave, and the square wave, she will note the similarity, due to the presence of the strong fundamental at 440 Hz in all three cases. In Chapter 2 we will revisit the absence of the even harmonics in the line spectrum of the square wave.

Before moving on we should comment on the little spikes in Figure 1.3.2, which are barely discernible next to the large spikes that represent the harmonics. Later in Chapter 2 we will look at this spectrum plot again, using decibels, and we will see that these little spikes represent a myriad of frequencies generated by the sharpness of the square wave's edges.

The autocorrelation of the square wave is shown in Figure 1.3.3. Once again, even with the presence of the higher harmonics, the autocorrelation suggests that the pitch of the square wave is 440 Hz. Notice that as the lag time increases from zero, where the square wave correlates perfectly with itself, the autocorrelation decreases linearly to zero where there is no correlation. Then, at 1.13 ms, which is at the completion of the first step, the correlation starts to rise toward unity where the time lag is 2.27 ms. As an exercise you might try to reason why the autocorrelation takes the shape given in Figure 1.3.3. Do not worry if you cannot figure it out.

This completes our introduction to elementary waves and to the four tools that we will use in the rest of the book: the time domain plot, the line spectrum, the cumulative line spectrum, and the autocorrelation. The next chapter will delve into these tools in greater detail.

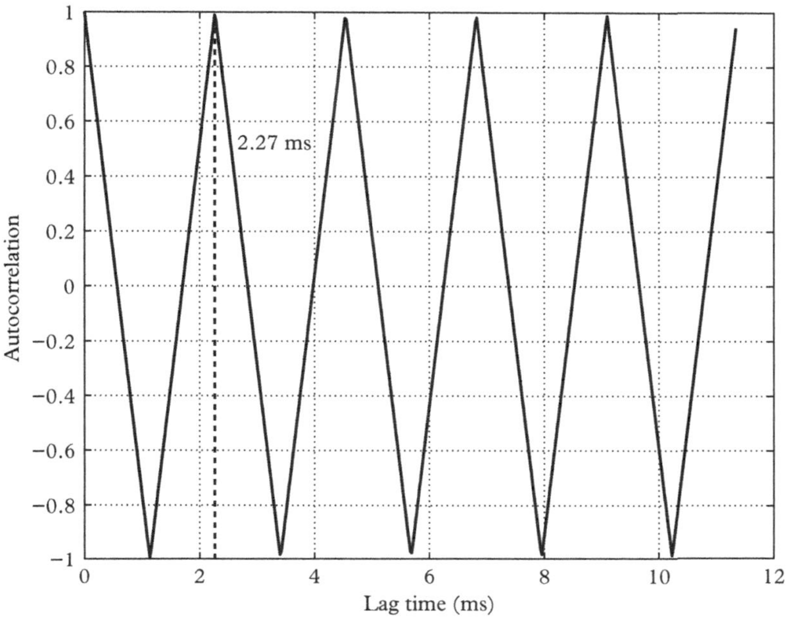

Figure 1.3.3 *Autocorrelation of a square wave*

1.4 Exercise

Study the Matlab script in the file FiguresPart1_Chap1.m. Run it and see if you can get the graphs in this chapter. Modify the script so that the amplitudes of the built-up wave *increase* with harmonic number. Compare the resulting graphs with Figures 1.2.3–1.2.6.

Look at the line spectrum of the square wave in Figure 1.3.2. Using pencil and paper sketch the cumulative line spectrum. Check your sketch by studying FiguresPart1_Chap1.m and finding the code that generates the built-up wave. Then find where CumLineSpectrum3.m is used and modify the code to compute and plot the cumulative line spectrum of the square wave.

2

The Basic Tools of Spectral Analysis

You have already been introduced to the line spectrum, the cumulative line spectrum, and the autocorrelation in the previous chapter, so why must we spend more time on them? The answer is that you do not have to. In fact, the musician-reader could perhaps skim this chapter and still understand a good bit if not all of the book. Therefore, this chapter, which is the longest in the book, is optional. However, reading it with a modicum of understanding will help you increase your ability to assimilate what comes after, so make a stab at it — you can always come back to it later.In any case, at least read this section before moving on.

This chapter will be the most difficult in the whole book. If there are parts that cause problems, please concentrate on the graphs and the verbiage rather than the equations (which are few and far between). If the graphs make sense but some of the mathematics does not, do not worry and keep trudging onward.

If you do decide to read this chapter, here is what you will come across in the following sections:

1. **The line spectrum as an approximation of a wave by a sum of sine waves.** We are going to continue the discussion of the square wave that started in the first chapter. We will try to approximate the square wave by a sum of sine waves. The coefficients that appear in the approximation will be used to construct a line spectrum. Therefore, the shape of a line spectrum for a wave will be presented as the result of an approximation of that wave by a sum of properly weighted sine waves.
2. **The Nyquist frequency.** In the line spectrum and cumulative line spectrum graphs, the maximum frequency on the frequency axis is the Nyquist frequency.We will make repeated references to it throughout the book.
3. **The frequency grid.** Since the line spectrum is a plot of a wave's strengths at various frequencies, the third section will spend some time describing the grid along the frequency axis. This will be important when we present the spectral centroid.
4. **Aliasing.** If the sound analyst tries to look for signal strength at frequencies greater than the Nyquist frequency then he/she will have to be aware of the phenomenon of aliasing, which is the topic of the fourth section.
5. **Removing average values.** The short fifth section discusses removing the average from a wave, and comments on signals with strength at zero frequency.
6. **Fine grid line spectrum.** The sixth section deals with the case where the frequency grid density afforded by the fast Fourier transform (FFT) is not sufficient and a finer grid is necessary at the cost of more computation time.

Spectral Analysis of Musical Sounds with Emphasis on the Piano. First Edition.
David M. Koenig, with comments by Delwin D. Fandrich.
Published in 2015 by Oxford University Press.

7. **Application to a noisy signal.** In the seventh subsection the tools and concepts already presented in this chapter will be illustrated with the spectral analysis of a noisy signal.
8. **The decibel and logarithmic axes.** The eighth section describes alternative ways of plotting the line spectrum. The decibel is introduced and illustrated. There is a short discussion of plotting the frequency logarithmically. The latter is not important, but using decibels is and it will appear throughout the book.
9. **The cumulative line spectrum.** This section will require a little thought. It is a simple extension of the line spectrum but since it is not as widely used it will likely be harder to read initially. However, in many ways it is a useful complement to the line spectrum. In fact, I will rarely present a line spectrum without the associated cumulative line spectrum.
10. **The spectral centroid.** Although sounding sophisticated, it is nothing more than a weighted average of the frequencies that appear in the line spectrum and is a compact measure of where the sound power is in a wave.
11. **The autocorrelation.** Although simple to compute, the autocorrelation is a bit difficult to interpret but it is worth our time because it is claimed to be an indicator of pitch. The autocorrelation, the line spectrum, and the cumulative line spectrum are the three main tools of analysis in this book.
12. **Maps.** In Part 3, when we take a detailed look at the piano with its 88 keys, an extension of the line spectrum, cumulative line spectrum, spectral centroid, and autocorrelation will be made. Instead of plotting, say, the percent accumulatedpower against frequency, we will also plot the percent accumulated power against the note number. Therefore, a three-dimensional plot (or map) will be generated that will show the percent accumulated power plotted over a plane representing the note number and the frequency. This approach will be applied to other instruments covered in Parts 1and 2 as well. This is all I will say about maps until Part 3.

Now that you have a big picture of the chapters contents, this is the point where you might want to skip on to the next chapter. If you are staying, let us start with the line spectrum.

Think of the line spectrum as providing a different view of a sound wave in a different domain; namely, the frequency domain rather than the time domain (or the audible domain). While the time domain plot of a wave shows how it changes its strength (amplitude) with time, the line spectrum shows how a sound wave's strength depends on different sine waves (with different frequencies and amplitudes) buried in the signal. In the next couple of sections the square wave will be approximated by a sum of weighted sine waves. From the approximation, the line spectrum for the square wave will be developed.

2.1 A Graphical Attempt to Approximate the Square Wave

Let us try to construct an approximation to the square wave by adding sine waves of various frequencies and amplitudes. Figure 2.1.1(a) shows how a single sine wave, with a period equal to that of the square wave, lines up with the square wave. This sine wave is a smooth but crude approximation to the square wave. Figures 2.1.1(b) and (c) show how sine waves with frequencies equal to the second and third harmonics of the square wave line up.

Figure 2.1.1(a) and (c) suggest that a sum of the first and third harmonics, with proper coefficients, could be used as a crude approximation of the square wave signal.

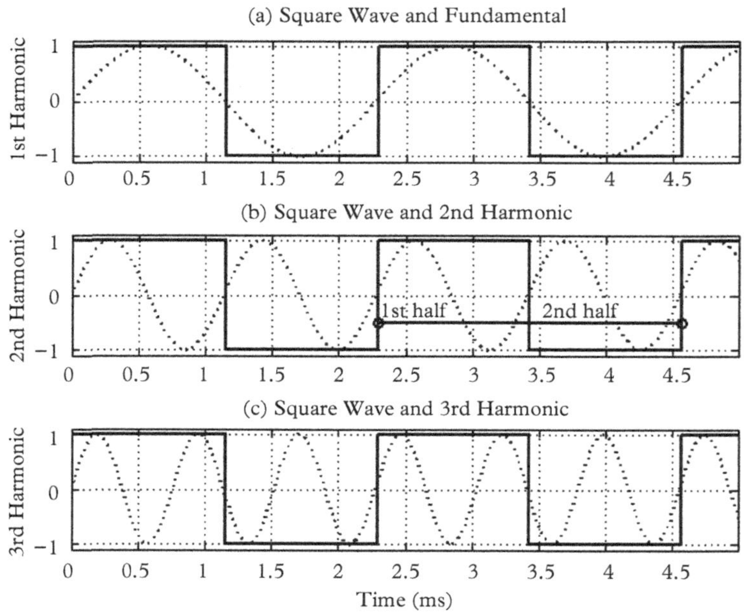

Figure 2.1.1 *A square wave along with (a) its fundamental, (b) its second harmonic, and (c) its third harmonic*

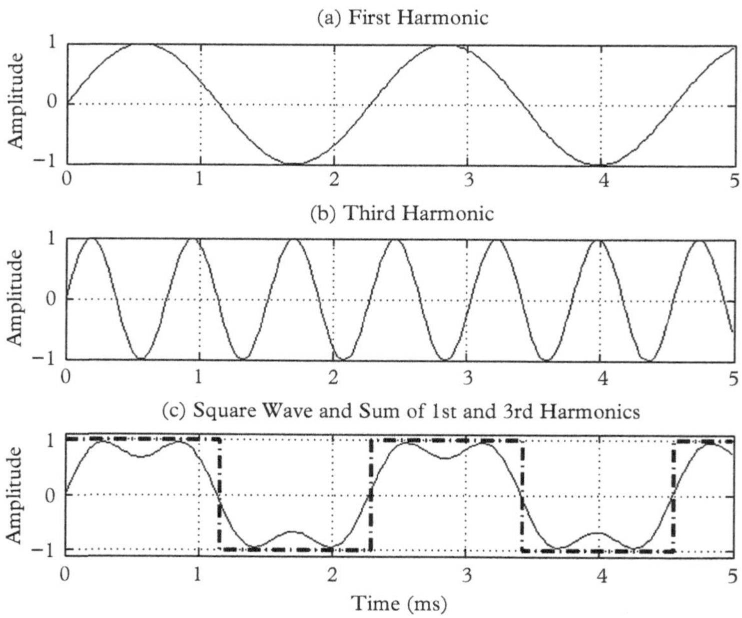

Figure 2.1.2 *(a) The first harmonic, (b) the third harmonic, and (c) the sum compared to the square wave*

On the other hand, look at the second harmonic in Figure 2.1.1(b). Note that the dotted curve has the same values during the first half of the square wave's cycle as it does during the second half. (The horizontal solid line with circles denotes the start and finish of one cycle.)Since the square wave is +1 on the first half of the cycle and –1 on the second half we would not expect the second harmonic to be useful in the approximation. This will be the case for all of the even harmonics. Therefore, this assertion is consistent with the structure of the line spectrum in Figure 1.3.2 where there is no signal strength at even harmonics.

Figure 2.1.2 shows how such a sum of the two sine waves, one with the frequency of the fundamental (440 Hz or 440 cycles s^{-1}) and one with the frequency of the third harmonic (1320 Hz), both with unity amplitude, add up to provide an interesting but inaccurate approximation to the square wave.

As mentioned, both the fundamental and the third harmonic have the same amplitude of 1.0. Figure 2.1.2 suggests that we might be able to improve the approximation by adjusting the amplitudes of the two harmonics and by adding more harmonics.

2.2 A Sum of Sine Waves as an Approximation to a Square Wave

In Section 2.1 we proposed that the square wave might be approximated by a sum of properly chosen and adjusted sine waves. Let's continue that analysis with the question "Can we find values for the coefficients $A_1, A_2, A_3, \ldots$ in Eq. 2.2.1 that will cause the sum of sine waves to behave like a square wave?" (There will be several equations in this section so if they bother you,concentrate on the graphs.) We answer that a square wave, $y(t)$, can be approximatedby the following sum:

$$\begin{aligned} y(t) &\cong A_1 \sin(2\pi f_1 t) + A_2 \sin(2\pi f_2 t) + A_3 \sin(2\pi f_3 t) + \cdots \\ \text{square wave} &\cong \text{fundamental} + \text{2nd harmonic} + \text{3rd harmonic} + \cdots . \end{aligned} \tag{2.2.1}$$

Note that Eq. 2.2.1 lines up with the word equation just below it. The "fundamental" is $A_1 \sin(2\pi f_1 t)$, which is a sine wave with a frequency of f_1 and amplitude (or strength) A_1. The "2nd harmonic" is $A_2 \sin(2\pi f_2 t)$, which is a sine wave with a frequency of f_2 where $f_2 = 2f_1$, etc. Therefore, the frequencies $f_1, f_2, f_3, \ldots$ are integer multiples of the fundamental frequency f_1, as in

$$f_2 = 2f_1, \quad f_3 = 3f_1, \quad f_4 = 4f_4, \quad \ldots,$$

and the coefficients $A_1, A_2, A_3, \ldots$ are related to the strengths of the harmonics that might show up in a line spectrum. The use of the symbol $\cong$ instead of the conventional equals sign means that the square wave, $y(t)$, is only approximately equal to the sum of the sine waves on the right-hand side of the $\cong$ symbol. Please make sure that you understand the three terms on the right-hand side of Eq. 2.2.1.

For this square wave, we choose, as before, the fundamental frequency to be $f_1 = 440$ Hz. The graphs in Section 2.1indicate that the coefficients of the *even* harmonics, $A_2, A_4, A_6, \ldots$, should be zero and that only those for the *odd* harmonics, $A_1, A_3, A_5, \ldots$, will be non-zero.

For example, Figure 2.2.1 shows how the following sum of a fundamental and three *odd* harmonics,

$$\begin{aligned} y(t) &\cong \frac{4}{\pi} \sin(2\pi f_1 t) + \frac{4}{3\pi} \sin(2\pi f_3 t) + \frac{4}{5\pi} \sin(2\pi f_5 t) + \frac{4}{7\pi} \sin(2\pi f_7 t) \\ \text{square wave} &\cong \text{fundamental} + \text{3rd harmonic} + \text{5th harmonic} + \text{7th harmonic}, \end{aligned} \tag{2.2.2}$$

approximates the square wave. Make sure you see the differences between Eq. 2.2.1 and Eq. 2.2.2 (the former contains all the harmonics while the latter contains only the odd harmonics; it also assigns values to the coefficients $A_1, A_3, A_5, \ldots$).

First, note in Figure 2.2.1 the presence of the ripples that result from the higher-frequency sine waves occurring in the sum. We see that trying to use sine waves to approximate a square wave is going to be a challenging task. Second, note that the coefficients (or amplitudes) of the sine waves in Eq. 2.2.2, namely

$$\frac{4}{\pi}, \quad \frac{4}{3\pi}, \quad \frac{4}{5\pi}, \quad \frac{4}{7\pi},$$

decrease with increasing harmonic number (which appears in the denominator). That is, each higher harmonic makes a smaller contribution to the sum.

If this sum is extended to include the 19th harmonic, then Eq. 2.2.2 becomes

$$y(t) \cong \frac{4}{\pi}\sin(2\pi f_1 t) + \frac{4}{3\pi}\sin(2\pi f_3 t) + \frac{4}{5\pi}\sin(2\pi f_5 t) + \cdots + \frac{4}{19\pi}\sin(2\pi f_{19} t). \qquad (2.2.3)$$

(Notice how the coefficients follow the sequence started above.) Figure 2.2.2 shows that the approximation provided by Eq. 2.2.3 is better but it suggests that no matter how many terms we use, a sum of sine wave terms is never quite going to replicate the sharp corners of the square wave.

This makes sense because, as was pointed out above, we are attempting to approximate a sharp-cornered function with a sum of functions that are smooth. A discussion of the ripple in

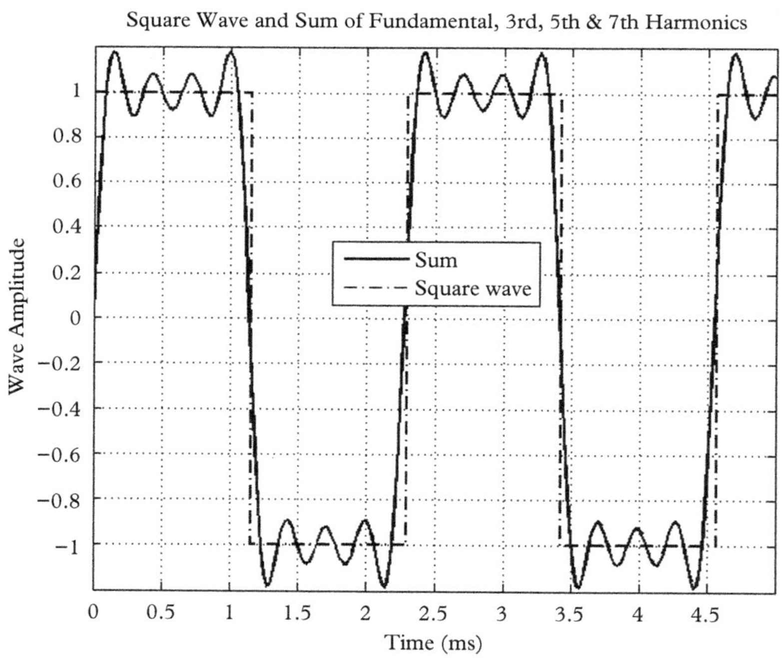

Figure 2.2.1 *A square wave and the sum of the first seven harmonics*

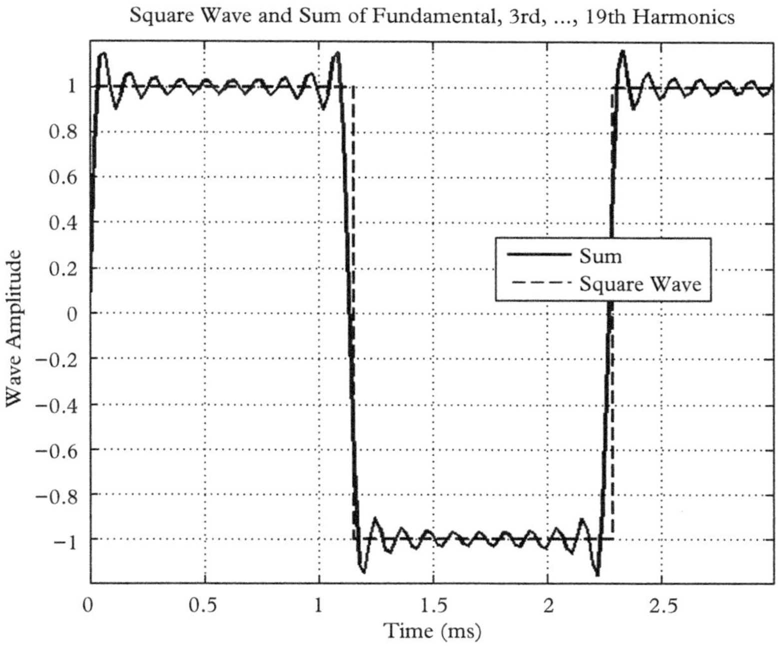

Figure 2.2.2 *A square wave and the sum of the first 19 harmonics*

the approximation is interesting but belongs in a more technical book (the reader can search for "Gibbs phenomenon" on the Internet to get some interesting discussions).

A Tentative Conclusion

We therefore conjecture that almost any type of wave can be approximated to "some extent" by a sum of properly weighted sinusoidal waves. Even if the approximation is not perfect we can gain insight into the nature of the wave by examining the distribution of signal power among the various frequencies ... and it is this insight that we are really after. Furthermore, the coefficients in the approximation can be used to construct the line spectrum.

Now, return to the square wave and Section 1.3, specifically Figure 1.3.2, where we plotted the absolute value of the coefficients obtained from the line spectrum; that is,

$$|A_1|, |A_2|, |A_3|, \ldots,$$

versus the frequencies

$$f_1, f_2, f_3, \ldots$$

That line spectrum in Figure 1.3.2 is re-plotted in Figure 2.2.3(b), and it can be compared with Figure 2.2.3(a) where the absolute values of the coefficients in Eq. 2.2.3, namely $|A_1|$, $|A_2|$, $|A_3|, \ldots, |A_{19}|$, are plotted against the coefficient index. Figure 2.2.3(a) and (b) are basically the same (as they should be).

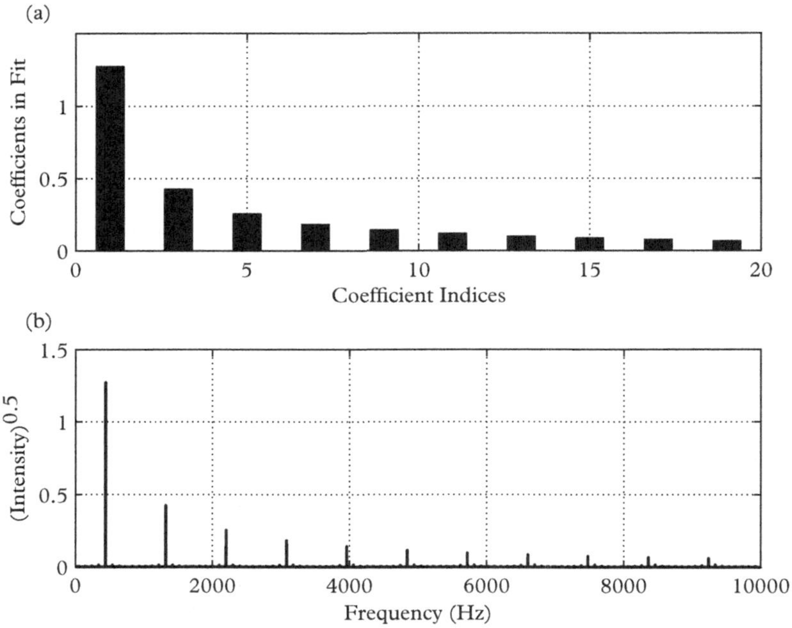

Figure 2.2.3 *(a) The coefficients in the square wave approximation, and (b) the line spectrum of the square wave*

Note that the coefficients in Figure 2.2.3(a) are zero for the even harmonics, and that this graph does not include the small amplitude "blips" that occur in Figure 2.2.3(b) in between the lines for the odd harmonics. We will talk about the "blips" later on in this chapter.

Partial Summary of Section 2.2

It appears that constructing a line spectrum is somewhat equivalent to fitting a sum of sine waves with properly chosen frequencies and amplitudes to the wave of interest and then plotting the absolute values of the coefficients of those sine terms against the frequencies. This kind of plot tells us about the frequency content of the signal. In general, the approximating sum will also contain cosines as well as sines, but this will not concern us here. Appendix 1 will look into this sum in more detail and discuss the idea of Fourier series approximation. However, all we need to agree on at this point is that we can think of a line spectrum as being made up of the coefficients from an approximation to the wave of interest containing a sum of sine waves.

2.3 The Folding Frequency and the Nyquist Interval

From the foregoing discussion, it appears that a signal (containing N samples with spacing of h seconds between samples) can be approximated by a sum containing $\frac{N}{2} + 1$ cosine and sine terms (where N must be an even number). The highest frequency used in the approximation is $\frac{F_s}{2}$, which is sometimes signified as f_{Ny}, the folding frequency or the *Nyquist frequency*, as in

$$f_{Ny} = \frac{F_s}{2} = \frac{1}{2h}.$$

Why this frequency is the highest will become apparent soon, but for the time being you should just accept it.

When the signal strengths are plotted against the frequencies, the frequencies range only over the following discrete values:

$$0, \frac{1}{Nh}, \frac{2}{Nh}, \frac{3}{Nh}, \ldots, \frac{1}{2h};$$

that is, from zero frequency to the Nyquist frequency, which is half the sampling frequency. This span of frequencies is often called the *Nyquist interval* and it is the interval that we will be interested in.

2.4 The Frequency Grid in the Line Spectrum Plot

If the wave being approximated has been sampled at a rate of F_s Hz(cycles s^{-1}) for a duration of T seconds then the sampling interval is

$$h = \frac{1}{F_s} = \frac{T}{N} \text{ seconds},$$

where N is the number of samples.

For example, if the sampling frequency F_s is 44,100 Hz and the duration T is 1.5s then the sampling interval is

$$h = 1/F_s = 2.2676 \times 10^{-5}\,\text{s} \quad \text{or} \quad 0.0227\,\text{ms}.$$

The number of samples is

$$N = T/h = 66{,}150.$$

The sum of sines and cosines in the approximation will contain terms for each frequency f_k, which is defined by

$$f_k = \frac{k}{Nh}, \qquad k = 0, 1, 2, \ldots, \frac{N}{2},$$

$$\text{or} \qquad f_k = 0, \frac{1}{Nh}, \frac{2}{Nh}, \ldots, \frac{1}{2h},$$

$$\text{or} \qquad f_k = 0, \frac{F_s}{N}, \frac{2F_s}{N}, \ldots, \frac{F_s}{2},$$

where $F_s = \frac{1}{h}$. Please make sure that all three ways of writing the frequencies make sense to you. Note that there are $\frac{N}{2} + 1$ frequencies, starting at 0.0 Hz and ranging up to and including $F_s/2$ Hz.

For the case of the square wave, only the terms in the sum that contain frequencies that are odd harmonics of the fundamental frequency, 440 Hz, will have non-zero coefficients.

These $\frac{N}{2}+1$ frequencies define a frequency grid that has values only at the $\frac{N}{2}+1$ discrete points and has a spacing between those points of $\frac{1}{Nh}$ or $\frac{1}{T}$ or $\frac{F_s}{N}$ Hz. In our current example,

$$F_s = 44,100\,\text{Hz},\ T = 1.5\,\text{s},\ h = 0.0000227\,\text{s} = 0.0227\,\text{ms},\ N = 66,150,$$

and the interval between frequencies in the frequency grid is

$$\frac{1}{T} = \frac{1}{1.5} = 0.6667\,\text{Hz}.$$

At the other extreme, consider the simple case where $N = 20$ with a sampling frequency of 1.0 Hz and therefore a sampling interval of 1.0s. Furthermore, assume that the samples are of a wave having two sinusoidal components with frequencies of 0.10 Hz and 0.25 Hz. Since $F_s = 1.0$ Hz, the folding or Nyquist frequency is 0.5 Hz. Therefore, as we will see in the next section, the two sinusoidal components in the wave can, in principle, be detected using this sampling rate because 0.10 Hz and 0.25 Hz are both less than 0.5 Hz. The time domain plot of the wave is shown in Figure 2.4.1. The twenty samples are shown as small circles. The actual continuous wave from which the samples have been taken is represented by the tiny dashes.

The line spectrum can be computed from the twenty samples by means of an FFT, and the result is shown in Figure 2.4.2.

There are two peaks, which just happen to be located at the frequencies of sinusoidal components in the wave because of the fortuitous choice of the sampling interval. In other words,

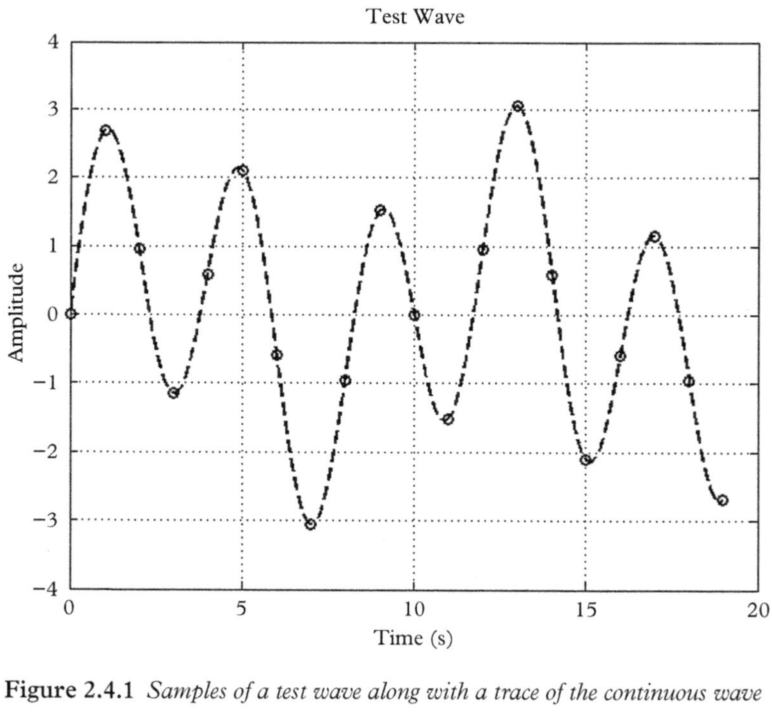

Figure 2.4.1 *Samples of a test wave along with a trace of the continuous wave*

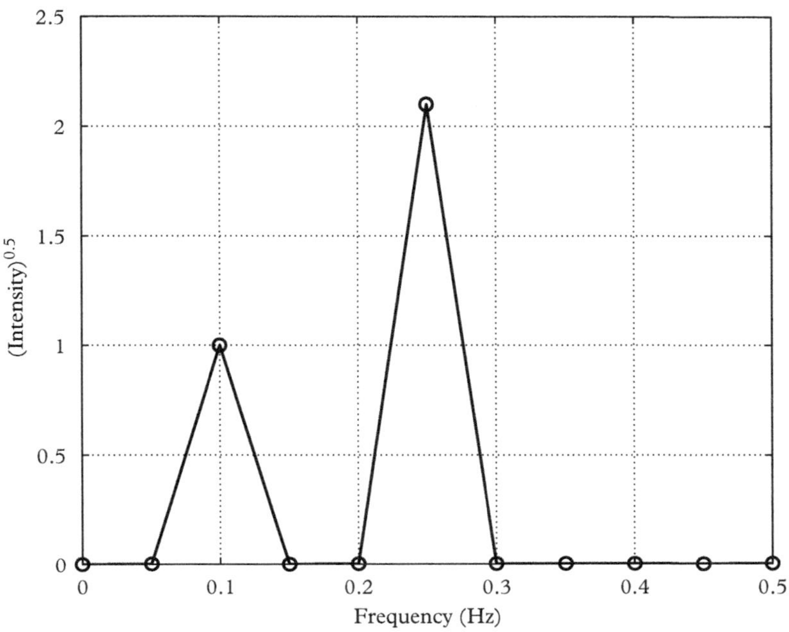

Figure 2.4.2 *Line spectrum of the sampled test wave*

an approximation to this twenty-sample wave would consist of just two sine waves with their associated coefficients and frequencies.

Table 2.4.1 shows the time grid t (column 1), the sample values y (column 2), the frequency grid f (column 3), and the line spectrum values Y (column 4). Note that the frequency grid ranges from 0.0 Hz to 0.5 Hz (the folding frequency) and that the frequency spacing is $F_s/N = 1/20 = 0.05$ Hz. There are $N = 20$ sample points and there are $N/2 + 1 = 11$ points on the frequency grid.

You might want to pause and think about these numbers in this section and how they have been developed. There have been two examples.The first example consisted of a large data set of 66, 150 samples with a sampling frequency of 44, 100 Hz over a period of 1.5s. The second example had only 20 samples with a sampling frequency of 1 Hz over a period of 19s.

2.5 Aliasing

Theoretically, the Nyquist frequency f_{Ny} is the highest frequency that can occur in the approximation developed in Section 2.2. If you are analyzing an unknown wave, the highest frequency that you can expect to identify using the line spectrum is f_{Ny}. If you suspect that the highest frequency component in the wave of interest is actually f_{max} then you should sample at a rate higher than $2f_{max}$. If you sample at a lower rate you risk experiencing aliasing, where frequencies in the wave greater than the Nyquist frequency show up as *aliased* lower frequencies in the Nyquist interval.

Table 2.4.1 *Time, sample values, frequency grid, and line spectrum for the sampled test wave.*

1	2	3	4
t	y	f	Y
0	0	0	0
1	2.6878	0.05	0
2	0.9511	0.1	1
3	−1.1489	0.15	0
4	0.5878	0.2	0
5	2.1	0.25	2.1
6	−0.5878	0.3	0
7	−3.0511	0.35	0
8	−0.9511	0.4	0
9	1.5122	0.45	0
10	0	0.5	0
11	−1.5122		
12	0.9511		
13	3.0511		
14	0.5878		
15	−2.1		
16	−0.5878		
17	1.1489		
18	−0.9511		
19	−2.6878		

To illustrate this, take a look at Figure 2.5.1. Here is a plot (the solid curve) of a sine wave having a frequency of 1.083 Hz and a period of 0.9231s. (See if this frequency makes sense to you by studying the graph. Does the solid curve really have a period of a little less than 1.0s?)

The wave is sampled at 1 Hz (one sample per second). The little black dots represent the samples. (Note that they are spaced apart by one second in the graph.) The sampling frequency of 1.0 Hz means that the Nyquist frequency is 0.5 Hz, which is the highest frequency that can be identified; the Nyquist interval therefore contains frequencies from 0.0 Hz to 0.5 Hz. But the wave being sampled has a frequency of 1.083 Hz, which is greater than twice the Nyquist frequency. What is going on here?

Because of this low sampling rate, the sampler (you) thinks she/he is sampling a wave with the frequency of the dashed curve in Figure 2.5.2, which has a frequency of 0.0833 Hz, and a

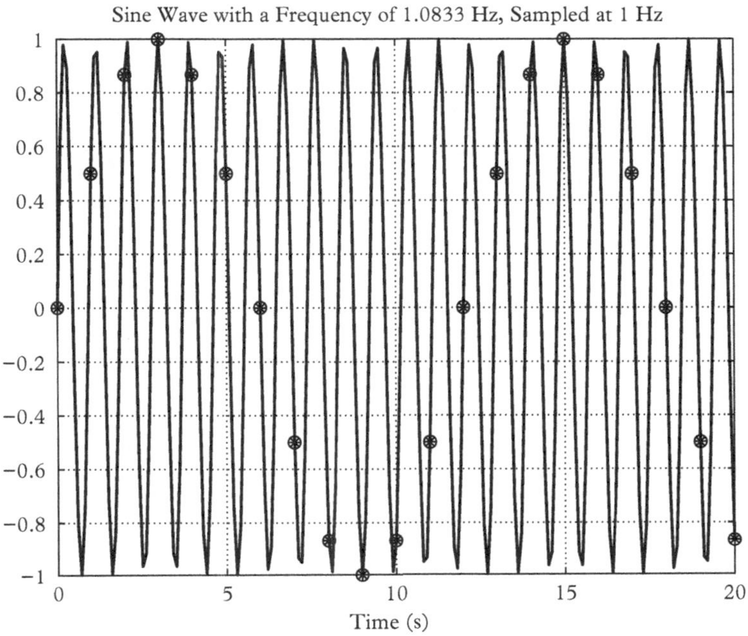

Figure 2.5.1 *Continuous trace and samples of a sine wave with a frequency of 1.0833 Hz*

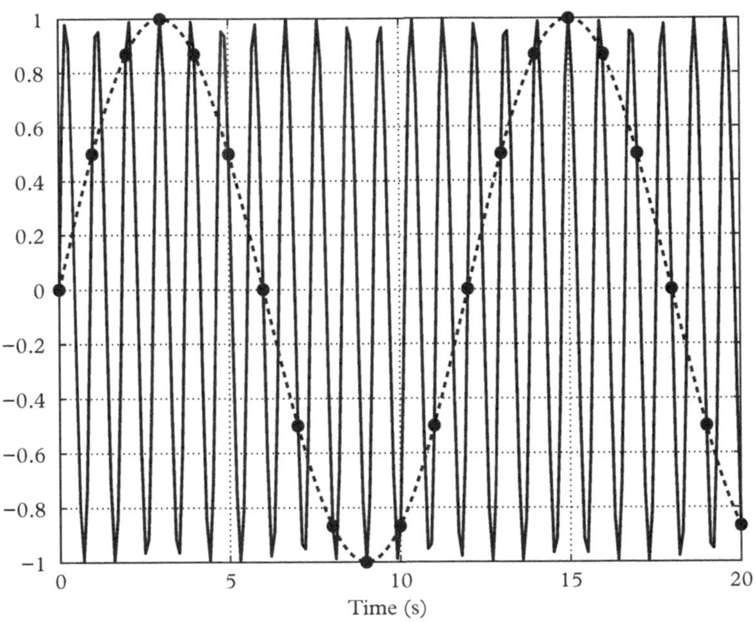

Figure 2.5.2 *The wave from Figure 2.5.1 as seen by the sampler (dashed curve)*

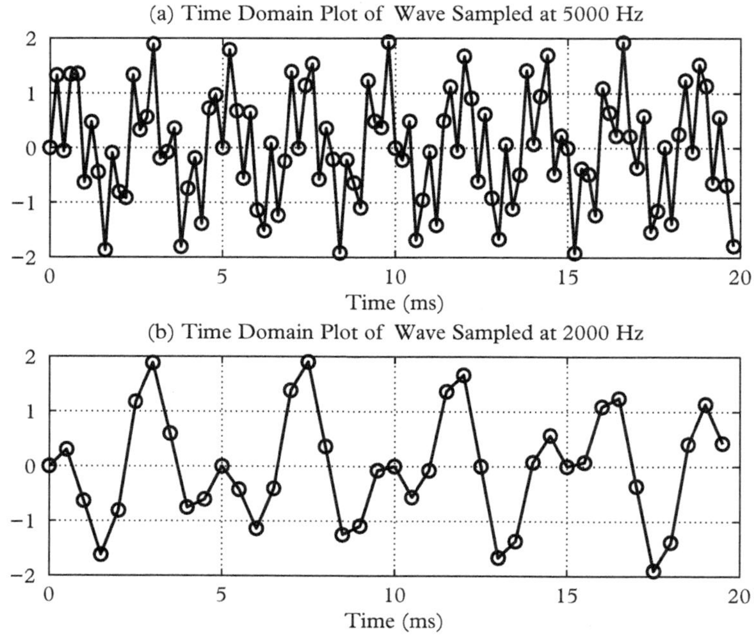

Figure 2.5.3 *Samples resulting from sampling at (a) 5000 Hz, and at (b) 2000 Hz*

period of 12.0s (verify that the period of the dashed curve in Figure 2.5.2 is indeed 12s). This low frequency does lie in the Nyquist interval. The sampler cannot see the true higher frequency wave. He/she only sees that wave's alias, which is a much lower frequency wave.

One more example! Figure 2.5.3 shows two plots of a sound wave containing two sinusoidal components: one with a frequency of 440 Hz (the A4 tone) and one with a frequency of 1760 Hz (the A6 tone). Figure 2.5.3(a) shows the result of sampling this sound wave at 5000 Hz. Since the Nyquist frequency for this sampling rate is 2500 Hz we should be able to identify both components.

Figure Figure 2.5.3(b) shows the result of sampling the same sound wave at 2000 Hz. Here, the Nyquist frequency is 1000 Hz so the sampled curve looks entirely different: it appears to contain the 440 Hz component plus a second component at 240 Hz, which is an alias of the component at 1760 Hz. These two sets of samples will have different sounds.

Figure 2.5.4 confirms this. Figure 2.5.4(a) shows the line spectrum of the result of sampling at 5000 Hz. Note that the frequency axis ranges from 0.0 Hz to 2500 Hz. There are two peaks at the expected frequencies: 440 Hz and 1760 Hz.

Figure 2.5.4(b) shows the result of sampling at 2000 Hz. Now the frequency axis only goes up to the Nyquist frequency of 1000 Hz and cannot show a component at 1760 Hz. Instead, there is a peak at 240 Hz, which is the alias of 1760 Hz. The sounds of these two waves areavailable in the file AliasSounds.mp3. The Matlab script in the file AliasSounds.m generates the graphs and the sounds.

Digital recordings of classical and popular music often use a sampling rate of 44,100 Hz, so the Nyquist frequency is 22,050 Hz. The highest note on a piano, C8, has a theoretical fundamental

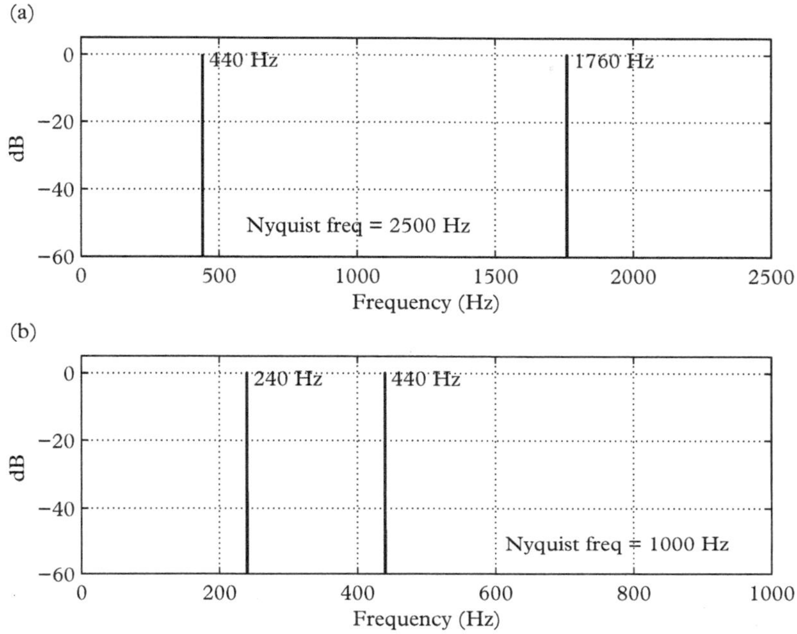

Figure 2.5.4 *Line spectra from sampling at (a) 5000 Hz, and from sampling at (b) 2000 Hz*

frequency of 4186 Hz. Some humans can hear frequencies up 20,000 Hz (although most adults have an upper limit of about 16,000 Hz). Later in this chapter we will take a look at the spectrum of the C8 note.

2.6 A Sine Wave with Zero Frequency

If you constructed a line spectrum for a wave of interest and found that there was signal strength at zero frequency, what would that mean? A zero frequency sine wave is a wave with zero frequency, that is, a constant. Therefore, if you observe signal strength at zero frequency your signal has an average value that is not zero. I have taken to subtracting the average from a signal before I analyze it just to make things a little simpler.

2.7 A Finer Grid in the Frequency Domain

FFTs can be used to efficiently calculate signal strengths at the discrete set of frequencies in the Nyquist interval noted in Section 2.3. If the FFT is used, then the spacing between the frequencies is $\frac{1}{Nh} = \frac{F_s}{N}$, which may be too coarse for the analysis at hand, especially if the number of samples N is low. Consequently, the line spectrum may have to be calculated over a finer frequency grid. If the FFT is not used to get the coefficients in the approximation and to construct the line spectrum, one is free to look for signal strength at *any* collection of frequencies.

The downside of not using the FFT is a significant increase in the time needed to calculate the coefficients (or signal strengths) unless some tricks are employed. We will have to use a fine frequency grid many times in the book, so it is fortunate that there are some Matlab tricks that one can use to address this computational concern. In any case, this issue should not concern the reader (but it is discussed in Appendix 1). When we do have to depart from using the FFT to have a finer spacing in the frequency domain we will employ the *nom de plume* of "fine grid DFT," where DFT refers to discrete Fourier transform, or "fine grid line spectrum."

We will make use of the fine grid line spectrum later in the book but, with these comments in mind, the reader might want to go back to Figures 1.1.6, 1.1.7, 1.2.3, and 1.2.5 and review the applications of the line spectrum. Make sure that the pictures of the line spectrum match your understanding of the time domain plots of the sound waves of interest.

The fact that the frequency resolution is $\frac{F_s}{N}$ suggests the downside of making the snapshots of the line spectrum as discussed in Section 1.2, especially Figure 1.2.6. To break the time axis into overlapping subintervals necessarily means the frequency resolution has to get poorer. Therefore, as you gain insight into how the line spectrum changes with time you lose detail with regard to the location of the lines in the frequency domain.

2.8 A Noisy Example—The Mystery Wave

To illustrate the use of the line spectrum for a different kind of wave, consider the three views of a signal in Figure 2.8.1. Note the different time scales in the three views.

Take a moment to study this wave. First, it is sampled at the same frequency as the previous signals, 44,100 Hz. Second, Figure 2.8.1(c) shows only a small portion of the wave. To show any

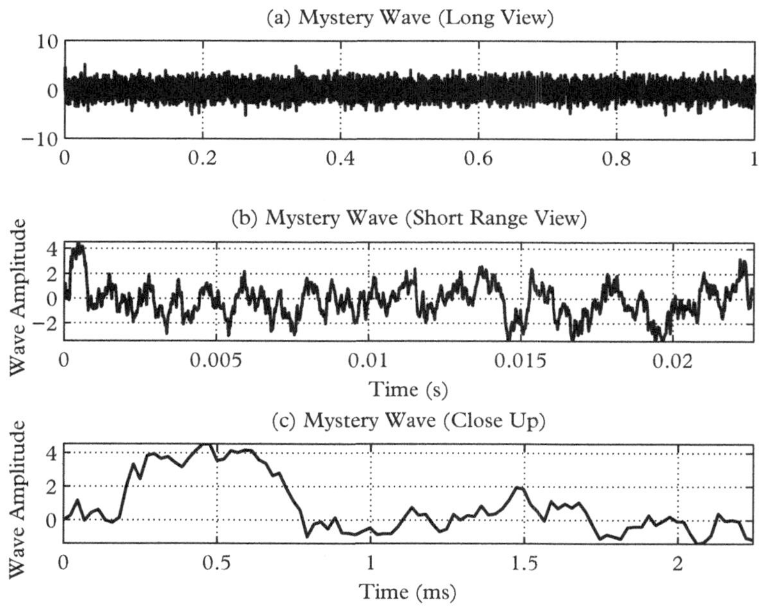

Figure 2.8.1 *Three views of the mystery wave*

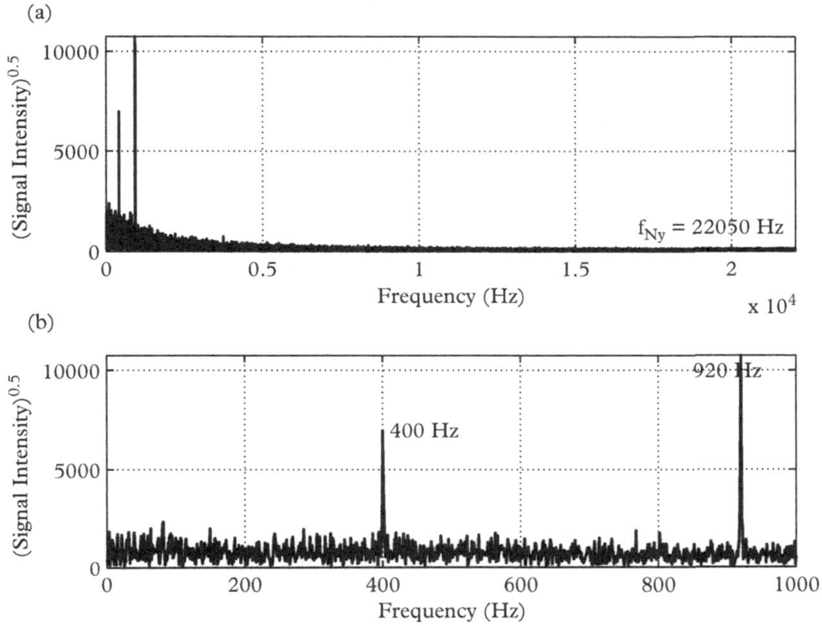

Figure 2.8.2 *Two views of the line spectrum for the mystery wave*

more, as in Figure 2.8.1(a), would make it difficult to perceive anything other than a blur. Third, it looks erratic and noisy. To get some insight into this signal we construct the line spectrum and present two versions of it in Figure 2.8.2.

Figure 2.8.2(a) shows the signal power over the entire Nyquist interval, from zero to 22,050 Hz, while Figure 2.8.2(b) takes a look at the interval between 0 and 1000 Hz.

We see a floor of hash suggesting a noisy background, which is consistent with the pictures in Figure 2.8.1. In Figure 2.8.2(a), the hash power tends to decrease with increasing frequency. In both plots in Figure 2.8.2 we see two lines at about 400 Hz and 920 Hz. Thus, it appears that the mystery signal is the sum of two sine waves and background noise. You can listen to the sound of the wave in Figure 2.8.1 in file MysterySignal.mp3, where you will hear mostly the soft buzz of the noise—the two sinusoidal waves with frequencies at 400 and 920 Hz are basically washed out.

2.9 Alternative Ways to Plot a Wave's Intensity in the Frequency Domain

Often, the absolute value squared of any coefficient in the sinusoidal approximations discussed above is considered to be proportional to the intensity of the signal at that frequency. That is, if

$$A_k, \; k = 0, 1, 2, \ldots, \frac{N}{2}$$

are the coefficients determined from, say, a line spectrum calculation based in turn on an FFT calculation, then

$$|A_k|^2, \ k = 0, 1, 2, \ldots, \frac{N}{2} \tag{2.9.1}$$

are the absolute value squared of the coefficients and they have units proportional to signal intensity. Whenever you see $|x|$ it means the absolute value of x; that is,

$$|5.1| = 5.1$$

and

$$|-5.1| = 5.1$$

and

$$|-5.1|^2 = 26.01.$$

So far we have been plotting *not* $|A_k|^2$, but rather $|A_k|$, and we have been using the label $\sqrt{\text{signal intensity}}$ or $(\text{intensity})^{0.5}$ on the y-axis to denote it. Plotting $|A_k|$ instead of $|A_k|^2$ emphasizes the lower-power components, and it is our preference.

Decibels and the Logarithmic World of Musical Sounds

Empirically, the human perception of sound intensity or power is more nearly proportional to the logarithm of the intensity rather than the intensity itself, so a new unit, the *decibel*, is often used to describe perceptual sound levels or sound level differences.

Sound levels in intensity can be converted to decibels by means of the following rather convoluted definition:

$$P_k = 20 \log_{10} \left(\frac{|A_k|}{\max\{|A_k|\}} \right), \tag{2.9.2}$$

where $\log_{10}(\ \ )$ is the base 10 logarithm of whatever is inside the parentheses and where P_k has units of decibels or dB. If the expression containing the logarithm bothers you, breeze through the next few sentences until you come to the discussion about Figure 2.9.1.

Equation 2.9.2 may be a little off-putting, so lets discuss it. The fancy term in the denominator,

$$\max\{|A_k|\},$$

is simply the maximum value of the absolute value of *all* the coefficients $|A_i|, \ i = 1, 2, \ldots, N/2$. The ratio

$$\frac{|A_k|}{\max\{|A_k|\}}$$

is a normalized or scaled version of the absolute value of the coefficient A_k and should be less than or equal to 1.0. Therefore, P_k is 20 times the base 10 logarithm of the normalized kth coefficient, $|A_k|$.

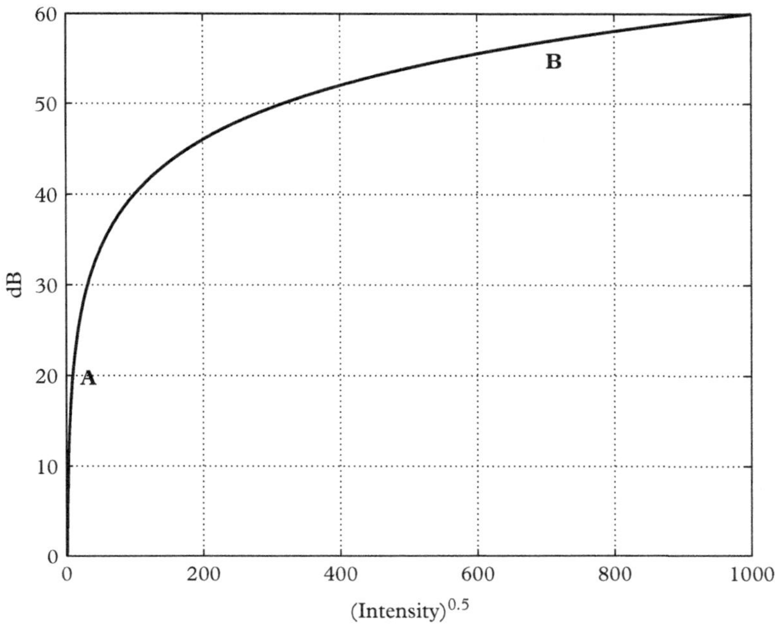

Figure 2.9.1 *Decibels versus the square root of the intensity*

If this is confusing, do not let it bother you and move on to the next figure. Figure 2.9.1 shows how changes in (intensity)$^{0.5}$ are related to changes in decibels.

At low intensities, near point A in Figure 2.9.1, a small change in (intensity)$^{0.5}$ is presented as a large change in decibels. The slope of the curve at this point is high. At higher intensities, near point B, the slope of the curve decreases. Thus, in a decibel-based line spectrum plot, low-power components are visually magnified relative to high-power components.

Consider an example to get a feel for the numbers. Lets assume that an FFT calculation has been carried out on a sampled wave and that the absolute value of the 23rd coefficient A_{23} has a value of 150.0. It is associated with the frequency f_{23}. Further, let's assume that $|A_{23}|$ is also the largest of all the coefficients. Next, assume that the 47th coefficient A_{47}, associated with the frequency f_{47}, happens to have an absolute value of 95.0. With that in mind, lets calculate the strength of these two components in decibels. The first component, the largest, has power P_{23} given by Eq. 2.9.2 as follows:

$$P_{23} = 20 \cdot \log_{10}\left(\frac{150.0}{150.0}\right) = 0.0\text{dB}.$$

The power is zero because $\log_{10}(1.0) = 0.0$.

The other component has power P_{47}, given by Eq. 2.9.2 as

$$P_{47} = 20 \cdot \log_{10}\left(\frac{95.0}{150.0}\right) = -3.97\text{dB}.$$

So, in this example, the decibel power associated with frequency f_{47} is "4 dB down" from the maximum component in the wave, which occurs at the frequency f_{23}. This example also attempts to demonstrate that a decibel is a *relative* measure. The intensity of the component having the frequency f_{47} is 4 dB down *relative* to the component having the frequency f_{23}.

For future reference, Eq. 2.9.2 can be turned around, or inverted, as in

$$\frac{|A_k|}{\max_i |A_i|} = 10^{\frac{P_k}{20}}. \tag{2.9.3}$$

(Do not worry if this inversion is confusing—it will not play an important part in what follows.)

Figure 2.9.2 shows the line spectrum of the mystery signal plotted in decibels.

The maximum value of the signal intensity occurs at 920 Hz, so the normalized intensity, using Eq. 2.9.2, is 1.0 at 920 Hz. Since the logarithm of 1.0 is zero, the peak at 920 Hz just touches the 0.0 dB line in Figure 2.9.2. The other peak at 400 Hz looks to be about 3 dB down from the maximum peak.

The background hash near the Nyquist frequency at point A on the right-hand side of Figure 2.9.2 is about 35 dB down from the peak at 920 Hz, so the hash near the Nyquist frequency at 22,050 Hz can be said to be 35 dB down from the maximum power in the signal.

Using Eq. 2.9.3, one can determine that when the decibel power is 40 dB down, it means that the absolute value of the coefficient A_k is 1/100th of the magnitude of the largest coefficient:

$$\frac{|A_k|}{\max_i |A_i|} = 10^{\frac{P_k}{20}} = 10^{\frac{-40}{20}} = 10^{-2} = \frac{1}{100}.$$

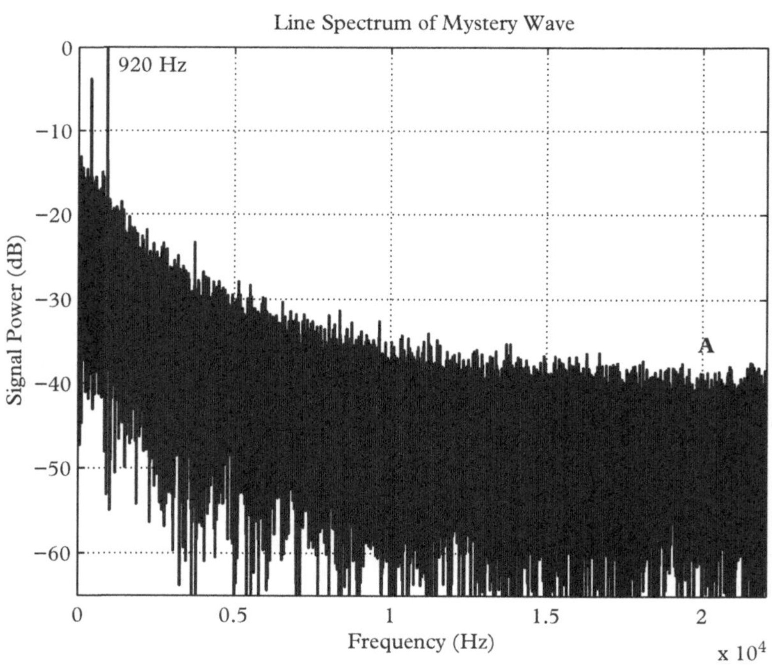

Figure 2.9.2 *Line spectrum of the mystery wave using decibels*

In the state of New York, the maximum allowable sound level is 82 dB for any motorcycle traveling at any speed. This is measured at a distance of 50 feet from the center of the lane in which the motorcycle is traveling. Most acoustics experts agree that a normal conversation in a quiet room has an intensity of about 60 dB relative to complete silence. The sound of breathing has an intensity of approximately 10 dB relative to complete silence. Thus, it is probably safe to say that if a sound wave's dominant harmonic component is assigned a reference intensity of 0.0 dB then any harmonic components more than 60 dB down from that cannot be heard.

As another example, consider the line spectrum of C8, the highest note on a piano, in Figure 2.9.3. The wave is sampled at 44,100 Hz, so the Nyquist frequency is 22,050 Hz.

Figure 2.9.3 shows the spectrum over the entire Nyquist interval using two different ways of plotting the intensity. There is a main spike near 4186 Hz and a small spike near 8372 Hz, which represents the second harmonic [not apparent in Figure 2.9.3(b)]. The intensity of the second harmonic is some 40 dB down from the fundamental. One also sees in both figures quite a bit of power at low frequencies. This below-the-fundamental signal intensity is called "hammer knock." We will spend more time on this phenomenon in Part 3.

We now return to the square wave introduced in Section 1.3 and studied in Section 2.1. Figure 2.9.4 shows the line spectrum of the square wave using decibels.

The fundamental at 440 Hz and its odd harmonics are still present, but now we see the presence of a huge number of additional components with amplitudes more than 40 dB down. These components could be considered part of the "noise" generated by the sharp corners of the square wave. Alternatively, and more correctly, they could be considered as aliases of the extremely high frequencies needed in the Fourier series (see Appendix 1) that attempts to approximate the square

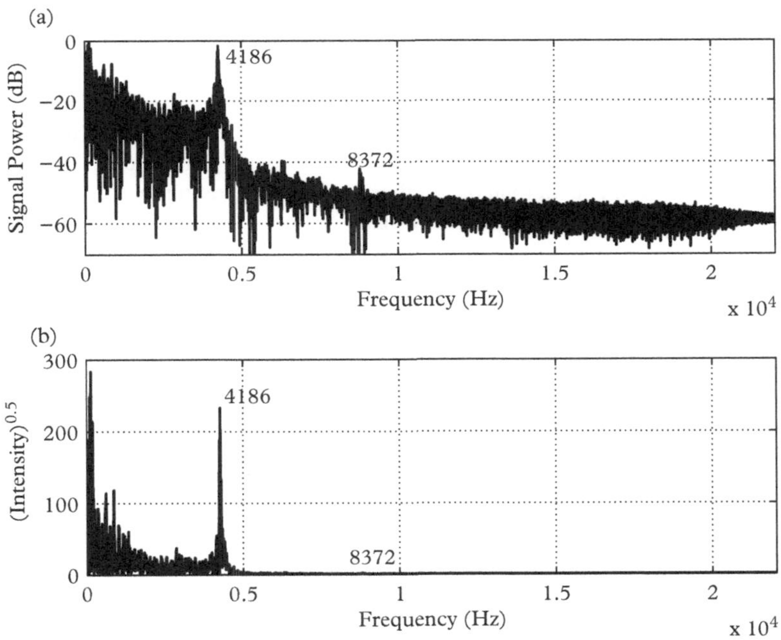

Figure 2.9.3 *Two views of the C8 note line spectrum: (a) in decibels, and (b) the square root of intensity*

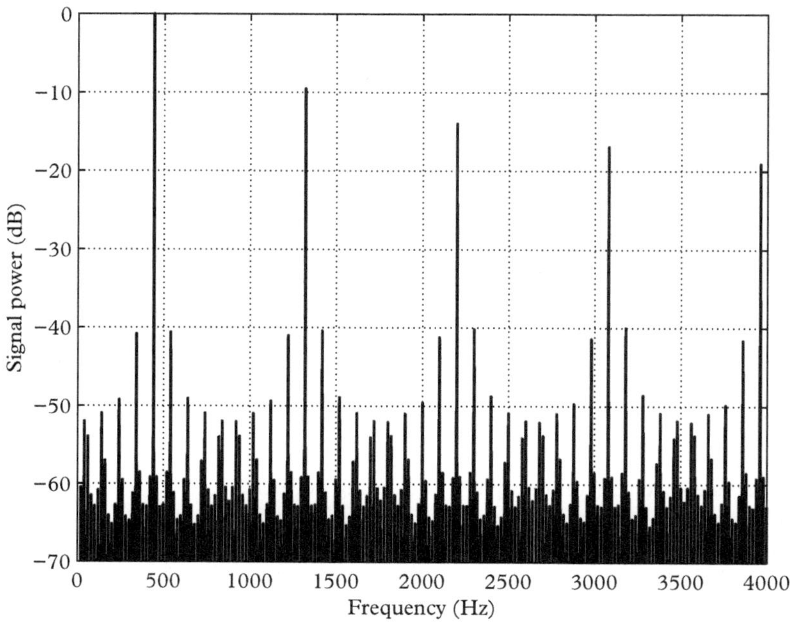

Figure 2.9.4 *Line spectrum of square wave using decibels*

wave shape. (Remember our earlier comments in this chapter about the large number of terms needed to approximate the square wave. Also, remember that the absolute values of the coefficients of those high-numbered harmonics are inversely proportional to the harmonic number.)

The reader might spend a few moments thinking about the differences between Figure 2.2.3 and Figure 2.9.4. Figure 2.2.3(a) comes from the signal that I generated as an approximation. I simply plotted the coefficients (a total of 19) that I used in generating that approximating signal. By definition, that figure is only going to show peaks or lines at the frequencies that I used in the approximation. Figure 2.2.3(b) comes from the FFT-based line spectrum for the actual square wave (with its square corners).

Similarly, Figure 2.9.4 results from the same approach that led to Figure 2.2.3(b). This approach samples the signal at 44,100 Hz, where the Nyquist frequency is 22,050 Hz. As we showed above, an approximation of a square wave in theory requires an unbounded number of harmonics and it will still never actually replicate the wave perfectly. Those harmonics required that are greater than the Nyquist frequency are going to become aliases and show up in the Nyquist interval—hence the huge number of little peaks in Figure 2.9.4.

Logarithmic Frequency Axis

The frequency axis is sometimes plotted logarithmically. For example, the spectrum for the square wave could be plotted as in Figure 2.9.5.

Note that the *y*-axis Figure 2.9.5(a) is in decibels and in Figure 2.9.5(b) it is $(\text{intensity})^{0.5}$. Since we are going to be interested in harmonics that occur at equally spaced intervals along a linear frequency axis, this manner of plotting will not be used *except* in Part 3, where it is helpful

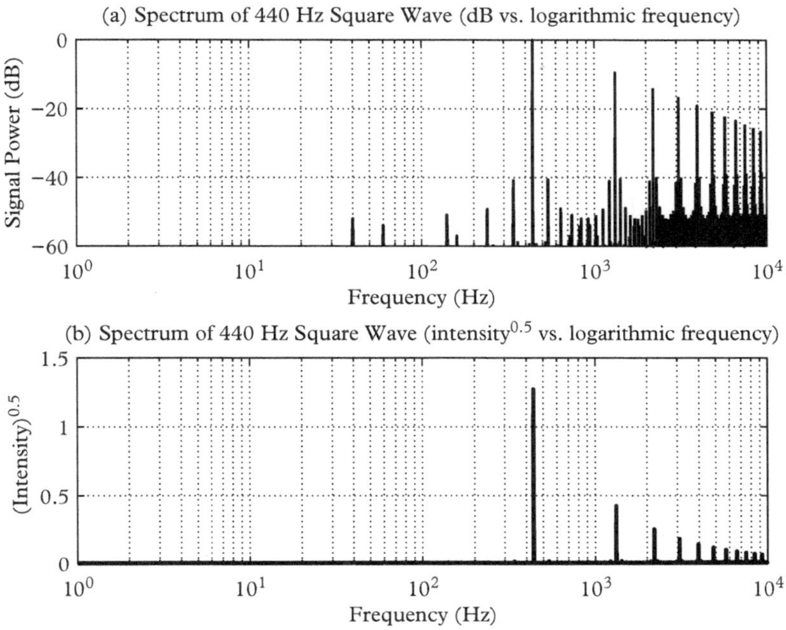

Figure 2.9.5 *Two views of a square wave using a logarithmic frequency axis*

in interpreting the piano maps because the fundamental frequency of the piano keys varies geometrically with the key numbers. However, many authors use logarithmic frequency axes, so the reader should not dismiss their use out of hand.

This section attempts to show how a musical sound is sensed logarithmically. This is the basis for our choice of logarithmic axes in some of the graphs. In Chapter 4, harmonics are discussed and the idea that the fundamental frequency of a note has a nonlinear relationship to its note number will further support our need to use logarithmic axes.

These last sections have been really tough. I hope you have made it through without too much pain. If you think this material is impenetrable in places I am sure you are not alone. You might want to reread it, but then again you might just want to keep going because the rest of the book will be "easy pickin's" by comparison. Warning! We still have to take a look at the cumulative line spectrum, the spectral centroid, and finally the autocorrelation.

2.10 The Cumulative Line Spectrum and White Noise

The Built-Up Wave

The cumulative line spectrum is a running sum of the line spectrum (using signal intensity units; that is, using the $|A_k|^2$). Figure 2.10.1 shows the line spectrum and the cumulative line spectrum for the built-up wave from Figure 1.2.4.

The x-axis in Figure 2.10.1(b) contains the frequencies in Hz. The y-axis gives the *accumulated* power in percent. Move along the frequency axis left to right. Figure 2.10.1(b) shows that until

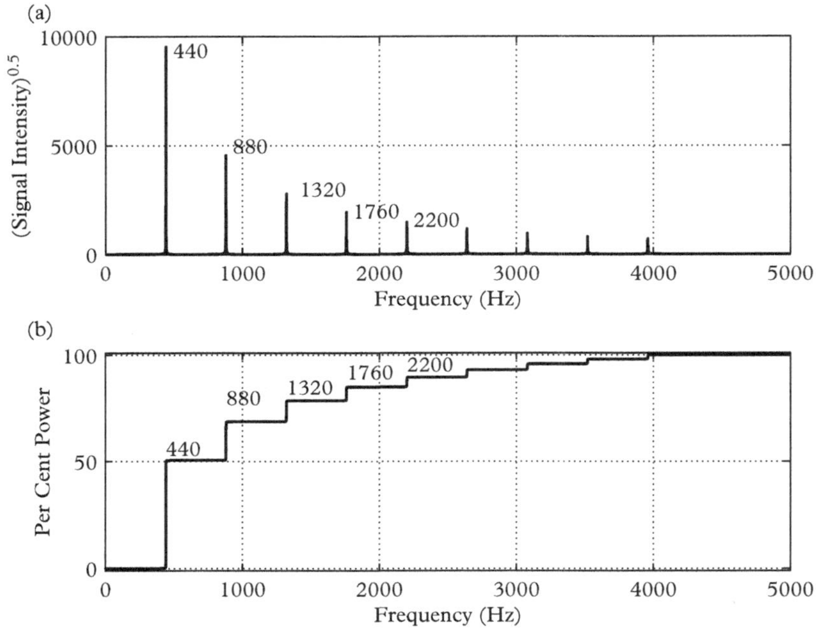

Figure 2.10.1 *Line spectrum (a) and cumulative line spectrum (b) of the built-up wave*

440 Hz there is no power accumulated. Then at 440 Hz there is a step of about 50%, indicating that there is a 50% contribution at that frequency. The graph is flat until 880 Hz, indicating that between 440 and 880 Hz there is no contribution. Then at 880 Hz there is a step of about 18%, indicating that the harmonic at 880 Hz contributes 18% to the total power of the signal. The graph shows more steps until 4000 Hz is reached, after which the curve is flat at 100% indicating that there is no more power at higher frequencies. This last comment should make sense to the reader because the built-up signal was designed to have only nine harmonics, the largest being $9 \times 440 = 3960$ Hz.

A 20-Sample Wave

To illustrate the details of the calculation, consider our simple case of 20 samples of a wave containing two sinusoidal components introduced in Section 2.4 and plotted in Figure 2.4.1. Table 2.10.1 shows the steps in the calculations.

The first and second columns contain the sample times t and the sample values $y(t)$. The third column contains the frequencies f from 0.0 Hz to the Nyquist frequency of 0.5 Hz. The fourth column contains the intensities $|A(f)|^2$ introduced in Eq. 2.9.1. Note that there are only two non-zero entries in the fourth column.

Lets briefly go through the steps of computing the cumulative line spectrum. Start at 0.0 Hz and clear an accumulator S. That is, set $S = 0$. (This quantity is shown in the fifth column.)There is no signal power at 0.0 Hz because we always assume that the average has been subtracted

Table 2.10.1 *Table of computations for the cumulative line spectrum of a sampled test wave.*

1	2	3	4	5	6
t	y	f	$\|A\|^2$	S	$\tilde{S}$
0	0	0	0	0	0
1	2.6878	0.05	0	0	0
2	0.9511	0.1	1	1	18.48429
3	−1.1489	0.15	0	1	18.48429
4	0.5878	0.2	0	1	18.48429
5	2.1	0.25	4.41	5.41	100
6	−0.5878	0.3	0	5.41	100
7	−3.0511	0.35	0	5.41	100
8	−0.9511	0.4	0	5.41	100
9	1.5122	0.45	0	5.41	100
10	0	0.5	0	5.41	100
11	−1.5122				
12	0.9511				
13	3.0511				
14	0.5878				
15	−2.1				
16	−0.5878				
17	1.1489				
18	−0.9511				
19	−2.6878				

from the wave before any calculations. At the first computed frequency on the frequency grid, $f_1 = 1/(N\,h) = 0.05$ Hz, add the intensity, or $|A_1|^2$ (from an FFT calculation), at that frequency to the accumulator:

$$\begin{aligned} S_1 &= S_0 + |A_1|^2 \\ &= |A_1|^2 = 0.0. \end{aligned}$$

In our example this intensity is zero because there is no signal strength at 0.05 Hz. In any case, S_1 is the accumulated signal power at f_1. At the next computed frequency on the frequency grid, $f_2 = 2/(N\,h) = 0.1$ Hz, add the intensity,1.0:

$$S_2 = S_1 + |A_2|$$
$$= |A_1| + |A_2| = 0.0 + 1.0 = 1.0.$$

Keep this going, up until the last computed frequency on the frequency grid, the Nyquist frequency, where

$$f_{\text{Ny}} = \frac{N}{2}\frac{1}{Nh} = \frac{1}{2h} = 0.5$$

and where the last bit of signal power is added to the sum:

$$S_{\frac{N}{2}} = S_{\frac{N}{2}-1} + \left|A_{\frac{N}{2}}\right|^2 = 5.41 + 0.0 = 5.41.$$

The accumulated powers at each frequency S_k reside in column 5. At this point, all of the available signal power has been accumulated and the sums can be scaled. We do this by dividing all of the sums in column 5, S_k, by $S_{\frac{N}{2}}$ and multiplying by 100. The result is $\tilde{S}$ and it resides in column 6. This is the accumulated power in percent.

The normalized accumulators, $\tilde{S}_0, \tilde{S}_1, \tilde{S}_2, \ldots, \tilde{S}_{\frac{N}{2}}$, are plotted against the frequencies $f_0, f_1, f_2, \ldots, f_{\frac{N}{2}}$. This plot is the cumulative line spectrum. Therefore, at each frequency, the cumulative line spectrum gives the percent power accumulated up to that frequency. By definition, the y-axis on the cumulative line spectrum plot will start at 0.0 and end up at 100.0.

For our 20-sample example the cumulative line spectrum is shown in Figure 2.10.2.

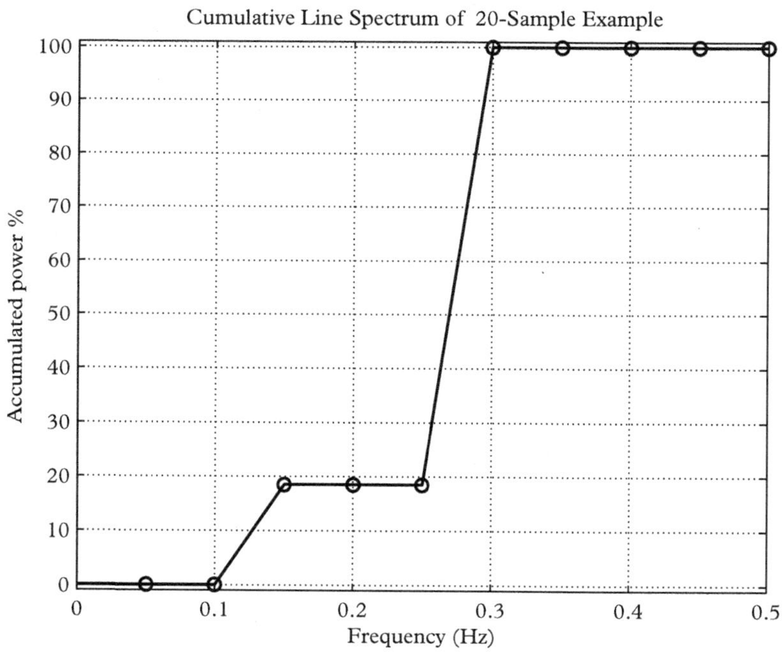

Figure 2.10.2 *Cumulative line spectrum of sampled test wave*

The reader should take a moment to study Figure 2.10.1 and Figure 2.10.2 to get a feel for how the line spectrum and the cumulative line spectrum relate. The cumulative line spectrum is helpful in deciding if there is any real intensity associated with what might appear as a thin isolated peak in a line spectrum.

The Mystery Wave

The cumulative line spectrum for the noisy mystery signal in Figure 2.8.1 is shown in Figure 2.10.3.

Figure 2.10.3(a) gives the cumulative line spectrum for the range of 0 to 1000 Hz. Four things should strike the reader. First, the power appears to grow linearly (and smoothly) with frequency as it accumulates the "hashy" background noise power. Second, there are jumps in the power at 400 Hz and 910 Hz where the two sine waves have their power. Third, as already mentioned, the cumulative line spectrum is much smoother than the line spectrum. Fourth, the graph in Figure 2.10.3(a) does not reach 100% because there is more power in the signal above 1000 Hz.

Figure 2.10.3(b) gives the cumulative line spectrum over the entire Nyquist interval. Here, one gets a better picture of how the power from the background noise accumulates—first quickly and then more slowly as the frequency increases. The jumps in power at 400 and 910 Hz are difficult to discern. This graph shows the important feature that the signal power comes from *all* of the frequencies in the Nyquist interval. This graph also shows that the background noise is not "white," which leads us to the next subsection.

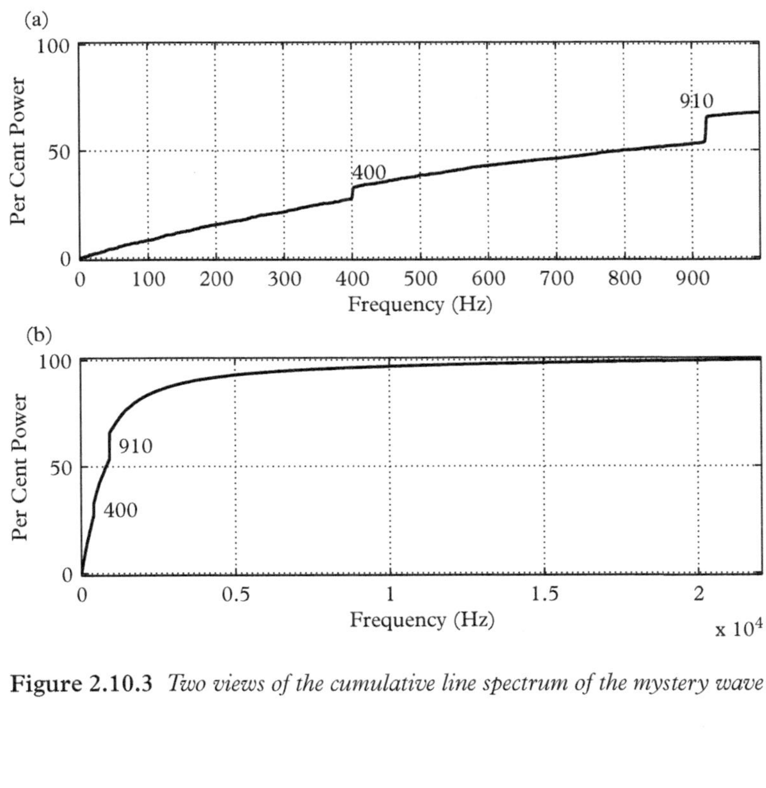

Figure 2.10.3 *Two views of the cumulative line spectrum of the mystery wave*

White Noise

Consider the two views (different time axis spans) shown in Figure 2.10.4 of a wave that has no apparent structure other than being busy and erratic.

The line spectrum of this wave is shown in Figure 2.10.5, where it is tempting to think that there may be a significant spike at 5126 Hz (or at several other frequencies). However, the spike does look narrow and there are other spikes that are also narrow, just not quite as high. The cumulative line spectrum, shown in Figure 2.10.6, addresses these questions nicely.

In Figure 2.10.6(a) the cumulative line spectrum of white noise consists of an apparently *smooth* ramp with unity slope. This supports the contention that white noise contains the same intensity, *on average*, at all the frequencies in the Nyquist interval. Figure 2.10.6(b) is a close-up of the cumulative line spectrum of white noise over a narrow range of frequencies, showing that the ramp in Figure 2.10.6(a) is not perfectly smooth. However, on a larger scale, the summing that takes place in the cumulative line spectrum does indeed smooth out the irregularities of the line spectrum and seamlessly accumulates power *as though* the line spectrum were perfectly flat.

As a thought exercise, think about a hypothetical white noise wave that has a perfectly flat line spectrum, i.e. its power is distributed perfectly equally across the Nyquist interval. Now compute the cumulative line spectrum for this wave. How would it look? It is important that the reader understands that the shape of the cumulative line spectrum would be a ramp. This is because as one moves along the frequency axis from left to right one accumulates more and more signal power linearly. Check out the answer in Figure 2.10.7. (To the engineer-reader, the ramp is the integral of the constant flat line.)

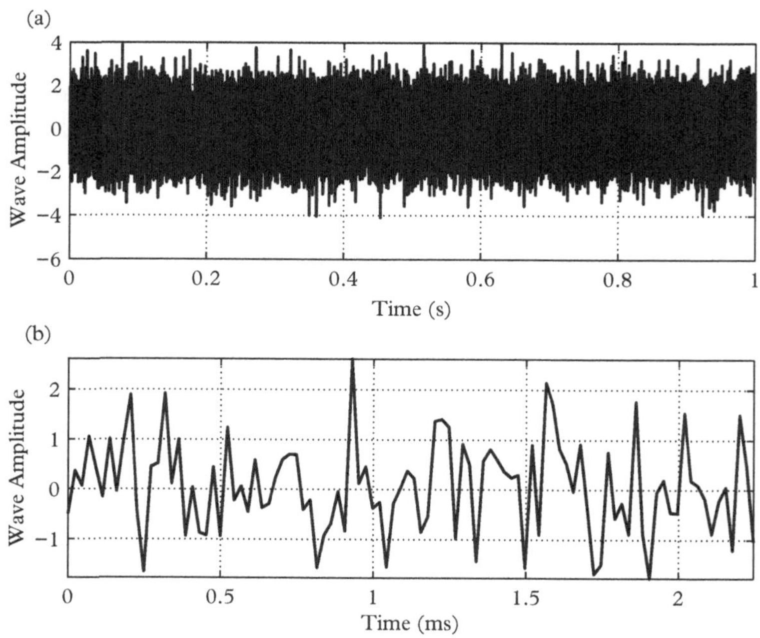

Figure 2.10.4 *Two views of the time domain plot of white noise*

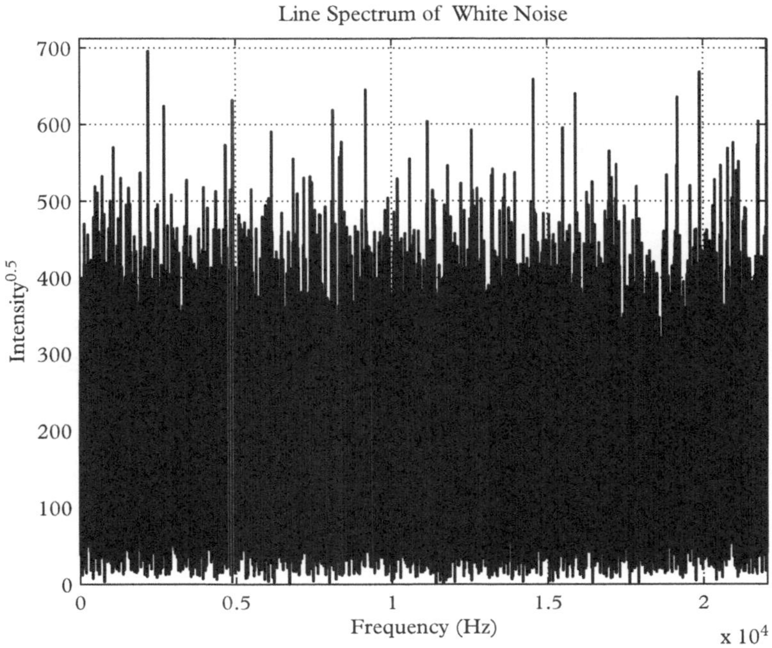

Figure 2.10.5 *Line spectrum of white noise*

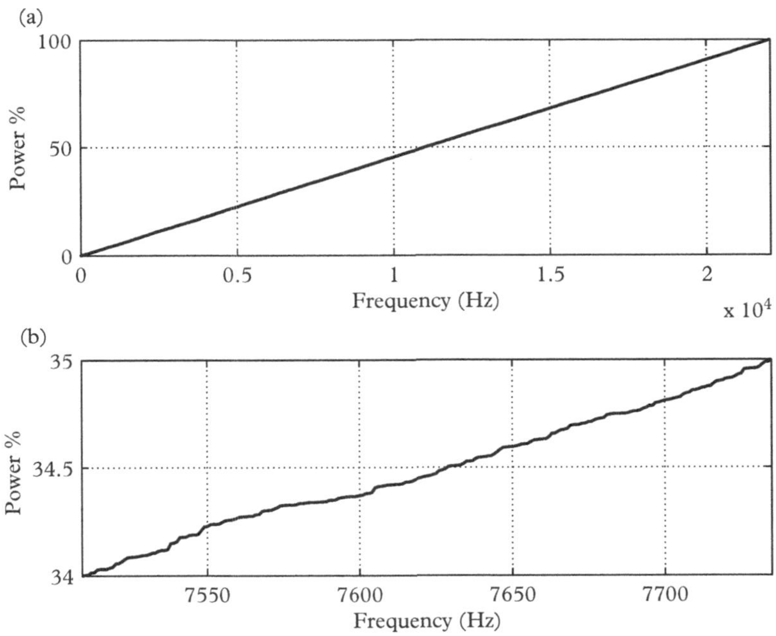

Figure 2.10.6 *Cumulative line spectrum of white noise: (a) over the Nyquist interval, and (b) over a narrow frequency interval*

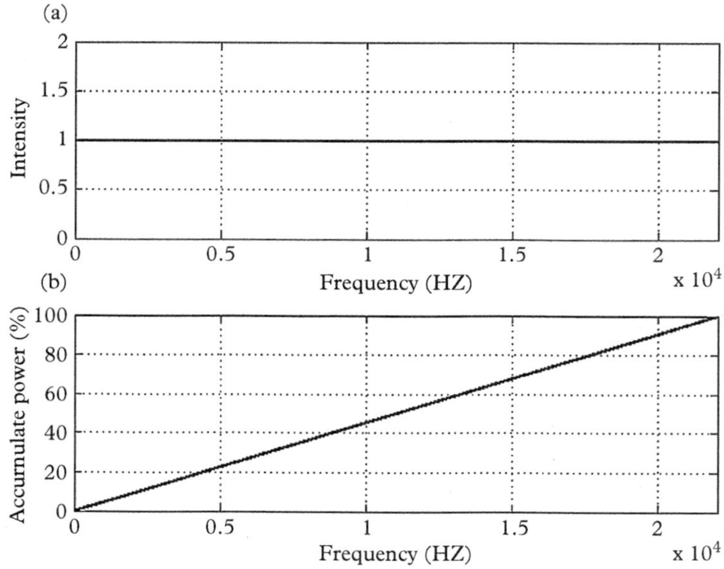

Figure 2.10.7 *(a) Line spectrum of hypothetical white noise, and (b) cumulative line spectrum of hypothetical white noise*

Since white noise is by definition an ideal random signal containing power distributed evenly (on average) over all the frequencies in the Nyquist interval, we can look at the mystery signal again. Figure 2.10.3 shows that the signal intensity from the background noise decreases with increasing frequency, and therefore the noise is not white.

Figure 2.10.6 can be useful in dispelling any concern about the small blips in the line spectrum in Figure 2.10.5 around 300 Hz and 810 Hz. Upon first glancing at Figure 2.10.5, one might think that those small spikes are representative of two more sinusoidal components lurking in the signal. Figure 2.10.6 shows that this is not the case. Therefore, the cumulative line spectrum can be a handy tool when the signals are noisy. As we will see later on, it also provides an easy way to decide where the signal energy is distributed among the partial frequencies.

The C8 Piano Note

Figure 2.10.8 shows the cumulative line spectrum of a piano C8 note that we mentioned earlier in Figure 2.9.3.

Note that only about 25% of the power comes from the fundamental at 4186 Hz, the rest coming from low frequency noise (which we earlier referred to as "hammer knock").

These sections on the line spectrum and the cumulative line spectrum have been mathematically and conceptually challenging. You have been introduced to some new tools that will need getting used to. The next section on the spectral centroid should be easier to read because it does not introduce any new concepts. Instead it builds on the components that make up the line spectrum and the cumulative line spectrum.

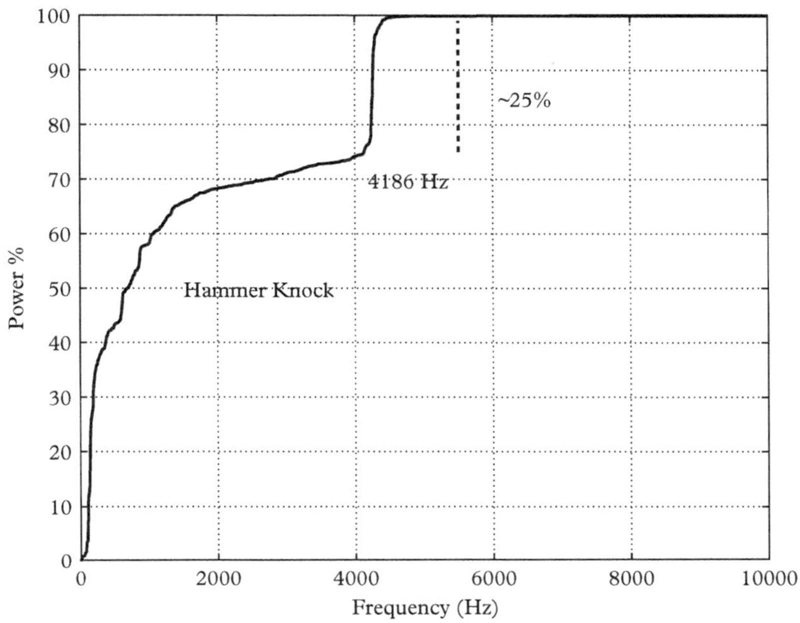

Figure 2.10.8 *Cumulative line spectrum of C8*

2.11 The Spectral Centroid

Throughout the book we will deal with spectra developed from observations and from simulations. You have been introduced to and hopefully are now somewhat familiar with the line spectrum and the cumulative line spectrum. Perhaps you are even impressed with the insight these tools can provide. To further enhance our arsenal I add another analytical weapon, the spectral centroid, which can compactly characterize a line spectrum. It will be used extensively in Part 3 when we talk about comparing pianos.

Once you have computed a line spectrum, the new tool, the spectral centroid, can be calculated from the frequencies and the signal strengths at those frequencies. The spectral centroid is nothing more than a weighted average frequency, where the weights are the signal strengths.

I could leave the definition in the previous paragraph alone and move on to the next section, but unlike with the line spectrum I am going to go into some of the computational details. If they bother you it is acceptable to pass on to the next section on autocorrelation. However, just remember that the spectral centroid is a weighted average of the frequencies that are used in the presentation of the line spectrum and that the weights are the signal strengths at those frequencies.

Our first guinea pig will be 20 samples of a wave consisting of two sinusoidal components. This wave was first studied in Section 2.4. Table 2.4.1 from that section is reprinted below as Table Table 2.11.1 with the addition of five more columns.

Column 6 contains the product of the signal intensity Y^2 (coming from the FFT calculation for the frequency f in column 3), i.e., $f\,Y^2$. The spectral centroid SpC is the sum of column 6, which is 1.203, divided by the sum of column 5, which is 5.41. The resulting ratio is 0.222 Hz, which is

Table 2.11.1 *Illustration of the spectral centroid calculation for the 20-sample wave.*

1	2	3	4	5	6
t	*y*	*f*	*Y*	*Y*^2	*f***Y*^2
0	0	0	0	0	0
1	2.688	0.05	0	0	0
2	0.951	0.1	1	1	0.1
3	−1.149	0.15	0	0	0
4	0.588	0.2	0	0	0
5	2.1	0.25	2.1	4.41	1.1025
6	−0.588	0.3	0	0	0
7	−3.051	0.35	0	0	0
8	−0.951	0.4	0	0	0
9	1.512	0.45	0	0	0
10	0	0.5	0	0	0
11	−1.512			5.41	1.203
12	0.951		**spec**	**cen =**	0.222
13	3.051				
14	0.588				
15	−2.1				
16	−0.588				
17	1.149				
18	−0.951				
19	−2.688				

the average frequency weighted by the strengths at each frequency. The two components that make up the wave have frequencies 0.1 Hz and 0.25 Hz with amplitudes of 1.0 and 2.1 respectively. The weighted average frequency of 0.222 Hz is between the two frequencies in the wave but closer to 0.25 Hz, because that component has a larger amplitude.

For the 20-sample example wave, Figure 2.11.1 shows the line spectrum along with the spectral centroid represented by the dashed vertical line.

Our second example for introducing the spectral centroid will be the built-up wave first presented in Section 1.2. Its line spectrum is re-plotted here as Figure 2.11.2, where the sampling frequency is 11,025 Hz and the number of samples is 44,100. The Nyquist frequency is therefore 5512.5 Hz. The spectral centroid, computed in the same way as for the 20-sample example above,

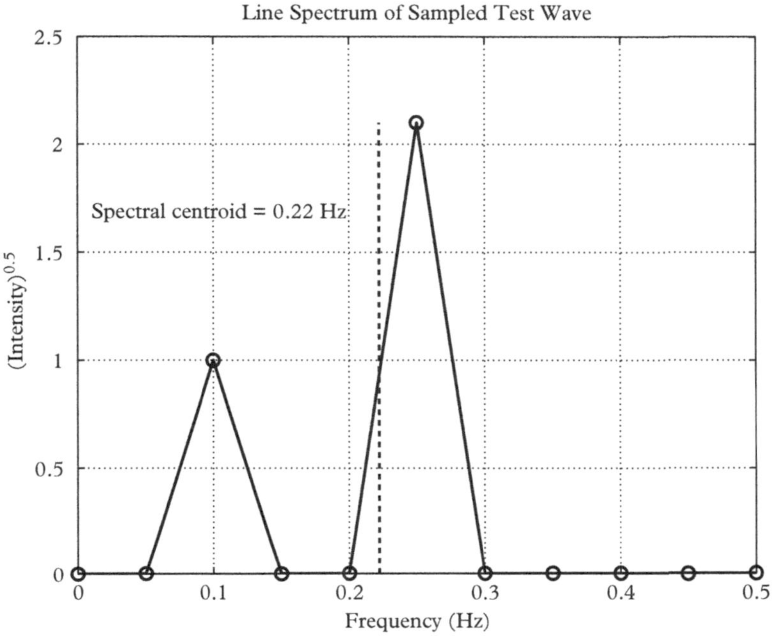

Figure 2.11.1 *Line spectrum of 20-sample wave showing the spectral centroid*

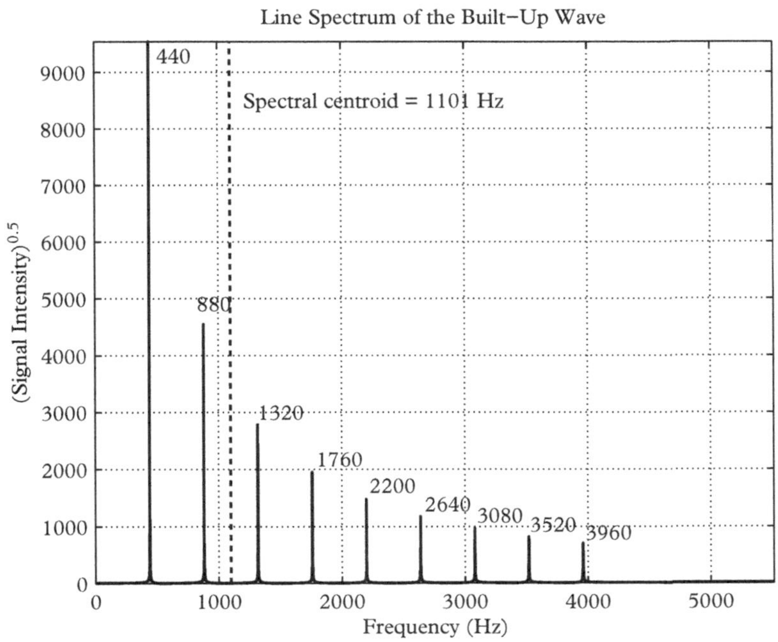

Figure 2.11.2 *Line spectrum of built-up wave showing the spectral centroid*

is 1100.6 Hz, represented by the vertical dashed line. Given the relative strengths of the nine harmonics in the wave, the location of the spectral centroid near the left of the spectrum should make sense.

The spectral centroid is a compact measure of how the signal strength is distributed amongst the harmonics. The line spectrum and the cumulative line spectrum also show how the signal power is distributed amongst the harmonics and they do a better job, but graphs are required. Sometimes it will be convenient to condense the information into a single number.

So, we have completed our discussion of the "unholy three": the line spectrum, the cumulative line spectrum, and the spectral centroid. Hopefully, you will become familiar with them as we proceed through the book. BUT ... we still have one more tool to deal with the autocorrelation.

2.12 The Autocorrelation

In the book *Why You Hear What You Hear*, the author Eric Heller claims that the autocorrelation is an indicator of a sound wave's pitch. To make this so, the analyst must choose the highest isolated autocorrelation peak that is associated with the smallest lag time and compute the frequency associated with the time lag. As we will see in the following example, this is sometimes difficult to do.

The 20-Sample Wave

The autocorrelation is probably more difficult to define and explain than it is to calculate, so we will do the latter first. Consider the 20-sample example wave that we have used in Sections 2.4 and 2.11.Table 2.12.1 shows the time of the sample, t, and the sampled values, $y(t)$, in the first two columns.

The third column shows the product of the samples, $y(t)^2$, with the sum of these products, 54.1008, at the bottom of the column. The fourth column shows the samples shifted in time by one sample, $y(t+1)$.

Look at columns 2 and 4 and note how the values line up. The fifth column shows the product of each sample with the corresponding sample shifted one sample in time, i.e. the product of the second and fourth columns, $y(t)\,y(t+1)$. The sum of these products, 8.0908, is shown at the bottom of the fifth column. Note that the second column, headed $y(t)$, has been extended by two zero samples (at the bottom of the column) to facilitate the multiplication between $y(t)$ and the shifted $y(t)$. The sixth column contains the samples shifted by one more sample, $y(t+2)$. The seventh column contains the products of each sample with the corresponding sample shifted two samples in time, $y(t)\,y(t+2)$, i.e., columns 2 and 6. The sum of these products, –33.78, is shownat the bottom of column 7.

There is an autocorrelation value for each shift, including the zero shift, and it is calculated by dividing the sum of the products by the sum of the third column, 54.1. For these three shifts of 0, 1, and 2, the autocorrelations are 1.0, 0.1496, and –0.62449, and they are shown in the last row of the table.

Note that the autocorrelation calculated for each shift is divided by the sum of the third column. This guarantees that the autocorrelation value for the samples with zero shift is unity. The autocorrelation for zero shift will be the largest value for all the shifts because the unshifted components always correlate exactly with themselves.

Hopefully, the reader could continue the calculation process for each successive shift. The autocorrelations for the case of carrying out ten shifts are shown in Figure 2.12.1. Because of

Table 2.12.1 *Illustration of the autocorrelation calculation for the 20-sample wave.*

1	2	3	4	5	6	7
t	*y(t)*	*y(t)y(t)*	*y(t+1)*	*y(t)y(t+1)*	*y(t+2)*	*y(t)y(t+2)*
0	0	0	0	0	0	0
1	2.6878	7.224269	0	0	0	0
2	0.9511	0.904591	2.6878	2.556367	0	0
3	−1.1489	1.319971	0.9511	−1.09272	2.6878	−3.08801
4	0.5878	0.345509	−1.1489	−0.67532	0.9511	0.559057
5	2.1	4.41	0.5878	1.23438	−1.1489	−2.41269
6	−0.5878	0.345509	2.1	−1.23438	0.5878	−0.34551
7	−3.0511	9.309211	−0.5878	1.793437	2.1	−6.40731
8	−0.9511	0.904591	−3.0511	2.901901	−0.5878	0.559057
9	1.5122	2.286749	−0.9511	−1.43825	−3.0511	−4.61387
10	0	0	1.5122	0	−0.9511	0
11	−1.5122	2.286749	0	0	1.5122	−2.28675
12	0.9511	0.904591	−1.5122	−1.43825	0	0
13	3.0511	9.309211	0.9511	2.901901	−1.5122	−4.61387
14	0.5878	0.345509	3.0511	1.793437	0.9511	0.559057
15	−2.1	4.41	0.5878	−1.23438	3.0511	−6.40731
16	−0.5878	0.345509	−2.1	1.23438	0.5878	−0.34551
17	1.1489	1.319971	−0.5878	−0.67532	−2.1	−2.41269
18	−0.9511	0.904591	1.1489	−1.09272	−0.5878	0.559057
19	−2.6878	7.224269	−0.9511	2.556367	1.1489	−3.08801
	0	0	−2.6878	0	−0.9511	0
	0	0	0	0	−2.6878	0
		54.1008		8.090817		−33.7853
		1		0.149551		−0.62449

the large sampling interval, the peaks in the autocorrelation would not be expected to line up with the periods of the sinusoidal components in the wave. However, due to our purposely fortuitous choice of the frequencies of the components in the wave, there is, in fact, a correspondence. The first peak resides at a time lag of 4.0s, which would correspond to a frequency of 0.25 Hz, while the second peak resides at a time lag of 8.0s, which would correspond to a frequency

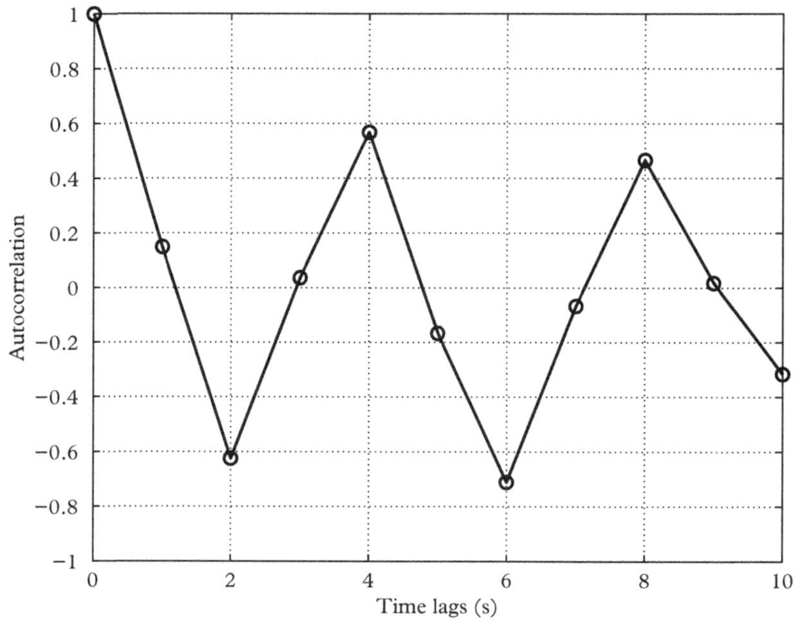

Figure 2.12.1 *Autocorrelation of 20-sample wave*

of 0.125 Hz. The actual frequencies of the components in the wave are 0.1 Hz and 0.25 Hz. Therefore there is a strong positive correlation of $y(t)$ and $y(t + 4)$ because the period of one of the sinusoidal components is

$$1/\,(0.25\,\text{Hz}) = 4\,\text{s}.$$

Also, $y(t)$ correlates strongly with $y(t + 8)$ because the other sinusoidal component in the wave has a frequency of 1/8s = 0.125 Hz. Note that the autocorrelation at the lag time of 2s is negative because $y(t)$ and $y(t + 2)$ tend to have different signs. This can be seen by looking at the second and sixth columns.

If we repeat the numerical experiment with the same wave when sampled at 44,100 Hz (originally plotted in Figure 2.4.1), the autocorrelation appears as shown in Figure 2.12.2.

The autocorrelation identifies a peak at a lag time of 3891 ms, which corresponds to a frequency of 0.257 Hz. It also identifies a peak at a lag time of 8s, which corresponds to a frequency of 0.125 Hz.

There are at least three comments to make. First, from a musical point of view there is no audible tone corresponding to these low frequencies. Second, both peaks have about the same height so there is no single, isolated highest peak to pick. Third, the first peak at a lag time of 3.891s points to a frequency that is not quite equal to that of one of the components in the wave. These results suggest that the autocorrelation may not always be an effective indicator of pitch, especially in this case because the frequencies are so low and because there is more than one dominant frequency.

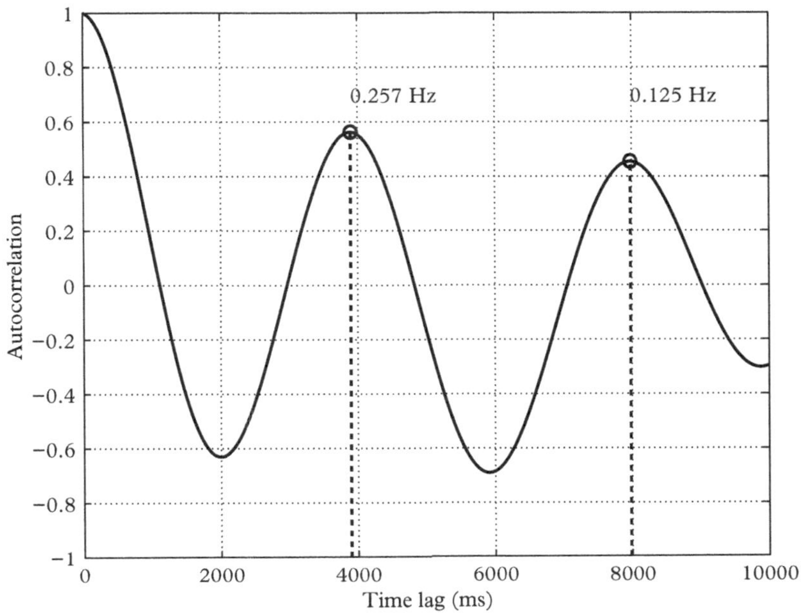

Figure 2.12.2 *Autocorrelation of example wave sampled at 44,100 Hz*

White Noise

Figure 2.12.3 shows the autocorrelation for white noise, which was introduced in Section 2.10. Each autocorrelation value is plotted as a small circle.

By definition, the samples of a white noise wave are completely independent of each other; therefore, there will be no lag time other than zero where the shifted white noise wave correlates with the non-shifted wave. Figure 2.12.3 shows that there is a lone autocorrelation point (or circle) at zero lag having a value of 1.0, while all of the other autocorrelation points are effectively zero. A close-up look at a small portion of Figure 2.12.3 is shown in Figure 2.12.4, where one sees that the autocorrelation values are not truly zero.

The Mystery Wave

An autocorrelation of the mystery wave introduced in Section 2.8 is given in Figure 2.12.5. None of the peaks are strong (except for the initial autocorrelation at a lag time of zero) and the locations of the peaks correspond to frequencies of 918.3 Hz and 450 Hz. In fact, the sinusoidal components have frequencies of 920 Hz and 400 Hz. Note that this is a simulated wave and the locations of the peaks will vary as the white noise, which is a random signal, varies. The weakness of the peaks is a consequence of the strength of the background noise.

A Fine Grid Estimator of the Autocorrelation

The problem of errors in estimating time lag associated with peaks in the autocorrelation can sometimes be addressed with a fine grid approach similar to that mentioned in Section 2.7. However,

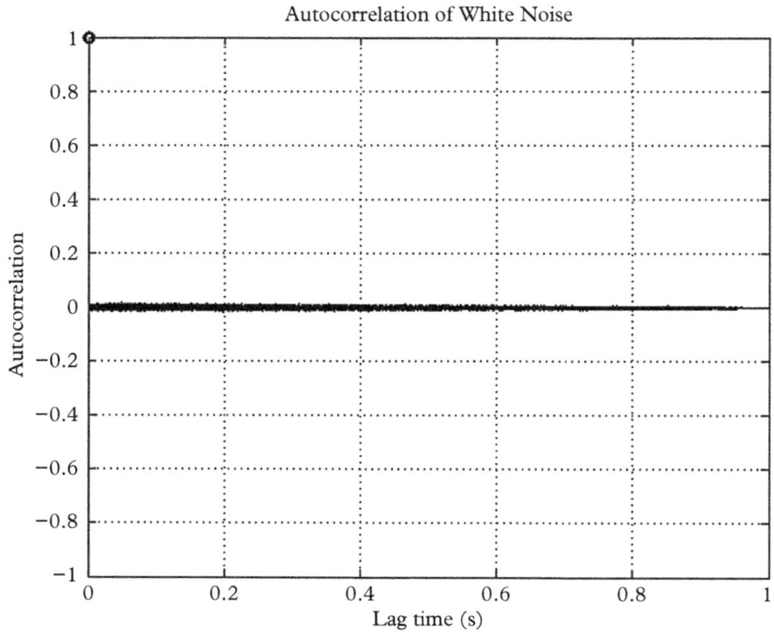

Figure 2.12.3 *Autocorrelation of white noise (big picture)*

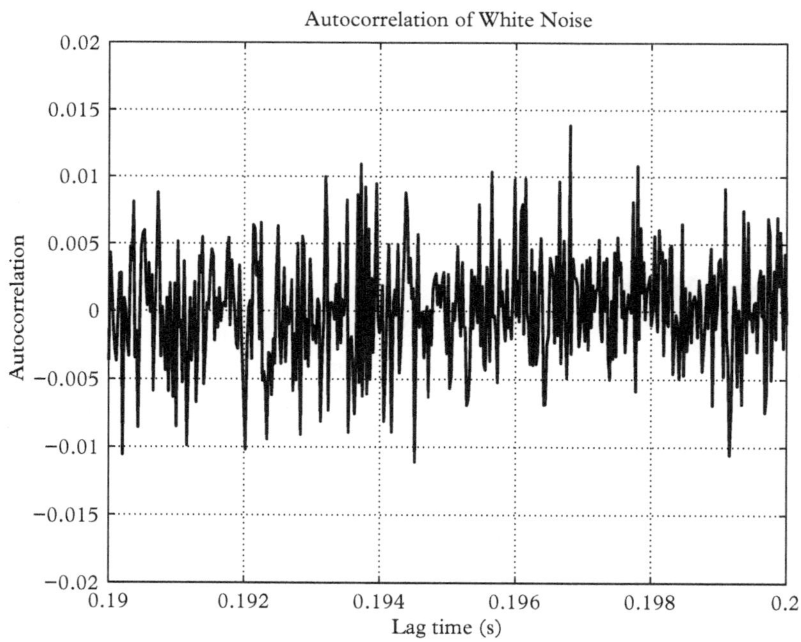

Figure 2.12.4 *Autocorrelation of white noise (small picture)*

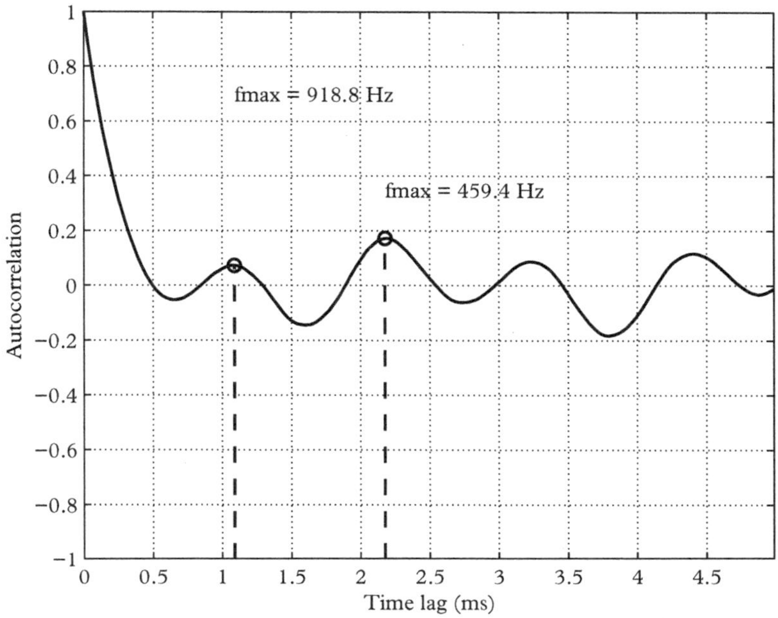

Figure 2.12.5 *Autocorrelation of mystery wave*

to do so requires an alternative definition of the autocorrelation based on the Wiener—Khinchin theorem that allows the autocorrelation to be computed via the inverse cosine Fourier transform. We will leave the details of the calculation toAppendix 1.

An Example Comparing Autocorrelation and Spectral Centroid

Consider two different waves with the line spectra shown in Figure 2.12.6(a) and (b). The first wave has signal power at three frequencies: 200 Hz, 400 Hz, and 600 Hz. The second wave has signal power at five frequencies: 200 Hz, 300 Hz, 400 Hz, 500 Hz, and 600 Hz. The spectral centroid (or average weighted frequency) is 400 Hz in both cases. However, the autocorrelations in Figure 2.12.6(c) show that the first wave has the strongest first peak at a lag time that corresponds to a frequency of 200 Hz while the second wave has a first peak at a lag time corresponding to a frequency of 100 Hz. The sounds of these waves clearly support these two estimates of pitch. The pitch of the second wave at 100 Hz does not come from signal power at 100 Hz, as indicated by the line spectrum. In fact, the pitches of both waves appear to be a result of the separation in the frequencies.

This example suggests that the average frequency sometimes may not give much insight. It also demonstrates that the line spectrum may not always be a good indicator of pitch. The sounds of these two waves are available in the file TwoWaves.mp3.

Now we are ready to read the rest of the book with a reasonable level of comfort.

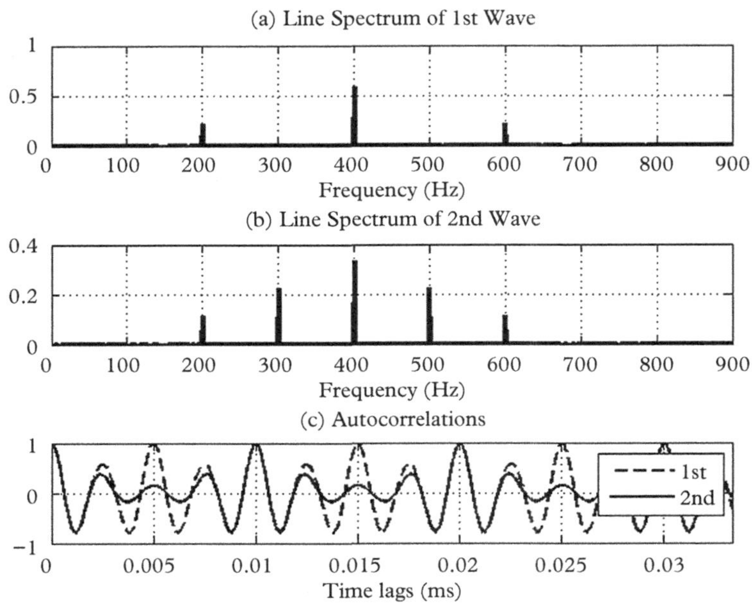

Figure 2.12.6 *(a) Line spectrum of first wave, (b) line spectrum of second wave, and (c) autocorrelations of first and second waves*

2.13 Exercise

Construct a Matlab script with the following code:

```
clear
close all
 Fs=44100;
 h=1/Fs;
 N=44100;
 t=(0:N-1)*h;
 D=0:.1:1;
 y = pulstran(t,D,'tripuls',0.1,-1);
soundsc(y,Fs)
 subplot(211)
 plot(t,y),grid
 title('Time Domain Plot of Wave')
 xlabel('time (s)')

[Y,f]=SimpleLineSpectrum2(y,h,0,1000);
subplot(212)
plot(f,Y),grid
title('Line Spectrum of Wave')
xlabel('frequency (Hz)')
```

Run this script, interpret the two graphs that result, and listen to the sound. What is a common name for the wave? Does the line spectrum make sense? Are the peaks where you would expect them to be? Does the sound make sense? By lifting code from the file FiguresPart1_Chap2.m, can you construct the cumulative line spectrum for this wave? How about the autocorrelation? If you cannot, can you estimate and sketch the shape of the two curves from what you know about the wave from the time domain plot and the line spectrum? Look for the script in the file named SawTooth.m to see the answers.

3

Analysis of Several Common Musical Sounds

This chapter looks at the time domain and spectral characteristics of 16 different kinds of sound *sources*. Although the chapter is relatively lengthy there are no new challenging technical concepts and both the musician- and engineer-reader might want to scan it before diving in. The sonic power of most of these sounds is distributed amongst a fundamental frequency and its harmonics. Only the bell chime, the clanging railroad connectors, the fire bell, and the train whistle do not conform to this structure.

3.1 The Tuning Fork

The tuning fork is designed to generate most of its sonic power at one frequency, in this case A4 or 440 Hz (the first A above middle C on the piano). The time domain plot of the wave is shown in Figure 3.1.1 (note the two different time scales).

Figure 3.1.1(b) shows a sinusoidal-like wave that has a period of about 2.3 ms, which is consistent with 440 Hz.

The line spectrum in Figure 3.1.2(a) shows a dominant spike in the neighborhood of 440 Hz with a nearly imperceptible contribution at around 120 Hz, perhaps due to the handling of the fork, to some resonance in the laptop surface, or to a characteristic of the microphone (Zoom H2). There is a slender spike at 880 Hz that is almost 40 dB down, and above 1000 Hz there is hash that is more than 60 dB down and cannot be heard. The source of the small peak at 2800 Hz is unknown. The cumulative line spectrum in Figure 3.1.2(b) supports the contention that effectively all of the sonic power comes from the fundamental frequency. The autocorrelation in Figure 3.1.2(c) presents the first peak at a time lag that is associated with the fundamental frequency.

A refinement of the line spectrum (using a fine grid DFT) in a small domain around 440 Hz is shown in Figure 3.1.3, which reveals a couple of things. First, the peak is near 440.8 Hz instead of 440 Hz (the dotted vertical line). The source of this deviation is not clear. Second, the shape of the peak is smooth with some small ripples (or lobes) instead of an isolated spike, as in Figure 3.1.2. This shape is an unavoidable consequence of the method used to calculate the spectrum and will show up repeatedly when we take close looks at the spectra of other sound waves later in the book (see Appendix 1 for a discussion of the source of the lobes). For example, the narrowness of the peak in Figure 3.1.3 depends on the length of the sample in time. For there to be a single line would require an infinite sample length and the source would have to be a pure sine wave.

Spectral Analysis of Musical Sounds with Emphasis on the Piano. First Edition.
David M. Koenig, with comments by Delwin D. Fandrich.
Published in 2015 by Oxford University Press.

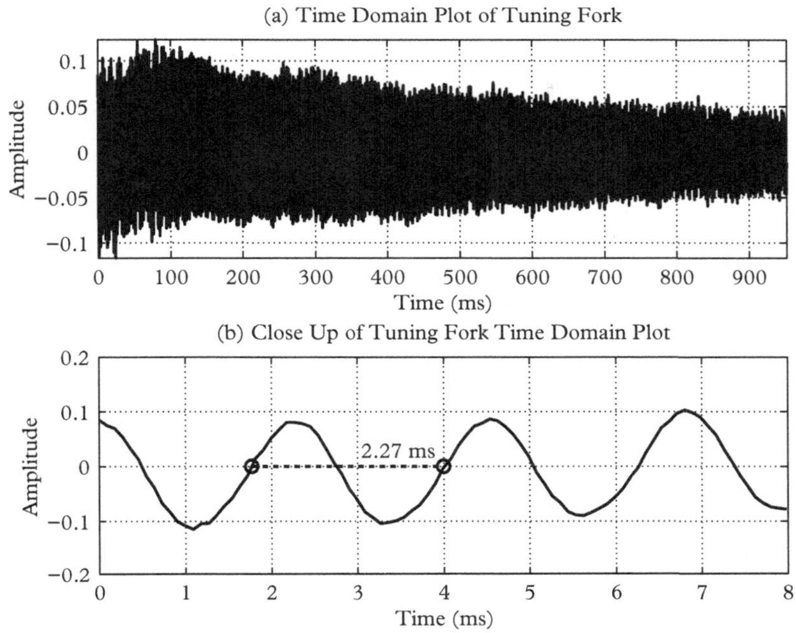

Figure 3.1.1 *Time domain plot of tuning fork: (a) big picture, and (b) small picture*

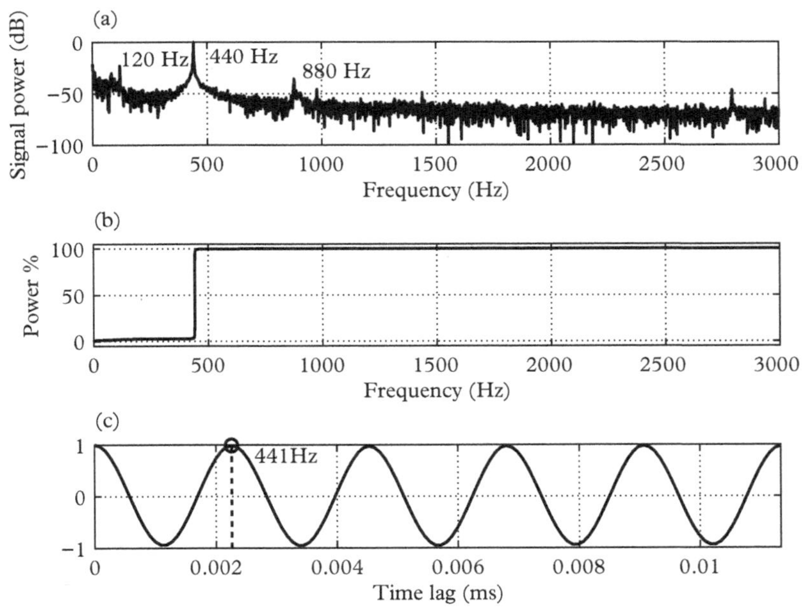

Figure 3.1.2 *Tuning Fork: (a) line spectrum, (b) cumulative line spectrum, and (c) autocorrelation*

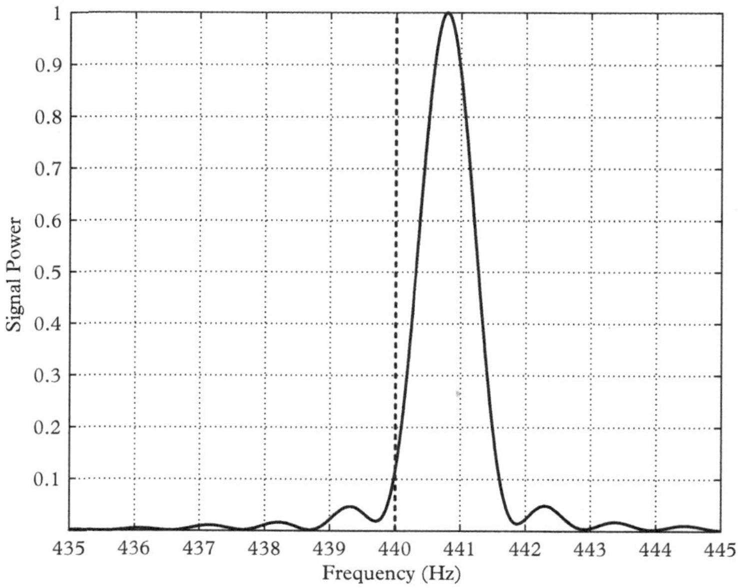

Figure 3.1.3 *Fine grid line spectrum of tuning fork centered at 440 Hz*

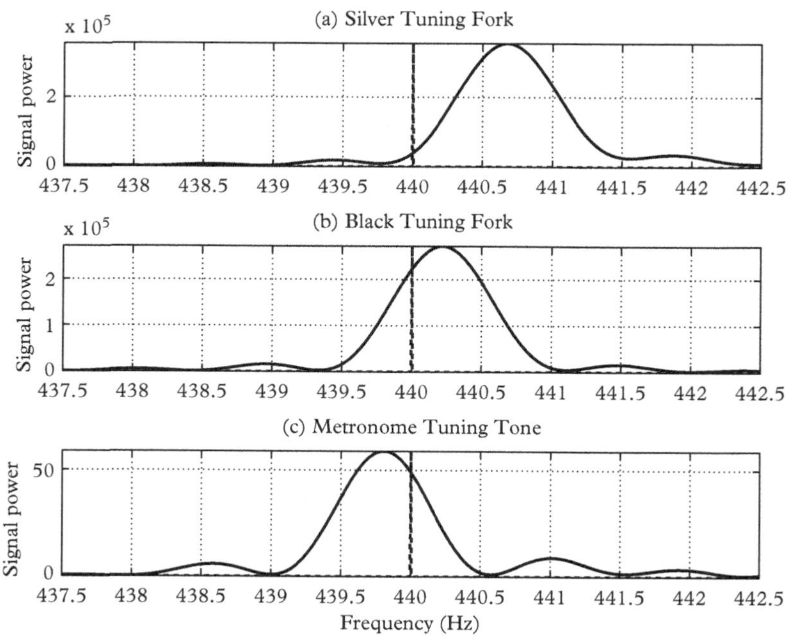

Figure 3.1.4 *Fine grid line spectra of (a) silver tuning fork, (b) black tuning fork, and (c) metronome tuning tone*

This tuning fork will be called the "silver tuning fork" because of its appearance. I also carried out similar tests on a second tuning fork, the "black tuning fork," and on the reference tone generated by a digital metronome. Figure 3.1.4 shows their fine grid spectra.Each is slightly displaced from the target of 440 Hz. Part of this discrepancy could be associated with the equipment used to collect the sounds, but since it was the same for all three tests I am tempted to conclude that the differences are real.

3.2 The Pitch Pipe, a Hum, and a "Laah"

A Lauren pitch pipe provides 13 tone choices from a C4 to a C5. After blowing into the pipe to generate an A4, I attempted to hum an A4. Figure 3.2.1(a) shows the time domain traces. A close-up look at the pitch pipe wave is shown in Figure 3.2.1(b). Note that in this graph one can discern the suggestion of a periodic component in the wave with a period of approximately 0.75 ms and with a frequency of approximately 1340 Hz. (Trying to deduce periodic components by looking at the difference in time between peaks is subject to error.)

As expected, there is a spike in the line spectrum in Figure 3.2.2(a) near 440 Hz. However, the third harmonic at 1320 Hz is dominant, as is verified by the cumulative line spectrum in Figure 3.2.2(b). It looks like almost 90% of the power comes from the third harmonic. Note the relatively "clean" distribution of power over all of the harmonics (only a little hash is in between or clustered about the harmonics and it is down by over 50 dB). The largest peak in the autocorrelation, Figure 3.2.1(c), appears at a lag time of approximately 2.25 ms, which

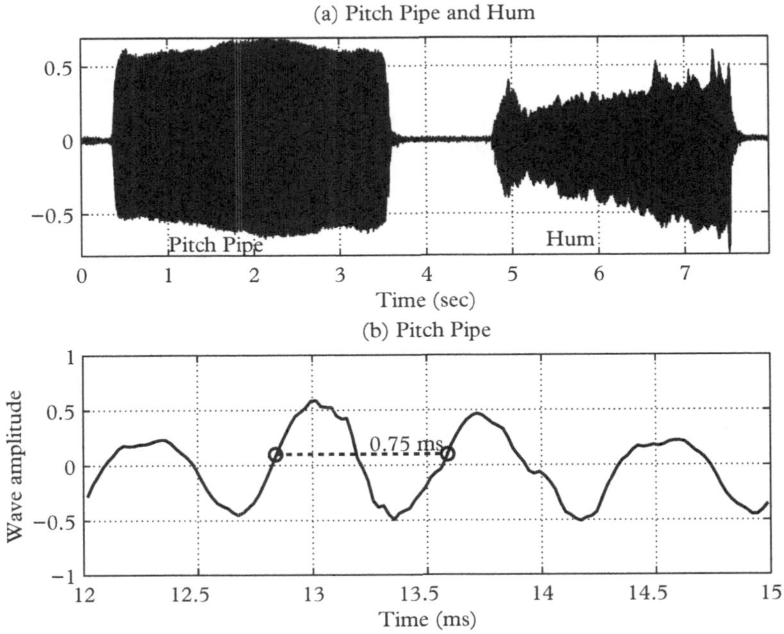

Figure 3.2.1 *Time domain plots of (a) the pitch pipe and a hum, and (b) a close-up of the pitch pipe*

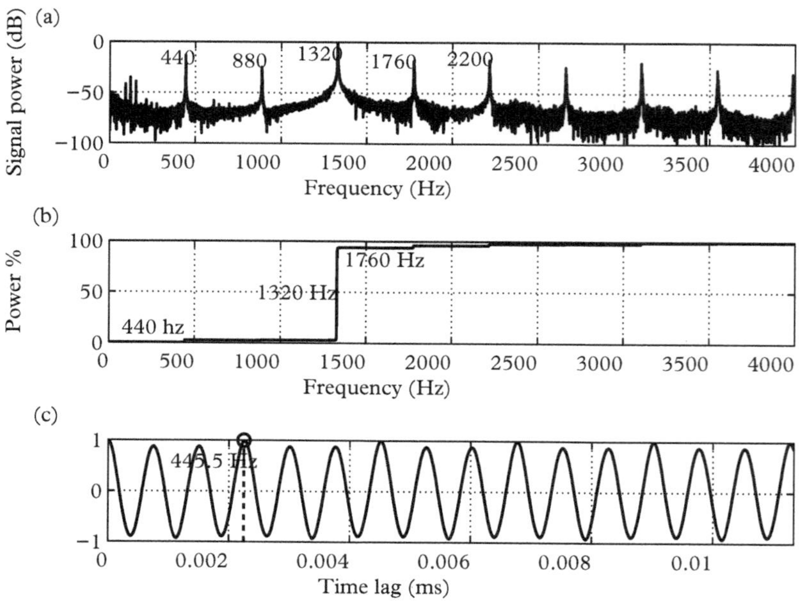

Figure 3.2.2 *(a) Line spectrum, (b) cumulative line spectrum, and (c) autocorrelation of pitch pipe*

corresponds to 445 Hz, but there are similar peaks at 0.75 ms and 1.5 ms, corresponding to frequencies of 1333 Hz and 667 Hz, which are fairly close to the harmonics. The autocorrelation in Figure 3.2.2(c) suggests no clear winner in a choice for pitch.

Using a fine grid DFT, Figure 3.2.3 gives a closer look at the area around 440 Hz for the pitch pipe, where the pipe appears to be more than 3 Hz sharp.

The line spectrum, cumulative line spectrum, and autocorrelation of my hum (the second wave in Figure 3.2.1) is shown in Figure 3.2.4. Compared to the pipe's spectrum, my hum shows more power at the second harmonic, more complex peaks, and a little hash at higher frequencies.

The peak near 440 Hz in Figure 3.2.4(a) for my hum suggests that I was perhaps able to duplicate the pipe's tone. However, there is something else going on. The hum's spectrum shows spikes at approximately 220 Hz, 440 Hz, 660 Hz, 880 Hz, and so on. Therefore, although I attempted to mimic 440 Hz, my hum's fundamental actually turned out to be near 220 Hz as evidenced by the spacing between the harmonics. That is, if I really was able to mimic 440 Hz then the harmonic spikes would have started at 440 Hz and would be separated by 440 Hz, but in fact they start at 220 Hz and are separated by 220 Hz. The cumulative line spectrum in Figure 3.2.4(b) shows that almost 60% of the power comes from 440 Hz. The autocorrelation in Figure 3.2.4(c) supports the contention that the pitch of my hum is closer to 220 Hz than to 440 Hz.

The audio for these two waves is available in the file PitchPipeAndHum.mp3. It appears that being able to aurally detect a few Hz deviation in frequency from target is not particularly easy; in fact, my hum was more than an octave off.

The line spectrum of the pitch pipe in Figure 3.2.5(a), followed by that of the hum in Figure 3.2.5(b), show how the harmonics shift and the frequency spacing halves as I stop blowing on the pipe and start humming in an attempt to mimic the pitch pipe.

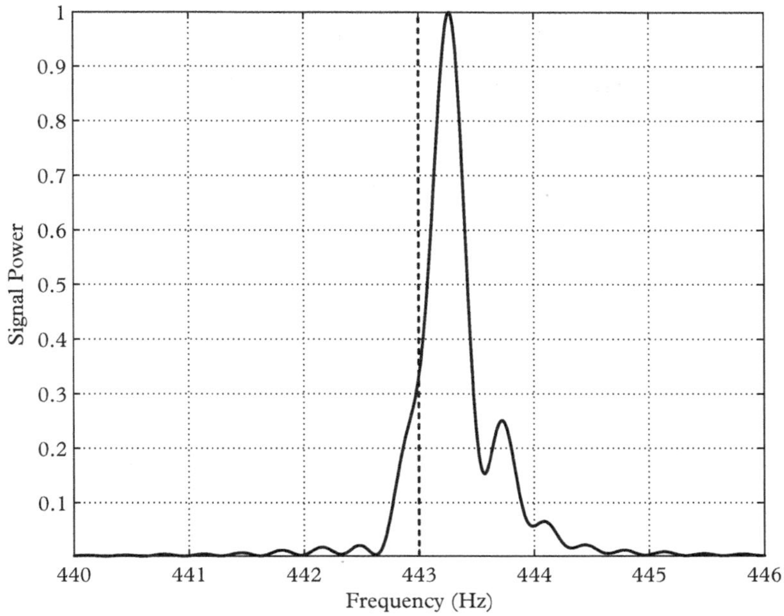

Figure 3.2.3 *Fine grid line spectrum of pitch pipe centered at 443 Hz*

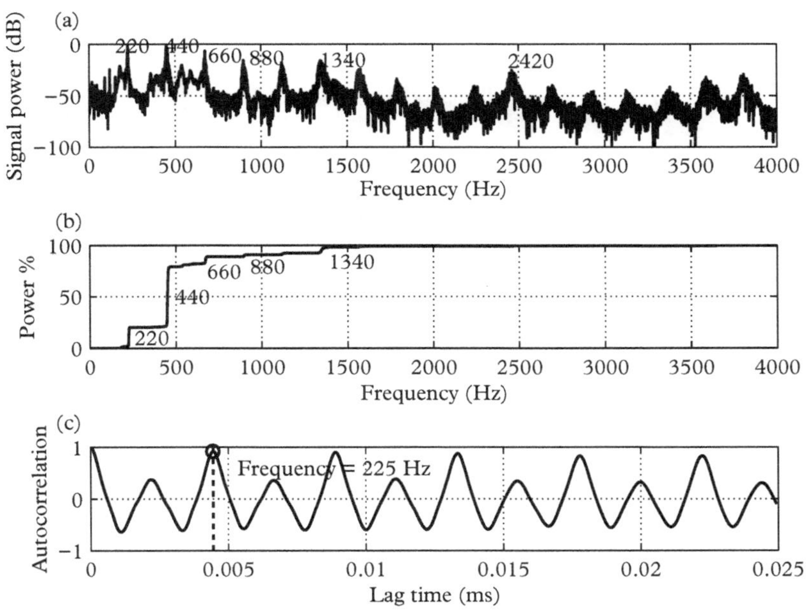

Figure 3.2.4 *Hum: (a) line spectrum, (b) cumulative line spectrum, and (c) autocorrelation*

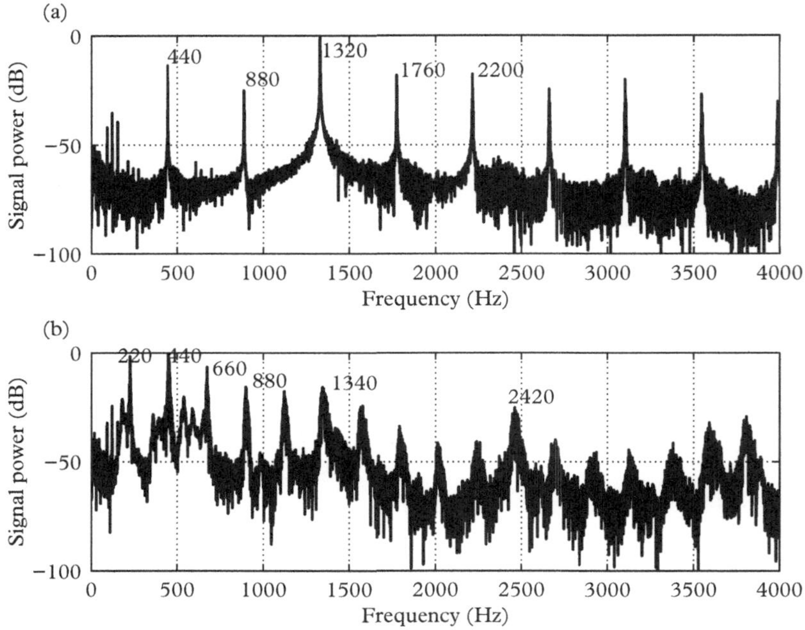

Figure 3.2.5 *Line spectra of (a) pitch pipe, and (b) hum*

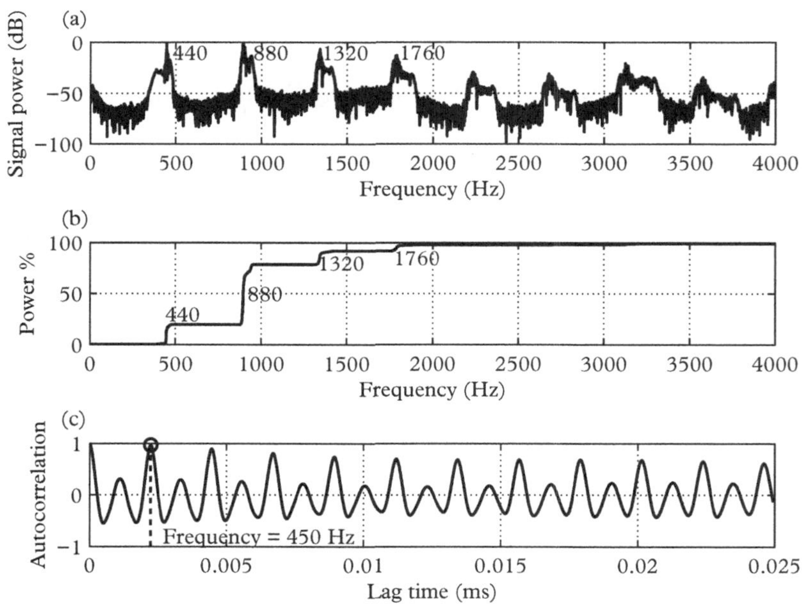

Figure 3.2.6 *Jenn's Laah: (a) line spectrum, (b) cumulative line spectrum, and (c) autocorrelation*

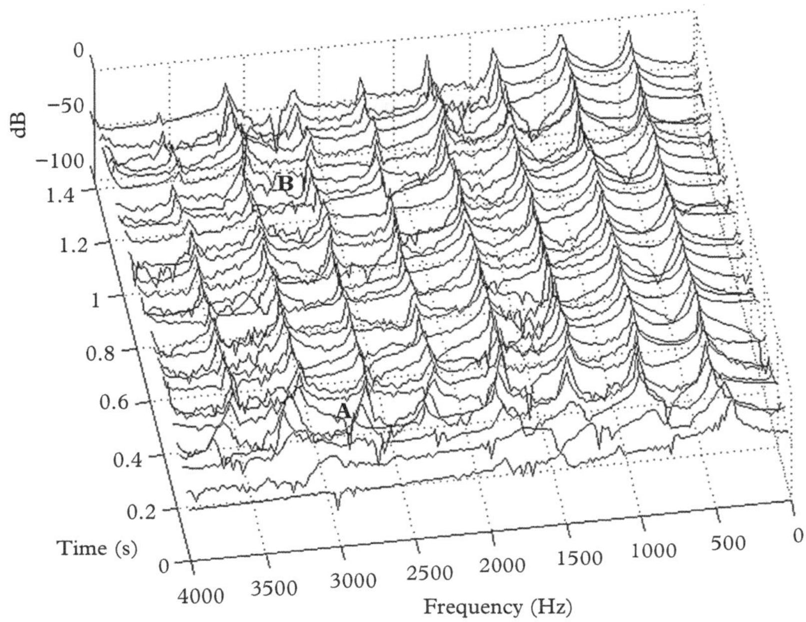

Figure 3.2.7 *Snapshots of the line spectrum of Jenn's Laah*

To compare my attempt at humming 440 Hz, I asked my daughter, Jennifer, to give it a try. The line spectrum in Figure 3.2.6(a) shows that the fundamental is near 440 Hz (actually nearer to 447 Hz) and the harmonics are spaced at approximately 440 Hz, so, unlike her father, she actually sang an approximation of an A4 tone. The cumulative line spectrum in Figure 3.2.6(b) shows that 20% of the power is at 440 Hz but nearly 60% is at 880 Hz. The autocorrelation in Figure 3.2.6(c) suggests that the pitch is 450 Hz.

Snapshots of the line spectrum are shown in Figure 3.2.7, where one can see a shift in frequency at points A and B, which can explain the broadening of the peaks in Figure 3.2.6(a). The audio for this "Laah" is in the file JennLaah.mp3.

3.3 A Door Chime, a Fire Bell, and Clanging Railroad Bars

All three of the waves in this section present many partials or overtones, but none has a harmonic structure.

Door Chime

The first wave in this section comes from the side door on my house (built in 1903). The line spectrum in Figure 3.3.1(a) has peaks at 688 Hz, 1712 Hz, 2985 Hz, 4475 Hz, 6164 Hz, 8067 Hz, 10,175 Hz, and 12,484 Hz, plus smaller peaks scattered in between these "major" peaks. There does not seem to be a fundamental with its harmonics. These spikes, which do not occur at integer multiples of some fundamental frequency, will be referred to as "partials" or "overtones" rather than harmonics. Note that the frequency axes are plotted in kilohertz (kHz) rather than Hz.

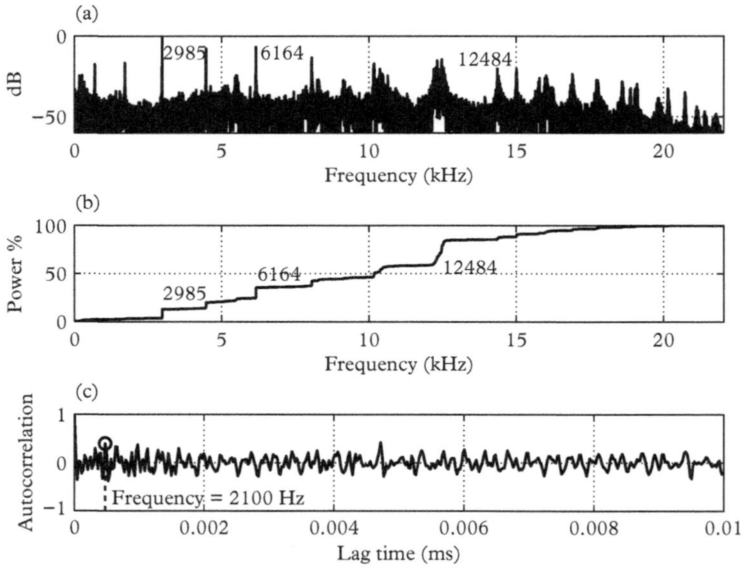

Figure 3.3.1 *Door chimes: (a) line spectrum, (b) cumulative line spectrum, and (c) autocorrelation*

The cumulative line spectrum, Figure 3.3.1(b), suggests that most of the power (~25%) comes from the peak near 12,484 Hz and that there is significant sonic power above 16,000 Hz, which is practically above the human hearing range (the highest key on the piano has a fundamental frequency of 4186 Hz). The highest peak in Figure 3.3.1(a) at 2985 Hz contributes 10% of the total power.

The autocorrelation in Figure 3.3.1(c) does not produce a strong isolated peak. I picked the maximum peak at about 4.7 ms, which corresponds to a frequency of 212 Hz. The first peak, not the strongest, occurs at a lag time of about 0.47 ms, which corresponds to a frequency of 2100 Hz. So, the autocorrelation does not appear to yield much insight for this sound wave.

A spectrogram (actually a series of snapshots of the line spectrum) of the door chimes is shown in Figure 3.3.2. At time zero there is power distributed sporadically across the entire frequency range (look at the back wall of the plot). As time progresses this band of power quickly diminishes and the various overtones develop. As time evolves further, one sees that the higher frequencies die out more quickly. The initial power distribution (near time zero) is likely due to the way the chimes are activated—by turning the handle quickly for approximately 0.5 s. The reader should compare Figure 3.3.2 to Figure 1.2.6 for the built-up wave, which has a similar structure. The sound of the chimes is available in the file DoorChimes.mp3.

Fire Alarm Bell

The second wave comes from an old fire alarm bell located in Burdett, NY. The bell is cast iron, three feet high, and three feet in diameter. The clapper is long gone but I was able to get a sound by striking it with a rock.

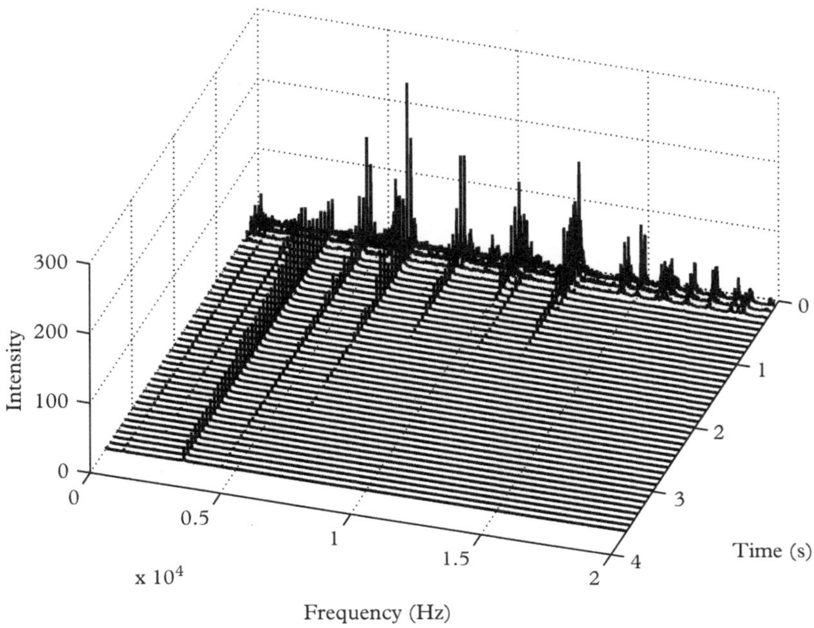

Figure 3.3.2 *Snapshots of door chimes' line spectrum*

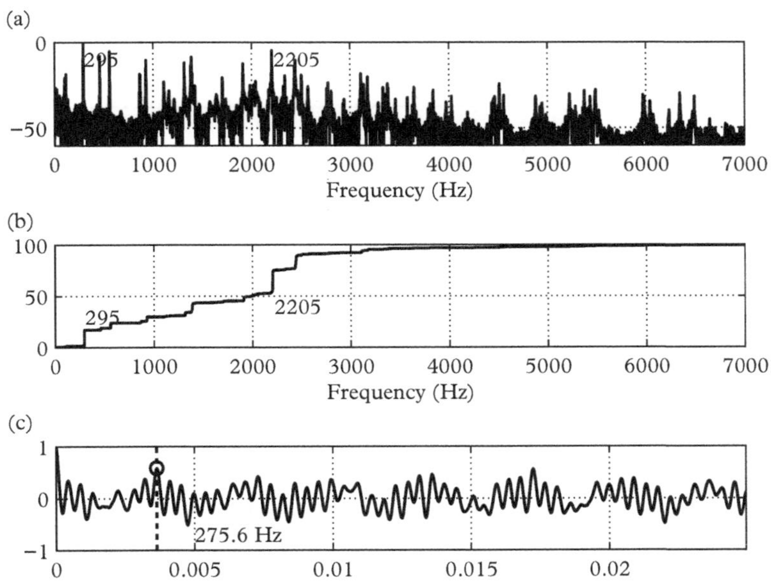

Figure 3.3.3 *Fire bell: (a) line spectrum, (b) cumulative line spectrum, and (c) autocorrelation*

The cumulative line spectrum in Figure 3.3.3(b) indicates that a large percentage of the sound appears at 295 Hz and 2205 Hz. The autocorrelation in Figure 3.3.3(c) shows that the highest and first semi-isolated peak occurs at a time lag corresponding to 276 Hz. This peak is only a little more isolated than that for the door chime. The sound of the fire bell and a pure sine wave of 276 Hz are available in the file FireBell.mp3.

Figure 3.3.4 *Railroad bars*

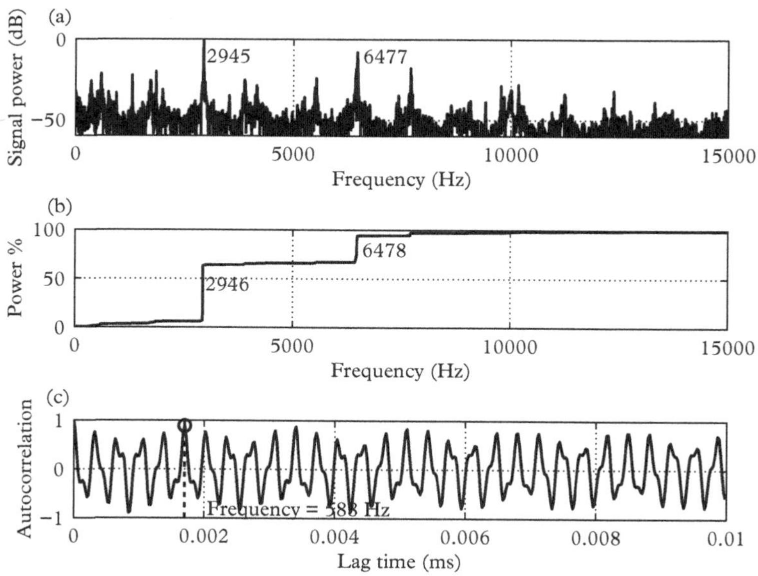

Figure 3.3.5 *Railroad bars: (a) line spectrum, (b) cumulative line spectrum, and (c) autocorrelation*

Railroad Bars

Figure 3.3.4 shows two railroad connectors that I found on the railroad tracks near my house. Figure 3.3.5 shows the line spectrum, cumulative line spectrum, and autocorrelation gathered when I clanged the bars together.

The line spectrum, Figure 3.3.5(a), and the cumulative line spectrum, Figure 3.3.5(b), indicate that the sound is dominated by peaks at 2946 Hz and 6478 Hz. Some of the peaks look as though they might be spaced harmonically, but close inspection shows that not to be the case.

The cumulative line spectrum shows that two frequencies exhibit strength: about 60% of the power resides at 2946 Hz and about 30% resides at 6478 Hz. As with the door chimes, it is difficult to pick a strong isolated peak in the autocorrelation plot shown in Figure 3.3.5(c). However, the frequency associated with the highest peak is 588 Hz. Upon listening to the clanging sound and a test signal having a frequency of 588 Hz, I have a difficult time associating the test signal with the sound of the clanging. These sounds are available in the file Clanging.mp3.

3.4 A Didjeribone®

The Didjeribone® is a cylindrical tube a little over a meter long and about 3 cm in diameter; the length is adjustable. Figure 3.4.1(a) shows the line spectrum for my attempt to get a tone. The fundamental is near 96 Hz, which is close to the G2 note. The odd harmonics are greater than the even harmonics. The cumulative line spectrum in Figure 3.4.1(b) shows that the first and fifth harmonics are the strongest. The sound of the Didjeribone is in the file digeridoo.mp3.

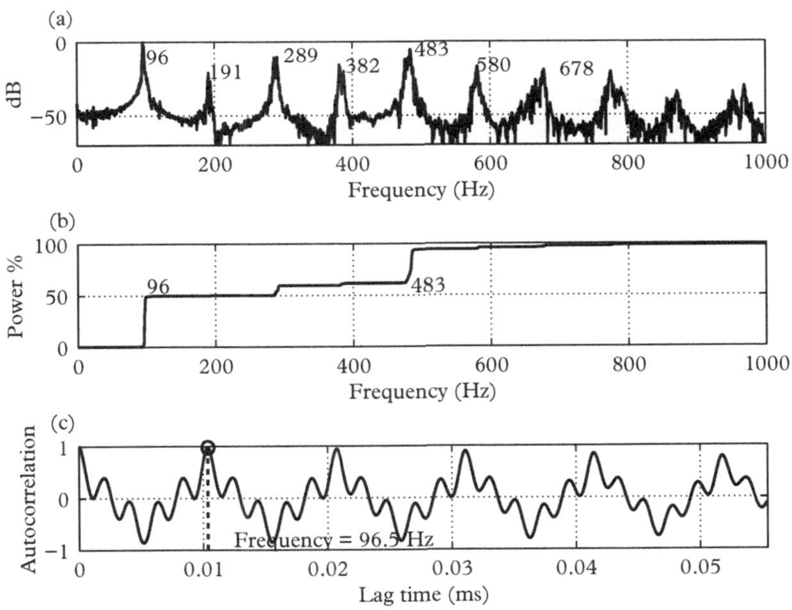

Figure 3.4.1 *Digeribone: (a) line spectrum, (b) cumulative line spectrum, and (c) autocorrelation*

The pitch according to the autocorrelation in Figure 3.4.1(c) is 96 Hz, which is consistent with the line spectrum.

We will return briefly to the Digeribone in Chapter 4, where we attempt to relate the observed harmonics with the physical dimensions of the instrument.

3.5 A Flute

A flute is a long tube closed at one end with a mouthpiece opening near the closed end. The flute player (in this case, a local music teacher) produces a sound from the flow of air across the mouthpiece opening. I gathered signals starting at B3 and ending at D7, all of which will be analyzed in Chapter 19 when we talk about cumulative line spectrum maps. The time domain trace of a typical note, A6, is shown in Figure 3.5.1.

The alternating peaks and valleys in the amplitude probably come from the flutist's natural tendency to add vibrato.

The cumulative line spectrum in Figure 3.5.2(b) shows that nearly 100% of the signal power resides at the fundamental frequency at 1804 Hz. Using the fine grid approach for the autocorrelation, Figure 3.5.2(c), yields a peak at a time lag that corresponds to 1804 Hz; therefore, the two spectra and the autocorrelation agree.

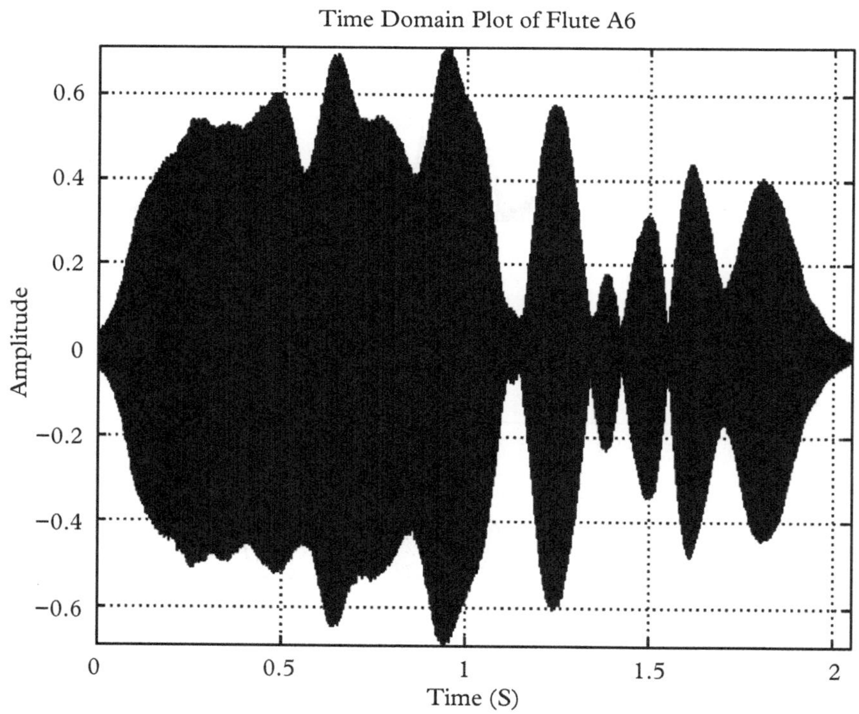

Figure 3.5.1 *Time domain plot of flute—note the lobes from the vibrato*

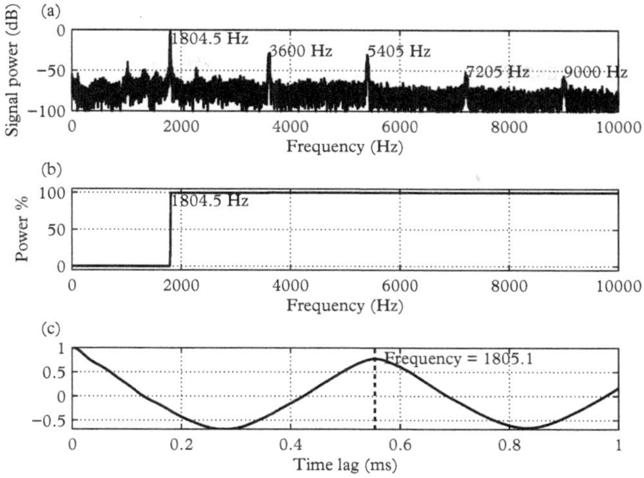

Figure 3.5.2 *Flute A6: (a) line spectrum, (b) cumulative line spectrum, and (c) fine grid autocorrelation*

3.6 A Cornet

Figure 3.6.1 shows the line spectrum, cumulative line spectrum, and autocorrelation of my attempt to play G4 (concert F4 at 349.2 Hz) on my WWII-vintage cornet. Both the cumulative line spectrum, Figure 3.6.1(b), and the autocorrelation, Figure 3.6.1(c), show that the first two harmonics contain most of the tone's strength and that the second partial dominates.

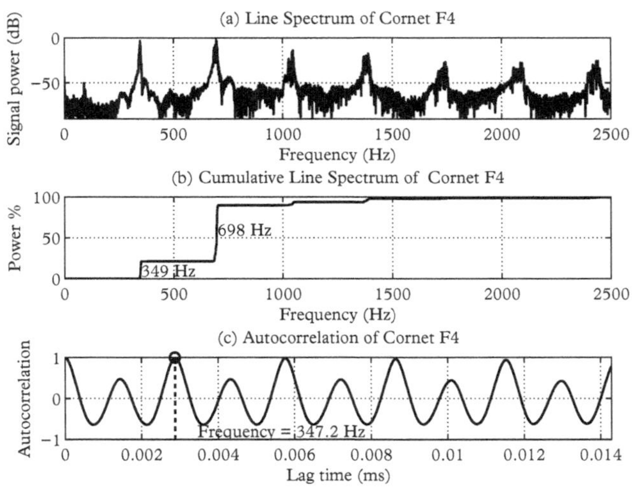

Figure 3.6.1 *Cornet: (a) line spectrum, (b) cumulative line spectrum, and (c) autocorrelation*

3.7 A Trombone

The trombone sound comes from a sophomore music major at Furman University. He played a 32-note chromatic scale, which will be studied in detail in Chapter 19 where the idea of cumulative line spectrum maps is presented. The theoretical fundamental frequency for E2 is 82.4 Hz. The line spectrum in Figure 3.7.1(a) shows peaks at 79 Hz, 158.5 Hz, 238 Hz, and so on, with the difference between the observed and the theoretical fundamental being approximately 3 Hz or 65 cents (see Appendix 3). The sixth harmonic is the strongest, while the fundamental (first harmonic) at 79 Hz is approximately 20 dB down and effectively does not appear on the cumulative line spectrum. However, the autocorrelation in Figure 3.7.1(c) shows that the pitch is indeed 79 Hz. This is an example of the so-called weak or missing fundamental, which we will reprise in Chapter 11.

The line spectrum, cumulative line spectrum, and autocorrelation for the highest note of the sequence, A4# (theoretical fundamental frequency of 466 Hz), are shown in Figure 3.7.2. The theoretical fundamental frequency for A4# is 466.2 Hz. Figure 3.7.2(a) shows peaks at 452 Hz, 903 Hz, 1355 Hz, and 1806 Hz. Therefore, this highest note is 16 Hz flat. The fundamental is dominant and the second harmonic is 8 dB down. The difference between the locations of the peaks is consistently 452 Hz. Compare this to the graphs for E2, where the fundamental is hardly noticeable.

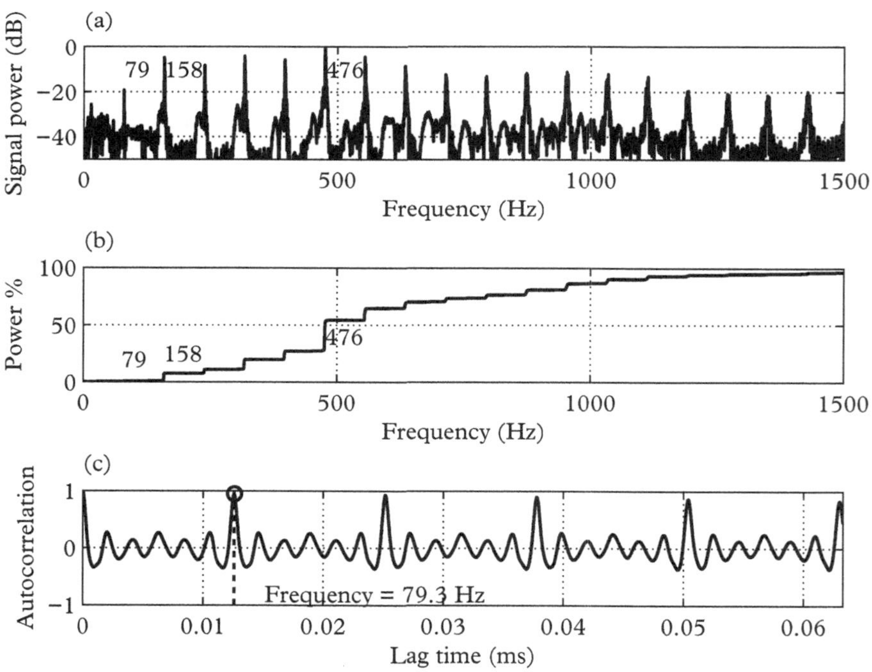

Figure 3.7.1 *Trombone E2: (a) line spectrum, (b) cumulative line spectrum, and (c) autocorrelation*

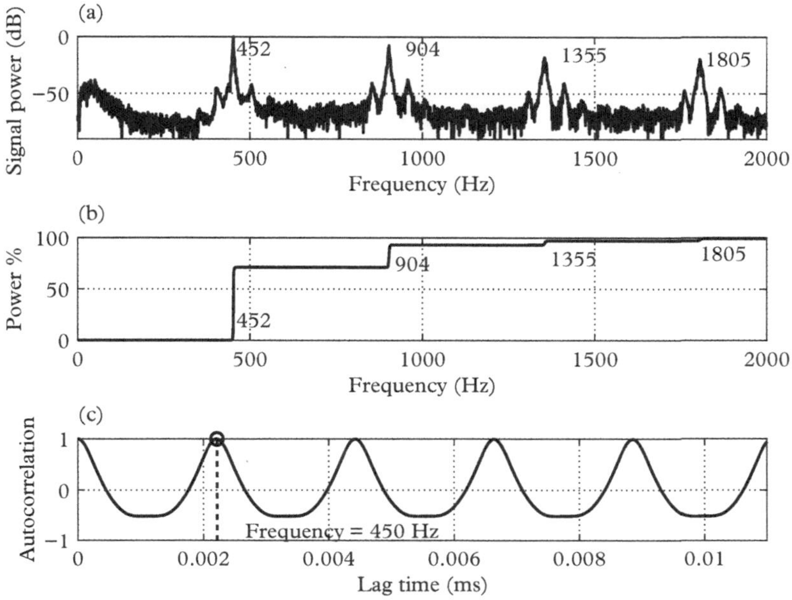

Figure 3.7.2 *Trombone A4#: (a) line spectrum, (b) cumulative line spectrum, and (c) autocorrelation*

3.8 A Bassoon

Figure 3.8.1 shows two views of a bassoon. A bassoon can be crudely described as two connected tubes with an attached mouthpiece. A mouthpiece with a reed is connected to the smaller tube having an inner diameter of 1.2 cm and a length of 89 cm. This tube is connected to a larger tube at the bottom having an inner diameter of 4.4 cm and a length of 134 cm. The top of the larger tube is open.

A local orchestra member was kind enough to play F2. The spectra for the F2 note are shown in Figure 3.8.2. The fundamental is 86 Hz, as evidenced by the spacing between the harmonics; however, the cumulative line spectrum shows that the fifth harmonic is dominant. Furthermore, the autocorrelation tells us that the estimated pitch is 86 Hz.

3.9 Drums and Cymbals

A professional drummer allowed me access to his drum set one night at a local jazz club where I gathered the following sounds.

Figure 3.9.1 shows the bass drum, where there is a fundamental at 81.5 Hz, which is close to the E2 tone.

The floor tom spectrum is shown in Figure 3.9.2, where there is a fundamental at 110 Hz (A2 tone) and partials at 175.5 Hz, 252 Hz, 322 Hz, and 397 Hz. The first peak in the floor

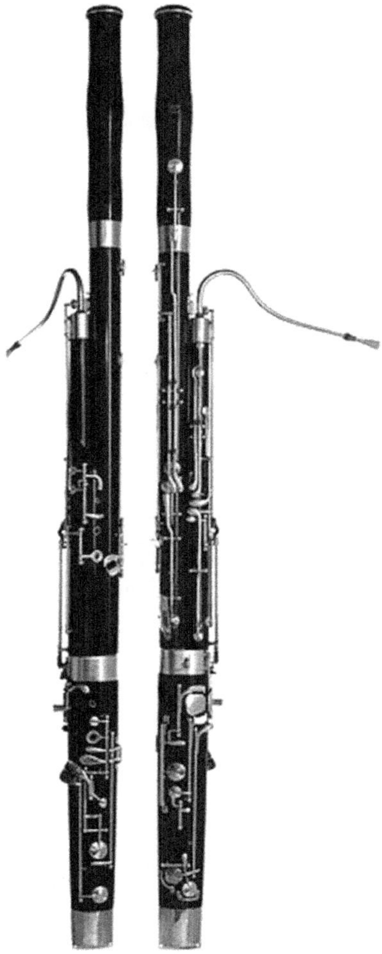

Figure 3.8.1 *Two views of a bassoon*

tom's autocorrelation plot is not the strongest but it gives a pitch that matches the first partial frequency. The partials are close to the theoretical overtones for a circular membrane given on p. 78 in Olson's *Music, Physics and Engineering* as follows:

$$\frac{175.5}{110} = 1.59 = \frac{f_{11}}{f_{01}} \ (1.59)\,,$$
$$\frac{252}{110} = 2.29 = \frac{f_{03}}{f_{01}} \ (2.30)\,,$$
$$\frac{322}{110} = 2.92 = \frac{f_{11}}{f_{01}} \ (2.92)\,,$$
$$\frac{397}{110} = 3.60 = \frac{f_{03}}{f_{01}} \ (3.60)\,,$$

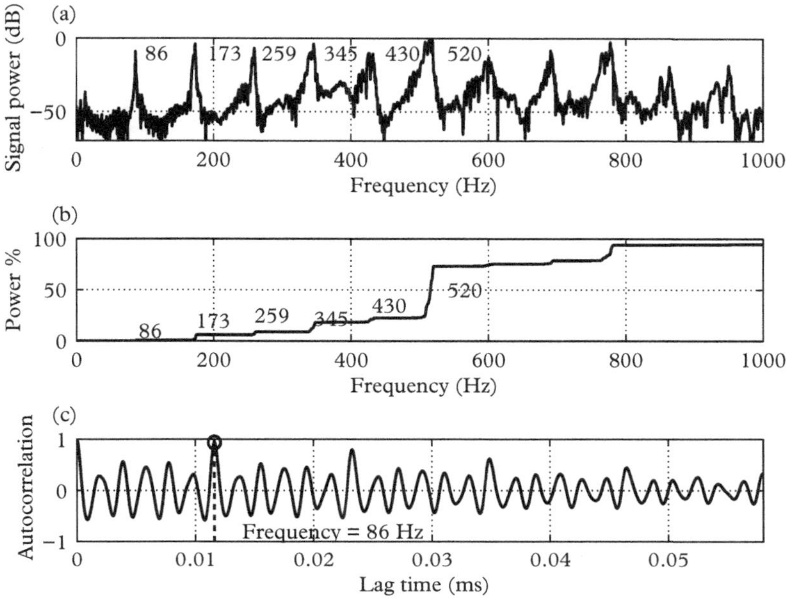

Figure 3.8.2 *Bassoon F2: (a) line spectrum, (b) cumulative line spectrum, and (c) autocorrelation*

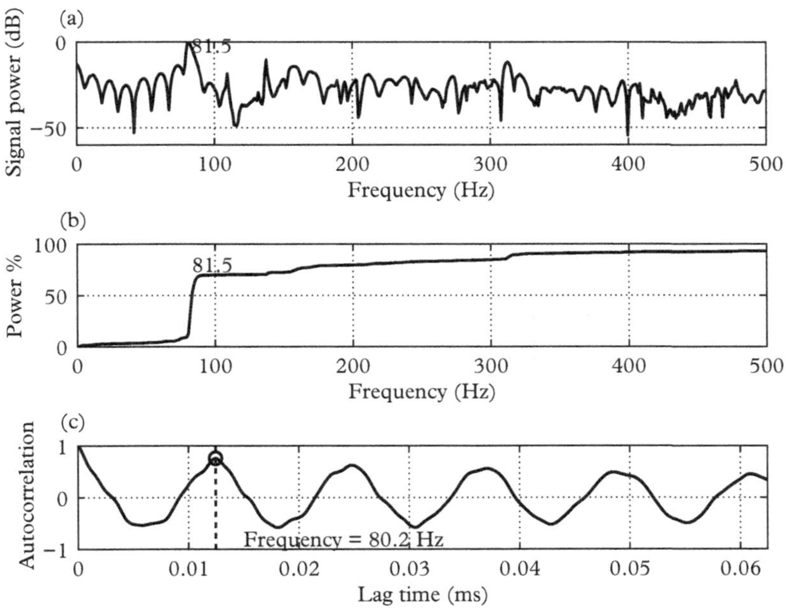

Figure 3.9.1 *Bass drum: (a) line spectrum, (b) cumulative line spectrum, and (c) autocorrelation*

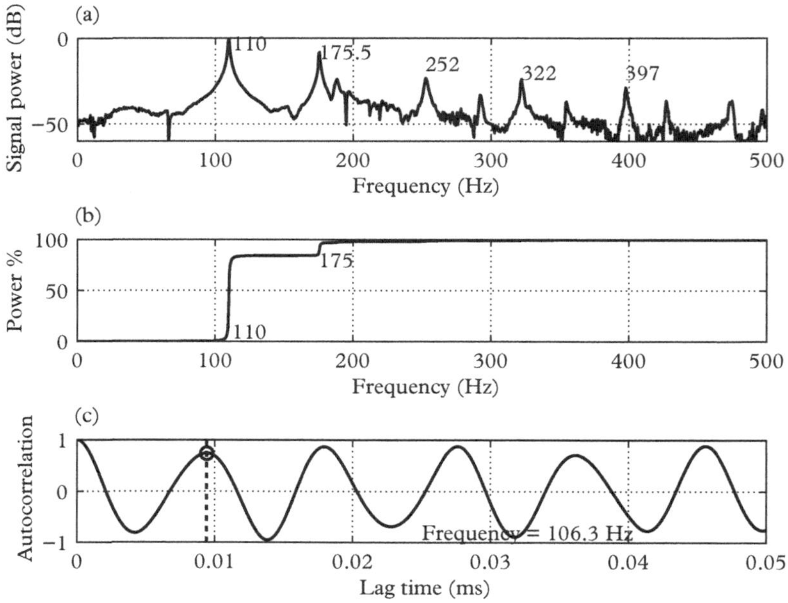

Figure 3.9.2 *Floor tom: (a) line spectrum, (b) cumulative line spectrum, and (c) autocorrelation*

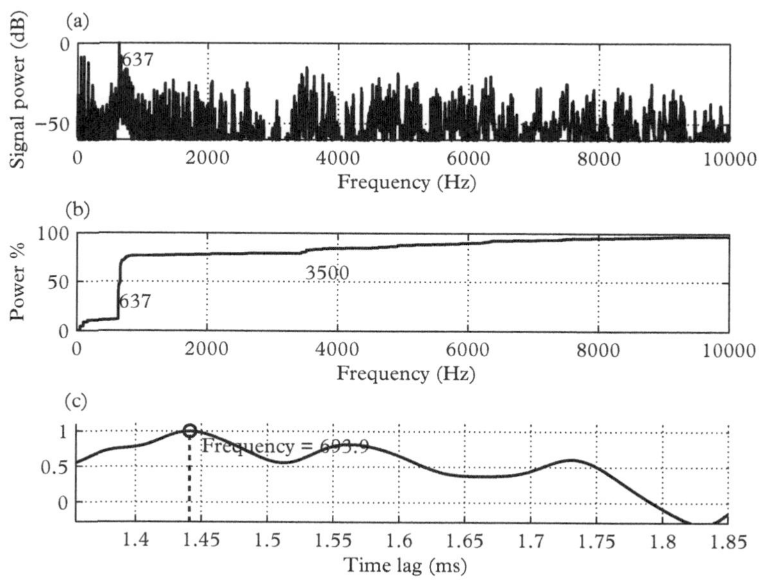

Figure 3.9.3 *Top hat cymbal: (a) line spectrum, (b) cumulative line spectrum, and (c) fine grid autocorrelation*

where $f_{01} = 110\,\text{Hz}$ is the fundamental and the other f_{ij} are the overtones associated with different modes of vibration. (The theoretical ratio values are given in parentheses.)

The longitudinal pressure variations in air resulting from the transverse vibrations of the circular drum membrane provide yet another example of the many ways sounds can be generated. We will revisit drumheads again in Chapter 7 when we deal with vibrating membranes.

Figure 3.9.3(a) shows the line spectrum for a top hat cymbal. There is a first partial or overtone, but there is no harmonic structure. Figure 3.9.3(b) shows the cumulative line spectrum, where the first partial at 637 Hz appears to be dominant. The fine grid autocorrelation in Figure 3.9.3(c) indicates that the pitch is 693 Hz, which is close to the peaks in the line spectrum and cumulative line spectrum.

3.10 Blowing over the Top of a Beer Bottle—The Helmholtz Resonator

If one blows over the top of an empty beer bottle shaped roughly like the one in Figure 3.10.1, a sound can easily be generated. The line spectrum, cumulative line spectrum, and autocorrelation of the result are shown in Figure 3.10.2.

The line spectrum in Figure 3.10.2(a) shows that there are two structures. The first consists of a fundamental at 186 Hz with harmonics at 371 Hz, 556 Hz, 744 Hz, 929 Hz, and 1115 Hz. The second consists of components at 1220 Hz, 1862 Hz, 2660 Hz, and 3480 Hz, which appear to be part of a harmonic structure with a fundamental of approximately 810 Hz. The cumulative line

Figure 3.10.1 *Generic bottle*

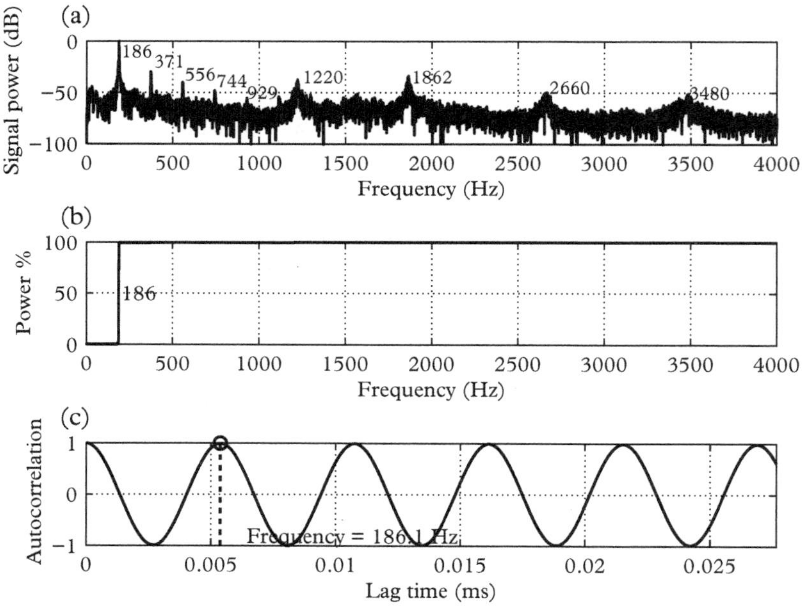

Figure 3.10.2 *Blowing over the top of a bottle: (a) line spectrum, (b) cumulative line spectrum, and (c) autocorrelation*

spectrum and the autocorrelation show that in spite of the interesting local structure, effectively all of the signal power is at 186 Hz.

The beer bottle is roughly approximated by a Helmholtz resonator model, which is an ideal object with a volume V, a neck of length L and an opening with a diameter D, somewhat like the drawing in Figure 3.10.3. The cross-sectional area of the opening is S.

The behavior of the Helmholtz resonator can be explained by a model consisting of a cylindrical plug of air in the neck having length L and diameter D. This plug is influenced by the mass of air in the body having a volume V. If the plug moves downward, the air in the body is compressed and its resulting increased pressure tends to provide a restoring force against the downward movement, just as a spring would act against the movement of an attached mass.

At the other end of the neck, the open end, the oscillatory movement of the plug radiates acoustical energy and experiences a resistance. Thus, the plug of air experiences two forces whenever it moves back and forth: the force of the body air against movement, and the acoustic resistance. Applying Newton's law of $F = ma$ allows one to develop a simple model for the behavior of the plug of air (see Kinsler and Frey, pp. 186–93).

Blowing across the top of the bottle can temporarily force this plug of air a little way down the neck, thereby compressing the air inside the body. That increased pressure now drives the plug of air out, but when it gets to its original position, its momentum takes it on outside the body a small distance. The rarified air inside the body then allows atmospheric pressure to push the plug of air back in. It can thus vibrate like a mass on a spring with the damping provided by the acoustic resistance (resistance to flow due to the viscosity of the air).

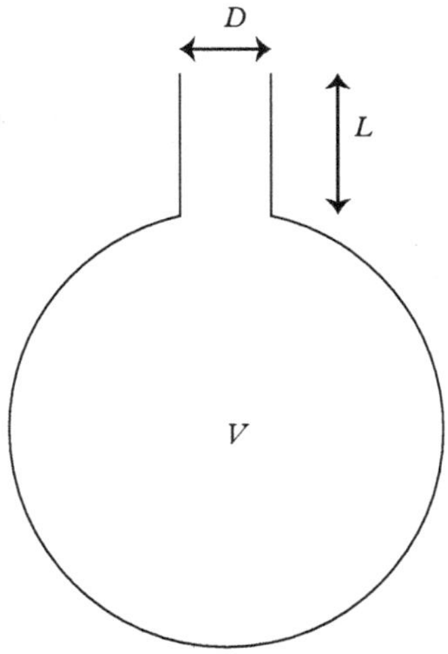

Figure 3.10.3 *Schematic drawing of a bottle showing neck diameter D, neck length L, and body volume V*

This simple model has a resonant frequency given (p. 193, Kinsler and Frey) by

$$f = \frac{c}{2\pi}\sqrt{\frac{S}{LV}}. \tag{3.10.1}$$

The opening of the top of the beer bottle has a diameter D of 0.021 m and a cross-sectional area S of 0.000346 m^2. The bottle neck length L is approximately 0.08 m. The speed of sound in air c is 343 m s^{-1}. The volume of the bottle is approximately 12 fl oz, or 0.000355 m^3. When these numbers are applied to Eq. 3.10.1, one gets f = 191 Hz, which is remarkably close to the observed fundamental frequency in Figure 3.10.2(a).

The fundamental and its harmonics depend on the dimensions of the neck and the volume of the bottle, while the higher overtones in the line spectrum depend on the shape of the bottle. The sound of me blowing over the top of a beer bottle is found in the file BlowingOverBottle.mp3.

3.11 A Train Whistle

I live near a railroad track. One day I ran along with the train as it slowly passed through our village and collected the sound of the whistle. The line spectrum over two frequency ranges is shown in Figure 3.11.1.

These line spectrum plots show the strongest peak at 931 Hz, along with two peaks at 422 Hz and 464 Hz. There is quite a bit of structure but nothing that appears to be part of a fundamental-with-harmonics grouping.

The cumulative line spectrum and the autocorrelation are shown in Figure 3.11.2. The autocorrelation shows many peaks. The highest occurs at a time lag (approximately 0.05 ms)

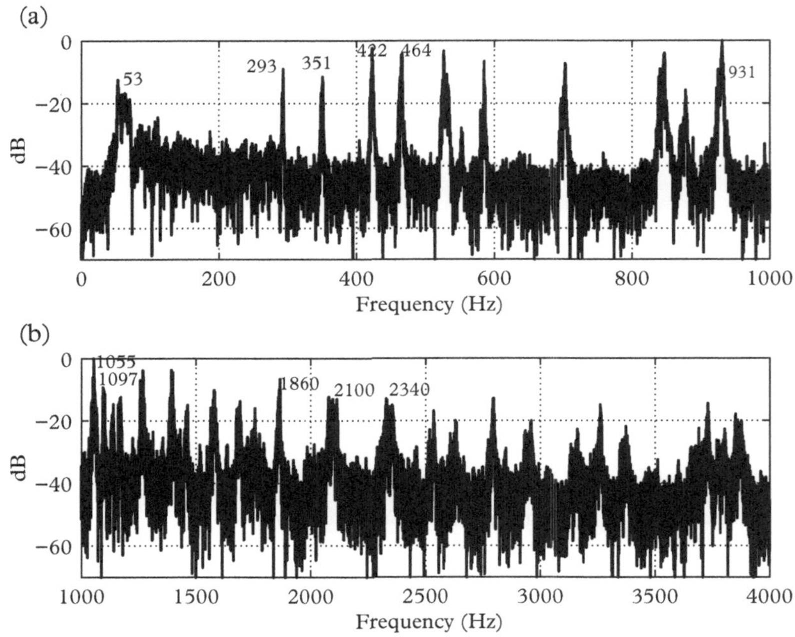

Figure 3.11.1 *Train whistle: (a) line spectrum 0–1000 Hz, and (b) line spectrum 1000–4000 Hz*

that corresponds to a frequency of 21.2 Hz, which shows no power in the line and cumulative spectra. The first peak is located at a time lag (approximately 0.002 ms) corresponding to 464 Hz, which matches one of the strong peaks in the line spectrum.

The time-varying nature of this wave suggests that the spectrogram (actually snapshots of the line spectrum) might be enlightening. Figure 3.11.3 shows that the signal power does change its distribution among the different frequencies with time. In particular, power at 846 Hz and 931 Hz (dotted lines in the intensity plane) show peaks at about 2.0 s, while the power appears to peak for 464 Hz at about 3.5 s.

3.12 Summary

The results of this chapter suggest that many familiar sounds have a consistent structure with signal power distributed discretely amongst a fundamental frequency and its harmonics. However, the everyday sound of a door chime, the clanging railroad connectors, the cymbals, the train whistle, and the fire bell have their own peculiarnon-harmonic structure. Consequently, it is difficult to mimic one of the non-structured sounds such as the sound of the chime (or cymbal) by humming or by going to a piano and picking out a single note that sounds like the chime. Furthermore, the author's attempt to mimic a pitch pipe by humming may sound fairly close, but was actually an octave off.

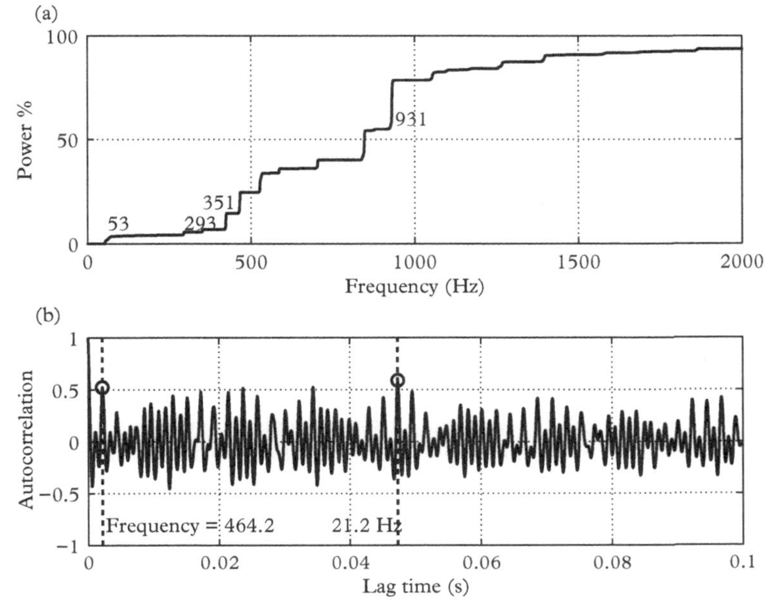

Figure 3.11.2 *Train whistle: (a) cumulative line spectrum, and (b) autocorrelation*

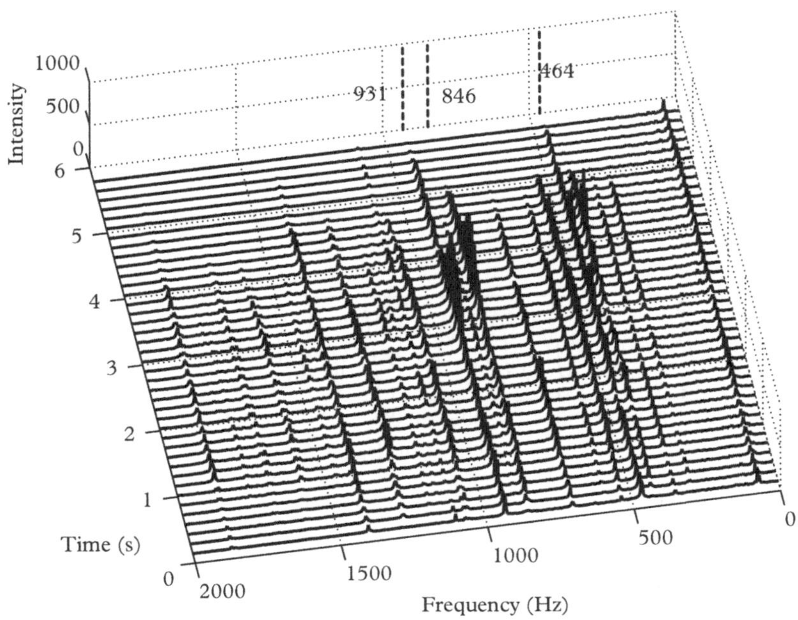

Figure 3.11.3 *Snapshots of train whistle line spectrum*

3.13 Exercises

Use a recorder such as a Zoom H2 (or your laptop computer) and record a common sound. The sound of an idling automobile engine or of someone whistling a tone might be two such candidates. You could also try rapping a wood stick against the side of a crystal wine goblet. Using a USB connector, read the wav file collected by the Zoom H2 into a computer having Matlab® and Audacity (if you do not have Matlab, try Octave). Use Audacity to pick out the wave and export it as a wav file to the computer. Read the file using the Matlab command

$$[\mathrm{y}, \mathrm{Fs}] = \mathrm{wavread\ (fname)}$$

With the data in the vector *y*, adapt parts of the Matlab scripts in the file FiguresPart1_Chap3.mto carry out the line spectrum and cumulative line spectrum. Does the spectrum show harmonics? If so, why? If you choose the idling engine, be warned that interpreting its spectrum will be quite a challenge. If you choose a goblet you might get something like Figure 3.13.1.

Does the goblet generate harmonics? Does the spectral centroid value make sense? Does the pitch value of 1050 Hz make sense? What note on the piano seems to be closest to the goblet's sound? How does the fundamental frequency of the piano note relate to the peaks in the line spectrum or the pitch from the autocorrelation?

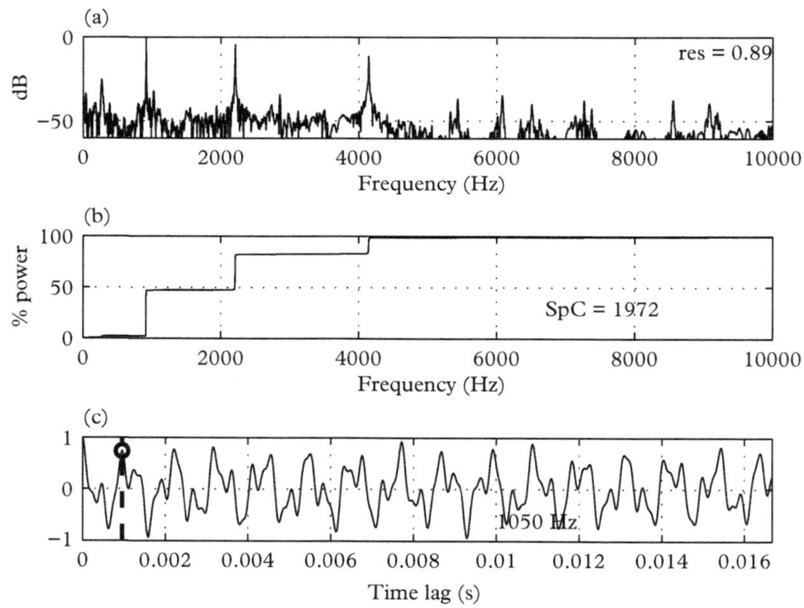

Figure 3.13.1 *Goblet: (a) line spectrum, (b) cumulative line spectrum, and (c) autocorrelation*

4

Harmonics in Musical Sounds

Where do the harmonics come from, and why do some sounds have them and others not? The pure sine wave and the tuning fork did not have any significant power at higher harmonics. The former's lack of power at higher harmonics was by design. In Section 3.1 we saw that the latter's second harmonic did appear in the line spectrum but it was approximately 40 dB down from the fundamental. A tuning fork is roughly similar to a vibrating bar with a mount in the center (the bottom of the "U"). Morse (p. 158) and Kinsler and Frey (p. 77) claim that it does exhibit harmonics, but that they damp out significantly more quickly than the fundamental. We will touch on vibrating bars in the next chapter.

One piano note, the highest, C8, was presented earlier and found to exhibit only one relatively attenuated harmonic besides the fundamental. However, we will discover in Part 3 that the lower-numbered piano notes indeed have a well-defined harmonic structure. The source of these harmonics will be discussed Chapter 5, which is devoted to the vibrating string. Suffice it to say for now that the piano string lengths, diameters, and tensions dictate which frequencies can be sustained, and they happen to be the fundamental and integer multiples of that fundamental.

The Didjeribone®, cornet, trombone, flute, and bassoon have harmonics because only waves with the proper wavelengths can repeat in their structures, which are ideally modeled by a tube with one open end and one closed end or by a tube open at both ends.

The bell chime and the clanging railroad connectors apparently have no physical structure to support harmonics.

In this chapter the idea that the harmonic frequencies of one note often line up with the fundamental frequencies of other notes is presented. This is followed by a discussion of open and closed tubes, which also exhibit harmonics because of their structure and which behave similarly to the above-mentioned musical instruments. The section closes with a discussion of the harmonics associated with a leaf blower.

4.1 Harmonics Associated with a Piano Note

Some musical instruments like the piano follow the equal tempered scale where the n th note's fundamental frequency is given by

$$f(n) = 440r^{n-49} \quad n = 1, 2, \ldots, 88 \qquad r = 2^{\frac{1}{12}} = 1.059463. \tag{4.1.1}$$

Spectral Analysis of Musical Sounds with Emphasis on the Piano. First Edition.
David M. Koenig, with comments by Delwin D. Fandrich.
Published in 2015 by Oxford University Press.

For example, the 49th note is A4 and has a fundamental frequency of 440 Hz. Therefore, A4#, which is the 50th note, would have a fundamental frequency of

$$f(50) = rf(49) = 1.05463 \times 440 = 466.1637 \text{ Hz},$$

and so on for other notes. There are, by definition, 100 cents between the n th and $(n-1)$th note in Eq. 4.1.1. We will take a closer look at cents in Part 3 and in Appendix 3.

Consider C3, the 28th note, which according to Eq. 4.1.1 has a fundamental frequency of 130.81 Hz. The *second* harmonic frequency of C3 is, by definition, a multiple of C3, as in

$$2 \times 130.81 = 261.62 \text{ Hz},$$

which is also the fundamental frequency of C4, the 40th note, as defined by Eq. 4.1.1.

The *third* harmonic frequency of C3 is

$$3 \times 130.8128 = 392.43 \text{ Hz},$$

which is "near" the fundamental frequency of G4 which, according to Eq. 4.11, is 391.99 Hz.

Table 4.1.1 and Figure 4.1.1 illustrate the relationship between the harmonics of C3 and the fundamental frequencies of several other notes in the equal temperament scale. The first column gives the note name. The second column gives the fundamental frequency of that note according to the equally tempered scale or Eq. 4.1.1. The third column gives the frequency of the closest C3 harmonic. The fourth column gives the harmonic number for the C3 note.

Table 4.1.1 *Relationship between the frequencies (in Hz) of the harmonics of C3 and the fundamentals of other notes*

Note Name	Equal Tempered Fundamental Frequency	C3 Harmonic Frequency	C3 Harmonic Number
C3	130.8128	130.8128	1
C4	261.6256	261.6256	2
G4	391.9954	392.4384	3
C5	523.2512	523.2512	4
E5	659.2551	654.064	5
G5	783.9909	784.8768	6
A5#	932.3275	915.6896	7
C6	1046.5022	1046.502	8
D6	1174.6591	1177.315	9
E6	1318.5102	1308.128	10
F6	1396.913	1438.94	11

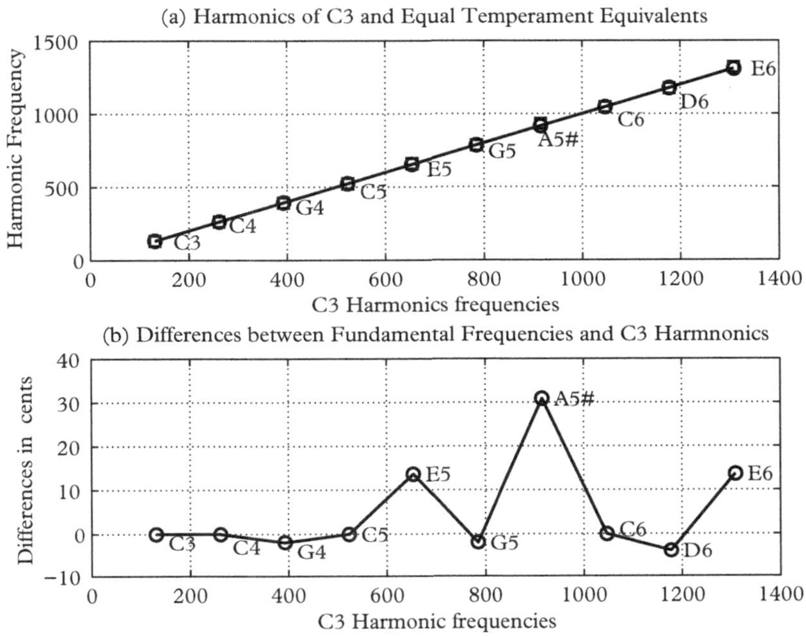

Figure 4.1.1 *Harmonics of C3 and its equal temperament equivalents*

For example, the G4 note in the third row of Table 4.1.1 has a fundamental frequency of 391.99 Hz, and the third harmonic of C3 has a theoretical harmonic frequency of 392.44 Hz.

Table 4.1.1 shows that the fundamental frequency of the notes an octave apart, namely C3, C4, C5, and C6, line up exactly with the second, fourth, and eighth harmonics of C3. There are the following coincidences for the C3 note:

1. The E5 note's fundamental frequency is about 5 Hz above the fifth harmonic of C3.
2. The G5 note's fundamental frequency is within 1 Hz of the sixth harmonic of C3.
3. The A5# note's fundamental frequency is approximately 16 Hz away from the seventh harmonic of C3.
4. The D6 note's fundamental frequency is within 3.0 Hz of the ninth harmonic of C3.
5. The E6 note's fundamental frequency is approximately 10 Hz away from the tenth harmonic of C3.

The same information is presented graphically in Figure 4.1.1. Figure 4.1.1(a) visually suggests that the harmonics of C3 and the fundamental frequencies of the selected notes are the same. Figure 4.1.1(b) shows the differences in cents. The greatest deviation is between the seventh harmonic of C3 and A5#, and it is approximately 30 cents.

The stave in Figure 4.1.2 shows the notes whose fundamental frequencies are close to the harmonics of C3.

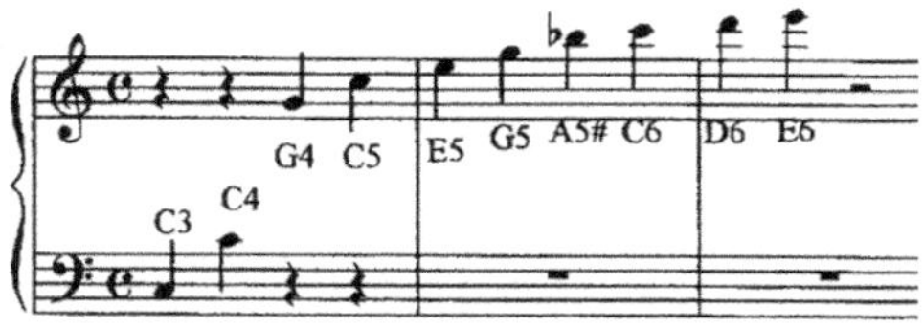

Figure 4.1.2 *Harmonics of C3 and its equal temperament equivalents*

Therefore, playing the C3 note, say by striking the piano key, generates a tone whose harmonic content approximately contains the fundamental frequencies of the above notes. That these harmonics do not match the C3 harmonic frequencies exactly is a consequence of the equal temperament scale, which is defined by Eq. 4.1.1. We will touch on this topic again in Part 3.

For the musician-readers, we should remark in passing that to find the harmonics of a particular note one uses the following musical intervals:

an octave 8ve (C3 to C4),
a major fifth M5 (C4 to G4),
a major or perfect fourth M4 (G4 to C5),
a major third M3 (C5 to E5),
a minor third m3 (E5 to G5),
a minor third m3 (G5 to A5#),
a major second M2 (A5# to C6),
a major second M2 (C6 to D6),
a minor second m2 (D6 to E6).

In summary, the intervals are: 8ve, M5, M4, M3, m3, m3, M2, M2, m2, . . .

Figure 4.1.3 shows the line spectra of some of these notes gathered from my Knabe grand piano with offsets so that they can be presented on one graph.

The top line in Figure 4.1.3 shows C3. The next line down shows C4. Note how the harmonics of C4 line up with some of those of C3—especially how the fundamental of C4 lines up with the second harmonic of C3. Next, the fundamental of G4 lines up with the third harmonic of C3. This continues until A5# is reached, where the seventh harmonic of C3 lines up with A5#'s fundamental. The fundamental of C6 should line up with the eighth harmonic of C3. However, because of where the hammer strikes the C3 string (at approximately ⅛ th of the string's length), there is no strength at that frequency.

The analysis of this section suggests that beyond the ninth harmonic the coincident notes are separated by no more than one note. For example, Table 4.1.1 shows that the 10th and 11th harmonics are separated by approximately one note, as in E6 to F6. Therefore, if one plays a C major chord with C3 as the root note, then improvised notes above E6 may not clash since they are aligned somewhat with the harmonics of the root note.

We will revisit the piano in Part 3, where the concept of cents and Eq. 4.1.1 will come up again.

Next, let us take a look at tubes, which are components of several of the musical instruments studied above and which can be used to generate harmonics.

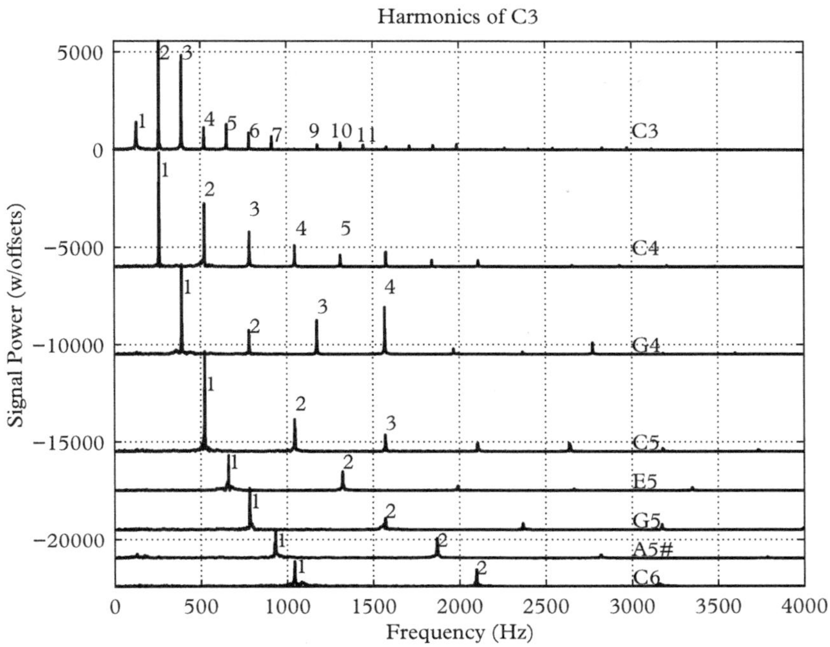

Figure 4.1.3 *Line spectra of C3 and its equal temperament equivalents*

4.2 A Tube Closed at One End and Open at the Other

Consider an idealized case where there is a tube of length $L = 1$ m that is open to the environment (ambient pressure) at one end and closed at the other. Assume that a pressure disturbance is somehow set up inside the tube at the closed end, perhaps because of a small mouthpiece or a small piston. Figure 4.2.1(a) shows how the resulting first harmonic or natural frequency is set up as a function of time. Note the times, 0.0 ms, 1.2 ms, 2.4 ms, 3.6 ms, 4.8 ms, and 6.0 ms, attached to the various curves. Also, remember that the y-axis in Figure 4.2.1 represents the longitudinal displacement along the axis of propagation even though it is plotted perpendicularly to the x-axis, which represents the length along the tube.

At the closed end (zero length) on the left of Figure 4.2.1(a), the longitudinal displacement has to be zero because there is no place for the air to move back and forth. At the open end ($x = L = 1$m), on the right side of Figure 4.2.1(a), the displacement pressure is zero (relative to the ambient pressure of the environment). It turns out that "the displacement pressure is equal to the negative rate of change of the longitudinal displacement with respect to distance." Using calculus, that last sentence would be equivalent to

$$p = -\frac{\partial y}{\partial x}.$$

Since the equation may be foreign to some readers and since the phrase in quotes is an equivalent way of presenting the concept, the reader can take his/her pick. Unfortunately, the quote is a bit off-putting, too.

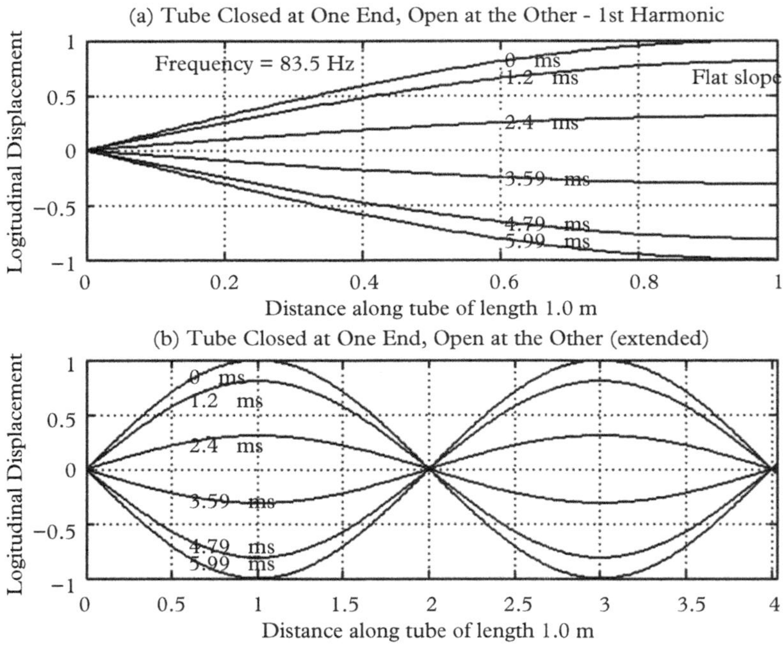

Figure 4.2.1 *First harmonic at various times for a tube closed at one end and open at the other*

In simple terms, this means that the slope at the tube's open end ($x = L = 1$m) in Figure 4.2.1(a) has to be zero (or flat) because of zero pressure (relative to ambient) at that point. The longitudinal waves shown in Figure 4.2.1(a) satisfy this requirement—note the "flat slope" in Figure 4.2.1(a). These longitudinal waves have a wavelength λ equal to four times the tube length (so Figure 4.2.1(a) shows only one quarter of the sine wave). That is, $\lambda = 4L$. The fundamental frequency is 83.5 Hz (for the case where $L = 1$ m), so the wavelength of the wave is 4.0 m.

Why does the quarter of a sine wave shown in Figure 4.2.1(a) turn out to be the fundamental? Because it is a portion of the sine wave that has zero displacement at one end and zero slope at the other. Figure 4.2.1(b) shows the same wave extended for 4 m to give you a picture of the wavelength of $4L$, and it gives you an alternative picture of the slope of the wave at the end of the tube where $x = L = 1$.

In general, the fundamental frequency for this tube is given by

$$f_1 = \frac{c}{4L}, \tag{4.2.1}$$

where c is the speed of wave propagation (the speed of sound). Equation 4.2.1 falls out of the solution of the appropriate boundary value problem for a tube open at one end. Similar equations are derived in Appendix 2.

The *third* harmonic will also satisfy these boundary conditions and it is shown in Figure 4.2.2.

Note that the wave in Figure 4.2.2 has a frequency of 250.5 Hz and a wavelength of $4L/3$ m. It has zero longitudinal displacement at $x = 0$, and where there is a node at about $x = 0.66$ m (see

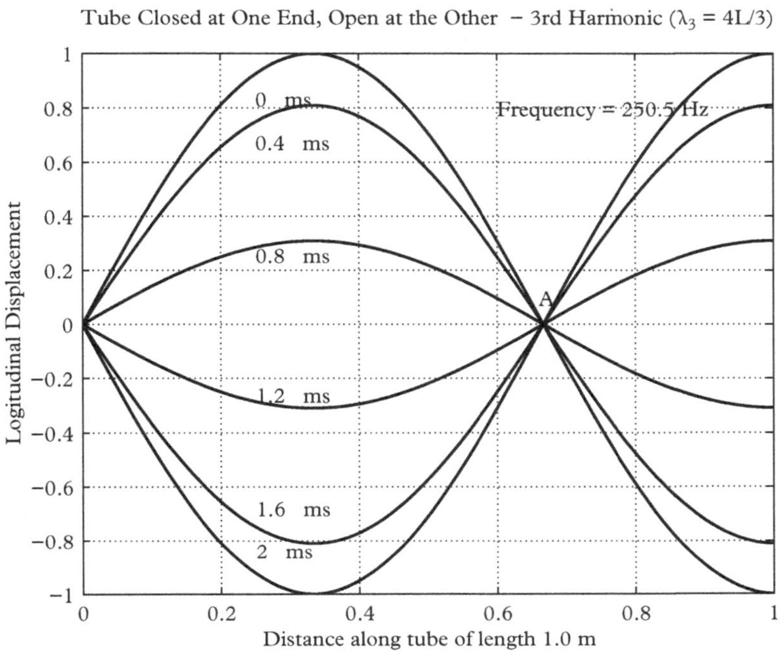

Figure 4.2.2 *Third harmonic at various times for a tube closed at one end and open at the other*

point A in Figure 4.2.2). The slope of the displacement with respect to distance is zero at $x = L$ (the open end of the tube on the right). Higher odd harmonics also satisfy these conditions.

The data from Figure 3.4.1, where I blew into one end of the Didjeribone® (having a length of 0.952 m) showed a fundamental frequency of 96 Hz. The theoretical fundamental frequency from Eq. 4.2.1 for a tube of that length would be

$$f_1 = \frac{c}{4L} = \frac{343\,\text{m/s}}{(4)(0.9525\,\text{m})} = 90\,\text{Hz}.$$

Figure 3.4.1 showed both even and odd harmonics, but the even harmonics are down from the odd harmonics by some 20 dB. So, even though I probably strained the ideal assumptions associated with the concept in this section by blowing into one end of the Didjeribone, the ideal model does roughly match the behavior of the Didjeribone. There is a similar match for the bassoon in Section 3.8 that you can develop for yourself.

4.3 A Tube Open at Both Ends

Consider a tube open at both ends. Figure 4.3.1 shows the wave shapes for the first three harmonics.

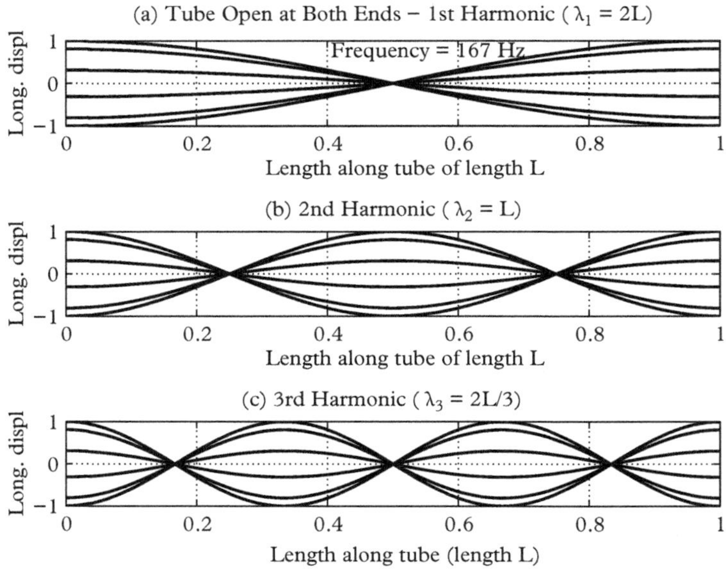

Figure 4.3.1 *First, second, and third harmonics at various times for a tube open at both ends*

At both ends of the tube the slope of the longitudinal displacement with respect to distance is zero because both ends are exposed to ambient pressure. For the first harmonic, the frequency and tube length are related as

$$f_1 = \frac{c}{2L}. \tag{4.3.1}$$

Equation 4.3.1 is derived in Appendix 2, where the results of solving the vibrating string problem are appropriated.

The idealized tube open at both ends allows for all the harmonics, while the tube closed at one end only allows for odd harmonics.

4.4 Another Look at the Cornet, Trombone, and Flute

Although it is not obvious, the cornet and trombone can be modeled by an open ended tube. Paraphrasing Professor John Askill in *Physics of Musical Sounds*, 2nd edition, when a mouthpiece and a flared bell are added to a long closed tube the high resonant frequencies become lower and the low resonant frequencies become higher such that they can be closely modeled by a tube open at both ends. I cannot find a good physical explanation for why this is so but I accept it as an empirical fact. In the next couple of paragraphs we will test this idea.

I carefully traced the path that the air takes on my cornet when all three valves are not depressed and came up with approximately 1.35 m. Using the open-at-both-ends formula in Eq. 4.3.1, the fundamental frequency is

$$f_1 = \frac{343 \text{ ms}^{-1}}{2 \times 1.35 \text{ m}} = 126 \text{ Hz},$$

which is in between B2 and C3 on the equal tempered scale (ETS). The cornet is a B♭ instrument, so when one plays C4 (no valves depressed) it sounds as though it has a fundamental frequency of A3#, which is 233 Hz. Therefore, it appears that the fundamental frequency of the cornet is approximately equal to a C3 on the ETS. The second harmonic of C3 is C4, which is the lowest note easily played on the cornet with no valves depressed. The next note playable with no valves depressed is G4, which is the third harmonic of C3. The third note playable with no valves depressed is middle C on the cornet or C5, and it is the fourth harmonic of C3. This sequence of playable notes with no valves depressed continues with E5, G5, A5#, and C6—all harmonics of C3. (As a youngster trying to learn the cornet I always wondered why I could get both A5# and C6 with no valves depressed. Unfortunately, when I was attempting C6 I often got A5#.)

In a similar manner I measured the length of the tube in the trombone in first position (when the slide is not extended) and got 2.85 m. After adding a correction for the flared bell of 0.06 m (well within the error of my measurements and therefore not likely to be significant), the following calculation yielded the fundamental frequency:

$$f_1 = \frac{343 \text{ ms}^{-1}}{2 \times 2.91 \text{ m}} = 58.9 \text{ Hz},$$

which corresponds approximately to A1# at 58.3 Hz, a note that cannot be comfortably played on the trombone. However, the second harmonic of A2# at 116 Hz is easily played. The next note obtainable with the slide in first position is F3 at 175 Hz, which is the third harmonic of A1#. The third note that can be played in first position is A3# at 233 Hz, which is the fourth harmonic of A1#. The trombone is also a B♭ instrument so these A# tones are C's in concert key.

The same approach can be taken for the flute. The lowest note played on the flute is B3 (with all holes closed), which has a fundamental frequency of 246.94 Hz. The length of the flute is approximately 27 in or 0.68 m. Applying Eq. 4.3.1, but solving for the length, gives

$$L = \frac{c}{2f} = \frac{343 \text{ ms}^{-1}}{2 \times 247 \text{ Hz}} = 0.694 \text{ m}.$$

Therefore, it appears that Eq. 4.3.1 is a fair approximation to reality.

4.5 A Triplet of Simulation Experiments

This subsection contains three simulation experiments where a tube open at one end will be stimulated with a sinusoid, white noise, and a step change. The resulting wave and its harmonics will be analyzed.

Experiment 1: Sinusoidal Forcing

To further support the ideas brought up in this chapter, consider the following mathematical experiment. We will solve the wave equation for the case of an ideal tube of length $L = 0.1$ m, open at one end. This tube is shorter than in previous simulations so as to get a higher natural frequency. At the closed end, the tube is forced with a sinusoid of frequency f_{CL} (imagine placing a piston at the closed end and disturbing the piston with short-amplitude sinusoidal displacements). The natural frequency of the tube is given by Eq. 4.2.1,

$$f_1 = \frac{c}{4L},$$

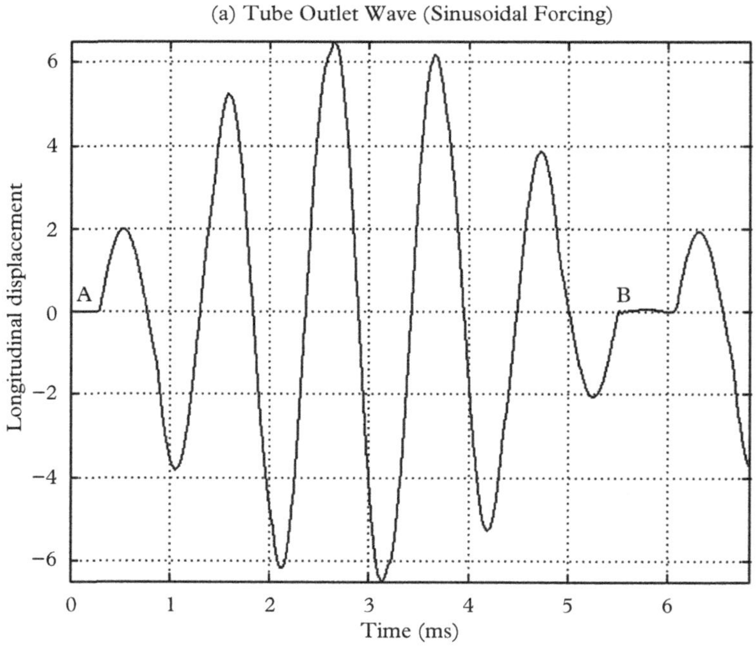

Figure 4.5.1 *Wave at outlet of open-ended tube for sinusoidal forcing at the closed end*

which for this experiment is 857.5 Hz. If the forcing frequency is an integer multiple of the natural frequency, then in principle the displacement amplitude will increase without bound because the model does not include friction. For the first simulation experiment, the forcing frequency was arbitrarily chosen to be 1.21 times the natural frequency, or 1037.5 Hz. Figure 4.5.1 shows the longitudinal displacement at the tube's open end, and Figure 4.5.2 shows the profile along the tube.

The flat spot in Figures 4.5.1 and 4.5.2 at $t = 0$ (point A in Figures 4.5.1 and 4.5.2) represents the time required for the forcing sine wave to reach the open end of the tube. This period of time is calculated from the length of the tube divided by the wave's propagation speed:

$$\frac{L}{c} = \frac{0.1 \text{ m}}{343 \text{ ms}^{-1}} = 0.29 \text{ ms}.$$

The other flat spot at about $t = 5$ ms (point B) represents the reflection and is twice as long.

Look at the back wall of Figure 4.5.2 where the position is 0.1 m for all time (point C). This is the point where the sine wave is generated and you can see the regular sinusoidal amplitude variation. At the other end of the tube, where the position is 0.0 m for all time, you can see the trace shown in Figure 4.5.1.

The line spectrum is shown in Figure 4.5.3. There are spikes at the natural or fundamental frequency of 857.5 Hz and also at its odd harmonics (2572.5 and 4287.5 Hz), which are diminished relative to the fundamental. There is also one spike at the forcing frequency of 1037.5 Hz (look at the two spikes close together at point A in Figure 4.5.3).

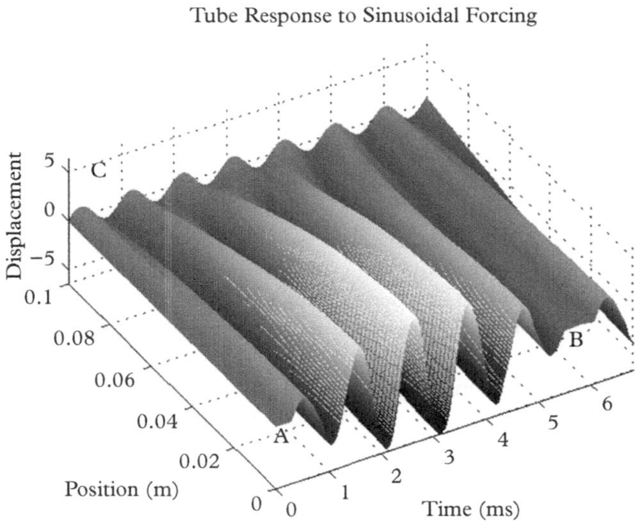

Figure 4.5.2 *Three-dimensional plot of wave at outlet of open-ended tube for sinusoidal forcing at the closed end*

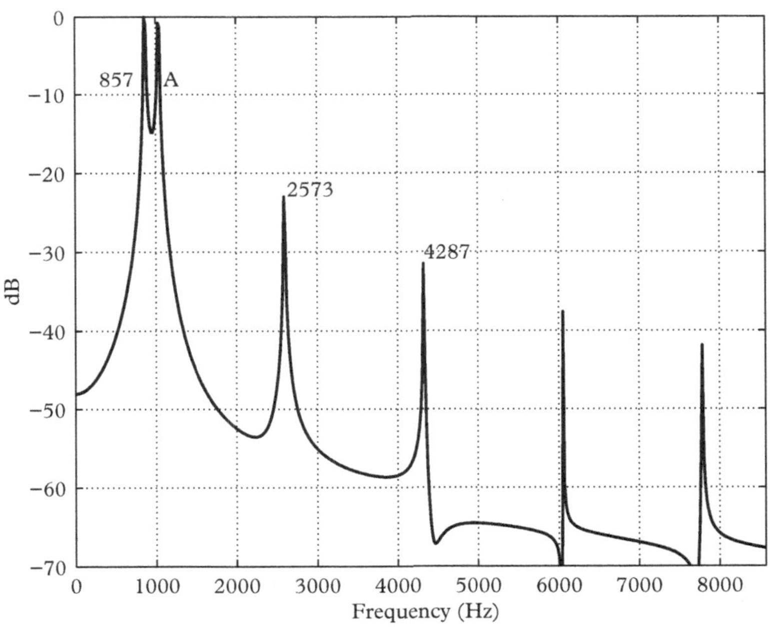

Figure 4.5.3 *Line spectrum of wave at tube outlet with sinusoidal forcing*

Experiment 2: Forcing with White Noise

We introduced white noise in Section 2.10 and stipulated that, on average, it contained a constant signal intensity at all the frequencies in the Nyquist interval. If we disturb the ideal tube with white noise at the closed end (with a randomly moving piston) then we should expect that only those waves with frequencies equal to the natural frequency of the tube and its harmonics will resonate. Figure 4.5.4 shows the time domain trace and the line spectrum of the longitudinal displacements at the outlet.

As with the case for sine wave forcing, there is a flat period in Figure 4.5.4(a) representing the time for the white noise to reach the end of the tube. After that, the response appears to be hash but if you look carefully there is a periodic trace buried in the noise. The line spectrum in Figure 4.5.4(b) shows what is buried in the apparent hash, namely the odd harmonics, which appear to have approximately the same strength.

These two simulation experiments suggest that the signal that one generates at the mouthpiece of a tube like a Didjeribone is probably a combination of a subconsciously chosen primary forcing sinusoidal signal along with parts of white noise. If your embouchure generated pure white noise then the Didjeribone would still produce a sound dominated by the natural frequency and its harmonics.

Experiment 3: Step Change Forcing

Figure 4.5.5 illustrates the case where the one-end-open tube is given a step disturbance at the closed end (imagine a piston at the closed end where $x = L = 0.1$ m). Note that the end of the

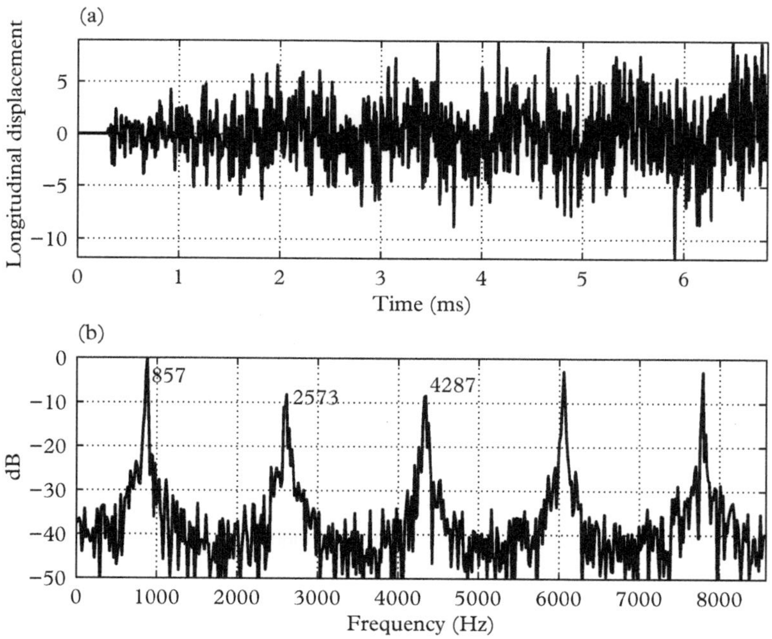

Figure 4.5.4 *White noise forcing of a tube closed at one end and open at the other: (a) longitudinal displacement, and (b) line spectrum of the outlet wave*

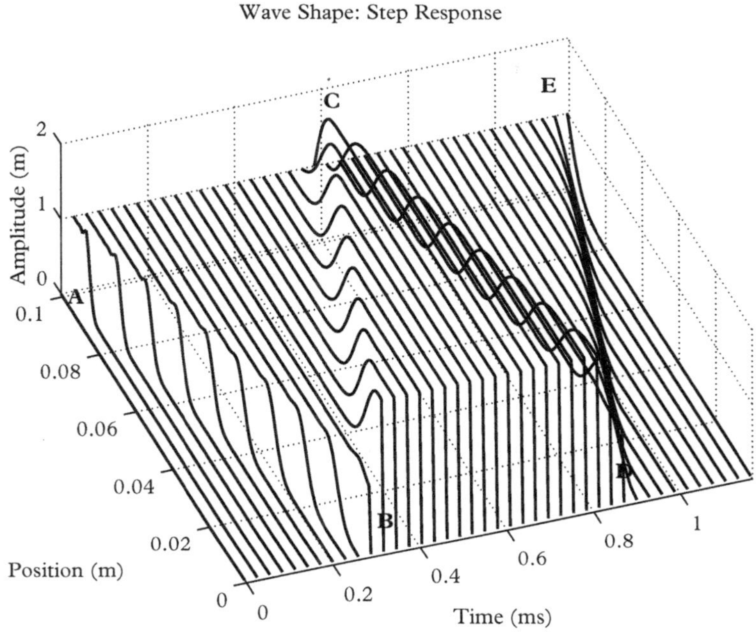

Figure 4.5.5 *Longitudinal displacement during step change forcing of a tube closed at one end and open at the other*

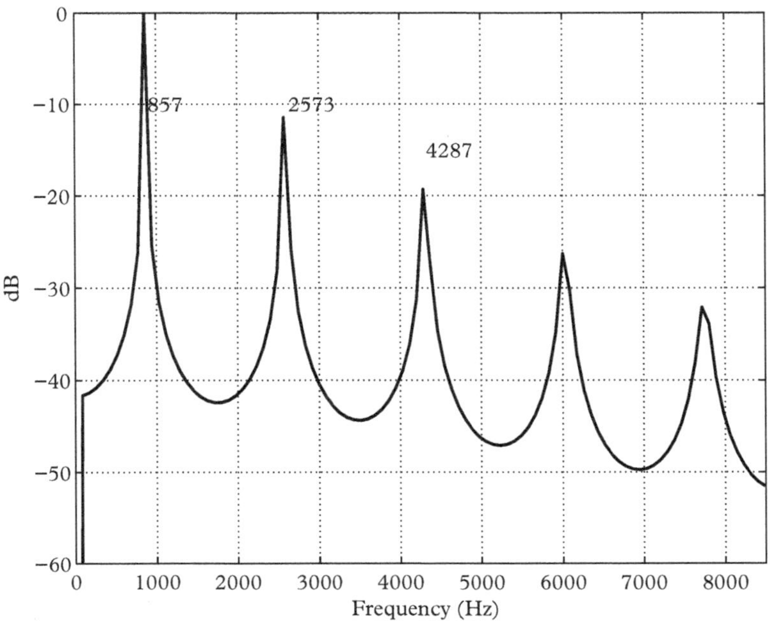

Figure 4.5.6 *Line spectrum of wave at tube outlet with step change forcing*

wave at position zero (the open end) is always flat because it is exposed to the environmental atmospheric pressure. The step starts at point A and evolves with time. At point B the wave has to flip upwards to maintain the zero slope (flatness) at the open end. At point C (the closed end) there is a reflection and the wave starts back towards the open end, which is at point D where there is another reflection and the wave moves toward the closed end at point E. If the reader has access to Matlab® then he/she can execute the scripts in the file FiguresPart1_Chap4.m and rotate the graph to get a better view of the transition from point D to E. Since the wave continues indefinitely, an echo-like response is implied.

The line spectrum of the signal at the tube outlet in Figure 4.5.6 shows that odd harmonics of the fundamental frequency are generated by the step change at the input.

These three experiments show that the response at the tube outlet will have harmonic components no matter what the stimulus is. The only difference is in the distribution of the power amongst the harmonics.

4.6 A Crude Example of Harmonics

To some folks the sound of a leaf blower is part of the music of the fall season, to others it is just an irritating noise. The line spectrum of my leaf blower is shown in Figure 4.6.1.

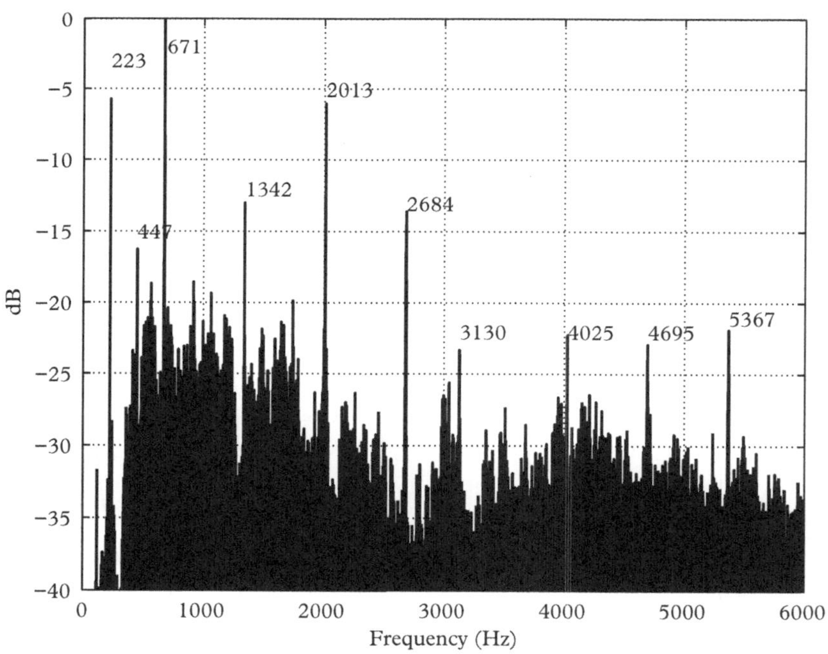

Figure 4.6.1 *Line spectrum of leaf blower*

Here is an interesting (to me, at least) presentation of some even and odd harmonics (or partials). The fundamental is at 223 Hz (–5.5 dB), which is close to the A3 note. The partials are

2nd	447 Hz,
3rd	671 Hz,
4th	917 Hz,
6th	1341 Hz,
9th	2013 Hz,
12th	2683 Hz,
14th	3130 Hz,
18th	4025 Hz,
21th	4695 Hz,
24th	5367 Hz.

Given that there is an electrically driven fan inside the blower this kind of harmonic structure is not surprising.

4.7 Summary

Harmonic frequencies of one note often line up with the fundamental frequencies of other notes, especially in musical instruments. Open and closed tubes exhibit harmonics because of their structure and behave similarly to several musical instruments. A tube open at one end with a stimulus at the closed end generates a harmonic response when the input is sinusoidal, white noise, or a step change.

4.8 Exercises

1. Find the code in the file FiguresPart1_Chap4.m that generates Figures 4.2.1 and 4.2.2 and make modifications so that the even harmonics are plotted instead of the odd harmonics. How are the graphs different and why?
2. Why do you think that Eq. 4.3.1 for an open-at-both-ends tube works for the cornet, trombone, and flute even though they are closed at one end?

Part 2

A Visual Analysis of Vibrating Objects (Strings, Membranes, Bars, Plates)

5 The Vibrating String

Part 2 deals with the components that make up many musical instruments. The first and probably the most studied is the vibrating string, which for our purposes is an integral part of the piano, the guitar, and the bass violin. We will spend considerable time on the plucked string, both idealized and real.

Adding an idealized soundboard to the flexible string shows how the resonant frequency of the soundboard influences the distribution of sound power. Disturbing the string at one end sinusoidally shows that the string behaves similarly to the tube that was studied in Chapter 4. The response of the string to a hammer strike is studied using two models and some actual data. The effect of stiffness in the string is studied next, both with a soundboard and without. A study of the force exerted by the string on the fixed ends leads to a comparison of stiff and flexible strings and an attempt to explain the differences. We complete our study of the vibrating string by looking at the longitudinal waves in a piano string from a Bösendorfer piano. In between the discussions of the simulations, there is a short look at a couple of real examples: strings on (1) a guitar, (2) a bass violin, and (3) in an experimental set-up.

5.0 Basic Parameters of the Flexible String

Consider a flexible, frictionless string of length L, stretched under tension $\tilde{T}$ between two fixed points. The string is not elastic. This means that when the string is disturbed from a straight line under no tension it does not have a natural tendency to return to the original undisturbed condition—tension is the only force on the string that can make it return. The string has a linear density of $\tilde{\rho}$ grams per meter. Under these conditions of non-zero tension, this string has a natural or fundamental or first harmonic frequency given by

$$f_{\text{nat}} = \frac{1}{2L}\sqrt{\frac{\tilde{T}}{\tilde{\rho}}}, \tag{5.0.1}$$

which is derived in Appendix 2. In words, Eq. 5.0.1 says that as the string length L increases the fundamental frequency f_{nat} decreases—a long string vibrates more slowly than a short string. It also says that the fundamental frequency will increase as the square root of the tension $\tilde{T}$ increases—a taut string will vibrate at a higher frequency than a slack string. Finally, a dense or heavy string will vibrate more slowly than a light string. The term "fundamental" is used because later on we will show that the motion of an ideal vibrating string is made up of the fundamental or first harmonic plus integer multiples of the first harmonic.

Spectral Analysis of Musical Sounds with Emphasis on the Piano. First Edition.
David M. Koenig, with comments by Delwin D. Fandrich.
Published in 2015 by Oxford University Press.

If the string has a length L of 0.51 m, a linear density $\tilde{\rho}$ of 6.8 g m^{-1} and is under a tension $\tilde{T}$ of 1370 N (or about 308 lbf) then Eq. 5.0.1 tells us that the fundamental frequency of the string is 440 Hz, which is equal to the fundamental frequency of A4, the 49th note on the piano (counting up from the lowest note) or A above middle C.

$$f_{\text{nat}} = \frac{1}{2L}\sqrt{\frac{\tilde{T}}{\tilde{\rho}}} = \frac{1}{2 \times 0.51}\sqrt{\frac{1370000}{6.8}} = 440\text{Hz}.$$

If plucked, this ideal string will vibrate (in a transverse direction, perpendicular to the string's axis when at rest) with a fundamental frequency of f_{nat}. Our model does not include damping so the vibrations will not decay with time. The string's natural period of vibration τ is the reciprocal of the fundamental frequency, as in $\tau = \frac{1}{f_{\text{nat}}}$. For this example string, the period is 1/440 or 2.27 ms. As we will see, the wave shape will repeat every τ ms.

The propagation speed c of disturbance waves along the example string (resulting from a pluck or a strike) is

$$c = \sqrt{\frac{\tilde{T}}{\tilde{\rho}}} = \sqrt{\frac{1370000}{6.8}} = 448.8\,\text{ms}^{-1}. \tag{5.02}$$

Since the string is 0.51 m long, it will take approximately $\frac{0.51}{448.5} = 0.0011364\text{s} = 1.136\text{ms}$ for a disturbance to propagate from one end of the string to the other. A round trip would require about 2.27 ms, which is the natural period of the string. The airborne longitudinal sound waves, discussed in Chapter 1, propagated at the speed of sound in air, which is 343 m s^{-1}.

Equations 5.0.1 and 5.0.2 can be combined to give the natural frequency in terms of the speed of propagation c and the string length L:

$$f_{\text{nat}} = \frac{c}{2L}. \tag{5.0.3}$$

Since the wavelength is related to the frequency by

$$\lambda = \frac{c}{f}, \tag{5.0.4}$$

we can combine Eqs. 5.0.3 and 5.0.4 to get

$$\lambda = \frac{c}{c/(2L)} = 2L. \tag{5.0.5}$$

So, the fundamental wavelength is twice the length of string.

Note that nowhere in these discussions has the diameter of the idealized string come up.

For future reference, we point out that the time required for a wave pulse to propagate from one end of a string of length L to the other end and back is

$$t_{\text{propagate}} = \frac{2L}{c}. \tag{5.0.6}$$

5.1 Plucking the Idealized Flexible String with Fixed End Points at 1/9th of its Length

The idealized plucked flexible string is studied because it can yield insight into a real string that might be part of a guitar or a piano. In this first section we will look at some simulated results for the case where the string is plucked at 1/9th of its length. The next two sections (Sections 5.2 and 5.3) deal with results for some real strings: a single piano wire, a single wrapped piano wire, a guitar string, and a double bass string. In Section 5.4 we return to the idealized string that is plucked at the center of its length.

We stipulate that the string is initially at rest so it traces a straight line in space between the two fixed end points (see Figure 5.1.1). Let the direction along this straight line be the longitudinal direction. We referred to it above as the string's axis and we will also call it the x-direction. Because the string is at rest, there is no movement in the transverse direction, which is perpendicular to the longitudinal direction. Any displacement in the transverse direction at any moment in time and at any position along the string will be symbolized by $y(x, t)$, where x is the position along the string and t is time.

Please note the difference in the transverse displacement of the string compared to the longitudinal displacement of the air molecules associated with the sound waves discussed in Chapter 1. Figures 5.1.1 and 5.1.2 show that $y(x, t)$ is the transverse displacement of the string at a point x along the string at time t.

Initially, at time $t = 0$, we have $y(x, t) = y(x, 0) = 0$ for $0 \leq x \leq L$, where $L = 51$ cm. In words, before the plucking, the initial value of the transverse displacement is zero all along the length of the string (where the position x varies from 0 to L), as is shown in Figure 5.1.2.

For this simulation, I chose the length, the tension, and the linear density such that the natural frequency of the string is 440 Hz. The propagation speed $c = \sqrt{\frac{\bar{T}}{\bar{\rho}}}$ is 0.449 m (ms)$^{-1}$. The period of the natural frequency is 2.27 ms. This string therefore has a natural or fundamental frequency equal to the idealized A4 note string for a piano.

By "plucking" we mean that the string is gently displaced to a new initial condition, symbolized by $y(x, 0)$ and shown in Figure 5.1.3, and then, at time zero, released.

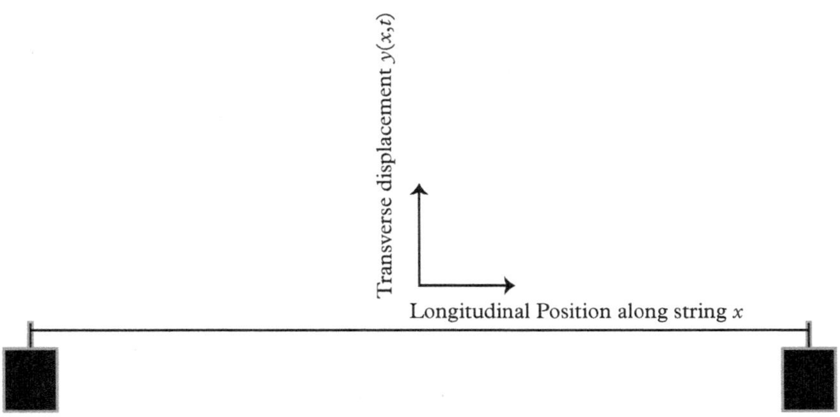

Figure 5.1.1 *Transverse and longitudinal displacements*

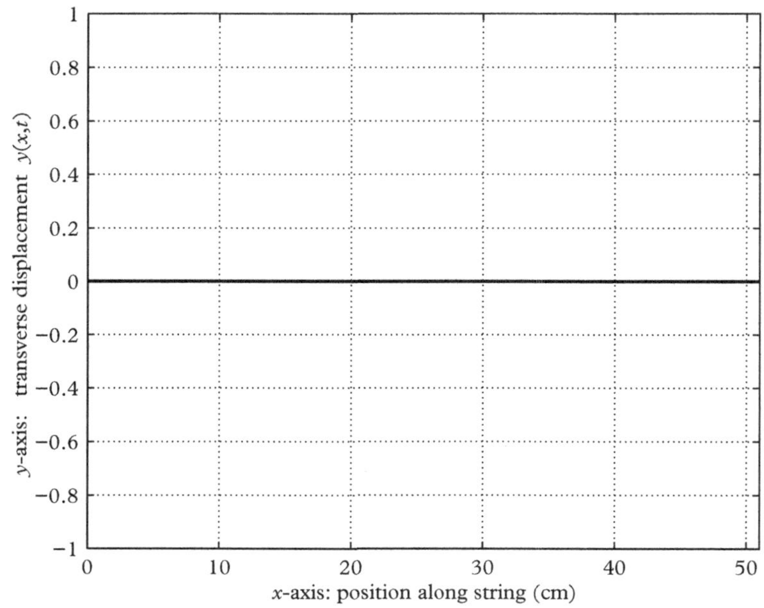

Figure 5.1.2 *Initial position of vibrating string (at rest before the pluck or strike)*

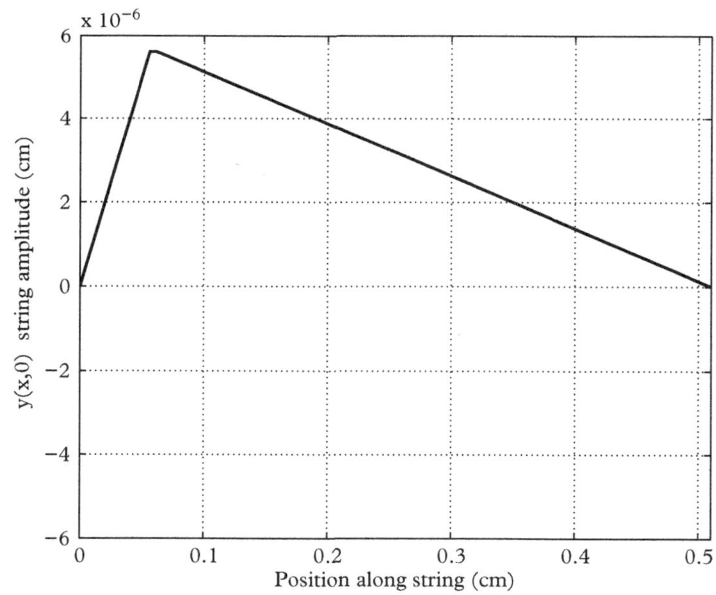

Figure 5.1.3 *Initial profile of vibrating string before being released (during a pluck)*

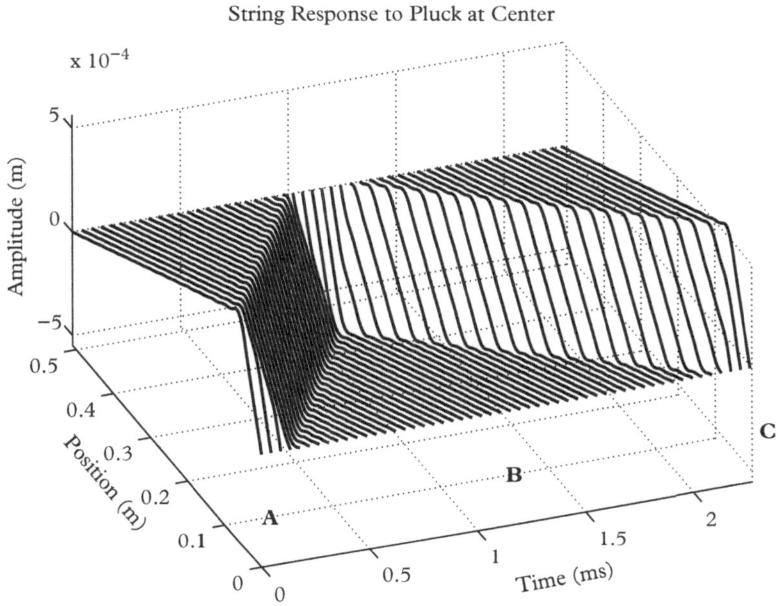

Figure 5.1.4 *Snapshots of the vibrating string response to a pluck at time A, the reflection at time B, and the return to nearly initial conditions at time C*

At the risk of over-repeating, the *x*-axis is the longitudinal position along the string. The *y*-axis is the string amplitude and it represents the transverse displacement of the string. Note that the "plucking point" is at 1/6th of the length, approximately where a hammer would strike a piano string. However, keep in mind that this is a pluck not a strike.

Upon releasing the string, the dynamic behavior can perhaps be best displayed on paper with a 3D graph showing snapshots of the string at various instants of time. Figure 5.1.4 shows snapshots plotted every 4.4 μs (microseconds). The total time span of the plot is 2.3 ms, which happens to be the period associated with this string's natural frequency of 440 Hz.

Since the string is frictionless, this behavior will repeat indefinitely. Figure 5.1.4 shows how the shape of the initial pluck propagates across the length of the string. The initial shape at time 0.0 is shown in Figure 5.1.4 at point A. As time progresses the shape of the wave evolves to a shape represented by point B, and then to a shape represented by point C, at which time the cycle repeats.

Upon close study of the *first* snapshot in Figure 5.1.4 *after* the initial condition, one will perhaps see a "kink" in the profile. A better picture is shown in Figure 5.1.5, where points A and B represent two snapshots of the string shape at times 0.0023 ms and 0.0046 ms. These snapshots show the splitting of the initial profile into two waves that will travel left and right to the fixed connections and then reflect back. The wave initially moving to the left has a much shorter distance to travel, hence the kink. Later on, we will look at a different kind of initial profile that will illustrate this better.

Alternatively, one can look at individual snapshots at various moments. Figure 5.1.6 shows three snapshots, all while the plucked peak is propagating to the right.

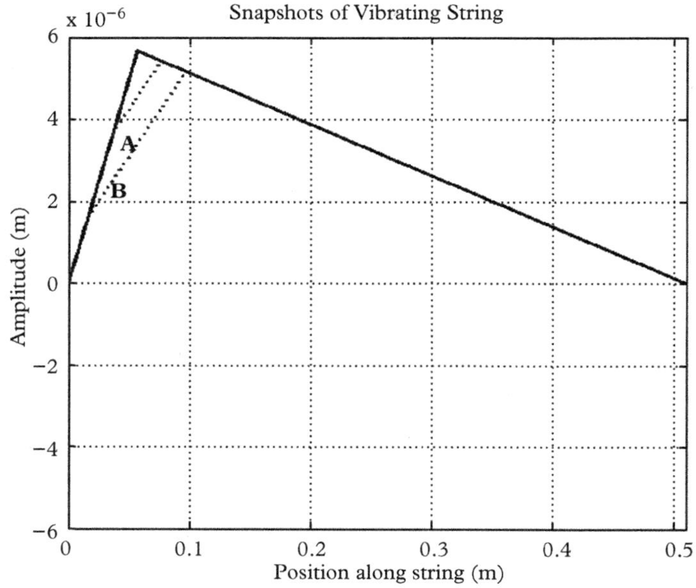

Figure 5.1.5 *Two snapshots of the vibrating string response showing the kink in the profile*

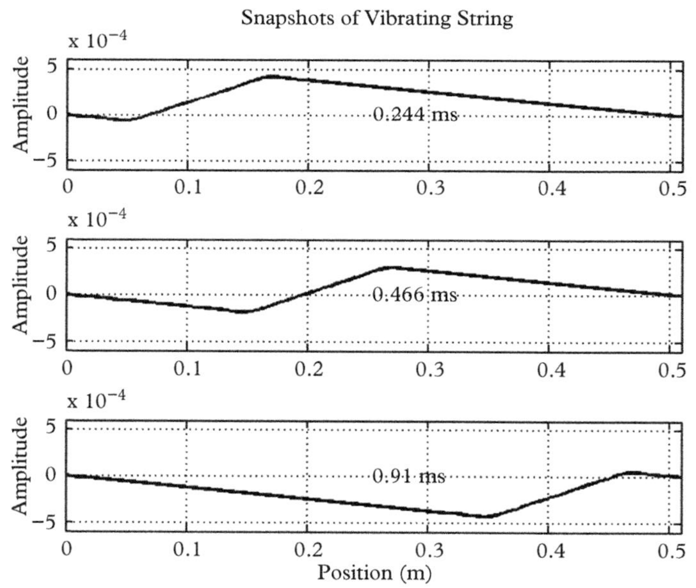

Figure 5.1.6 *Three snapshots of the vibrating string response showing how the peak propagates*

Close inspection of Figures 5.1.4–5.1.6 shows that at time zero the shape of the wave takes on the initial shape shown in the leftmost portion, point A of Figure 5.1.4. Instead of a smooth curve, note the sharpness of the corner at time zero in Figure 5.1.4. As time progresses and as the string is released, the peak of the transverse displacement of the wave appears to decrease in height and propagate longitudinally toward the rigid connection on the right, at $x = 0.51$ m. When the peak reaches the rigid connection at 0.51 m, it is reflected, and the now-negative trough appears to move from right to left until at 2.3 ms it returns almost to its initial shape. This makes sense because the propagation speed of a disturbance for this string is 448 m s^{-1}, or 0.448 m $(ms)^{-1}$, and since the length is 0.51 m it will take

$$\frac{0.51 \text{ m}}{0.448 \text{ m } (\text{ms})^{-1}} = 1.136 \text{ ms}$$

to travel the length of the string. By the end of 2.27 ms, the disturbance will have propagated back to the other end of the string. In 1 s, the wave will propagate back and forth along the string 440 times.

The reader can get the most insight from watching the video of the vibrating string that can be accessed from the website entitled "Response of a Flexible String to Three Initial Conditions" (no. 5). The file containing the Matlab script generating the figures used in Chapter 2, FiguresPart2_Chap5.m, also contains a short movie of the string movement. Appendix 2 discusses the mathematical methods used to generate the plotted data and explains how the initial sharp-edged profile evolves into the smoother shapes.

Gathering the String's Sound: Three Approaches

What kind of "sound" would emanate from this plucked string? The end points of the string are fixed, so, without a soundboard, the amount of sound energy transmitted to the surrounding air and then to our ears by just the vibrating string would be relatively small.

Method One—Averaging

However small the effect would be, let us assume that each element along the string displaces air particles vertically or transversely and generates local longitudinal sound waves. Therefore, at each instant of time, t, we will add up the transverse displacements, $y(x, t)$, at every "spot," x, along the string. In other words, at each instant of time, we would go to a graph like that in Figure 5.1.6 and take the average of the displacements over the entire length of the string, and obtain a value $S(t)$. (You may wish to skip the next sentence.) Effectively, we are carrying out the following:

$$S(t) = \frac{1}{L}\int_0^L y(x, t)\, dx \cong \frac{1}{L}\sum_k y(x_k, t)\, \Delta x.$$

The signal $S(t)$ represents a longitudinal sound wave that impinges upon our eardrums. Figure 5.1.7 shows a schematic of how the transverse string waves generate longitudinal sound waves in air.

Figure 5.1.8 shows what this wave $S(t)$ (normalized to have unit amplitude), however weak, looks like. This longitudinal wave is periodic (with a period equal to the natural period of the string) but it is not a pure sine wave. The line spectrum of this wave is given in Figure 5.1.9. The frequencies along with the number of the harmonic are shown in this spectrum.

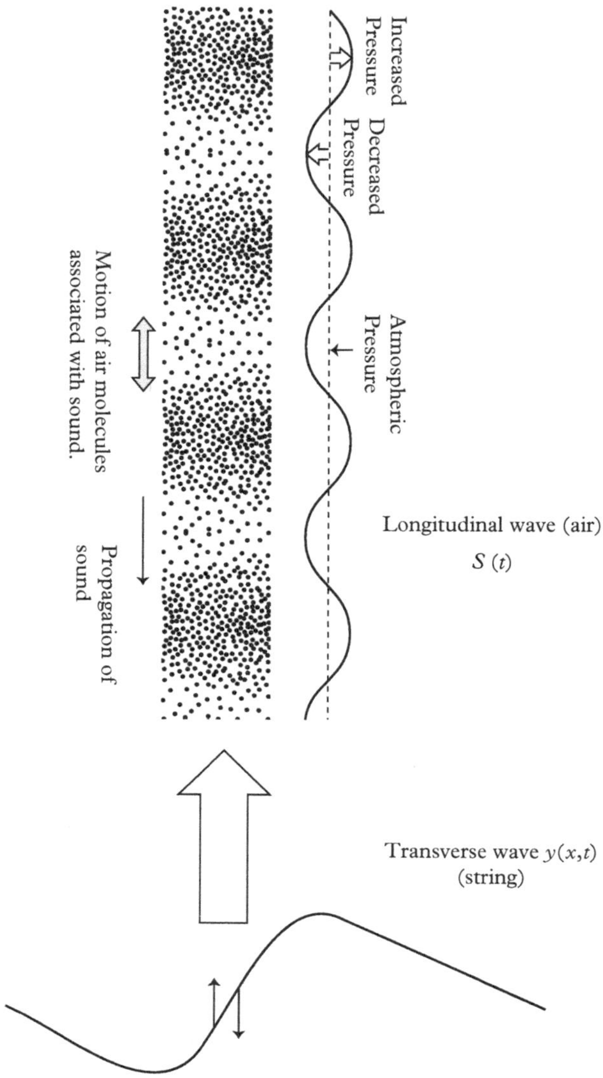

Figure 5.1.7 *The evolution of the transverse string vibrations into longitudinal vibrations in air*

At the natural frequency of 440 Hz there is a line, and there are lines near the odd harmonics of 1320 Hz, 2200 Hz, and 3080 Hz. The even harmonics are more than 60 dB down. Also, the ninth harmonic at 3690 Hz is lower than perhaps expected. Why?

In the summing process, any sinusoidal components in the wave with even harmonics would cancel each other just as they did when we analyzed the square wave in Chapter 1. A sine wave with a frequency equal to an even multiple of the fundamental frequency would have as many positive half cycles and negative half cycles, so there would be no net average wave creation.

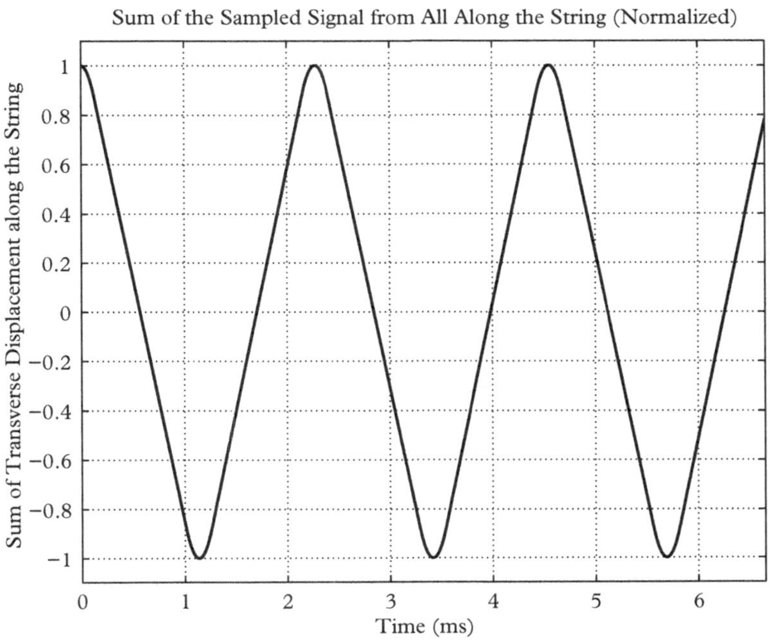

Figure 5.1.8 *The signal generated by summing the sampled transverse vibrations over the length of the string at every instant of time*

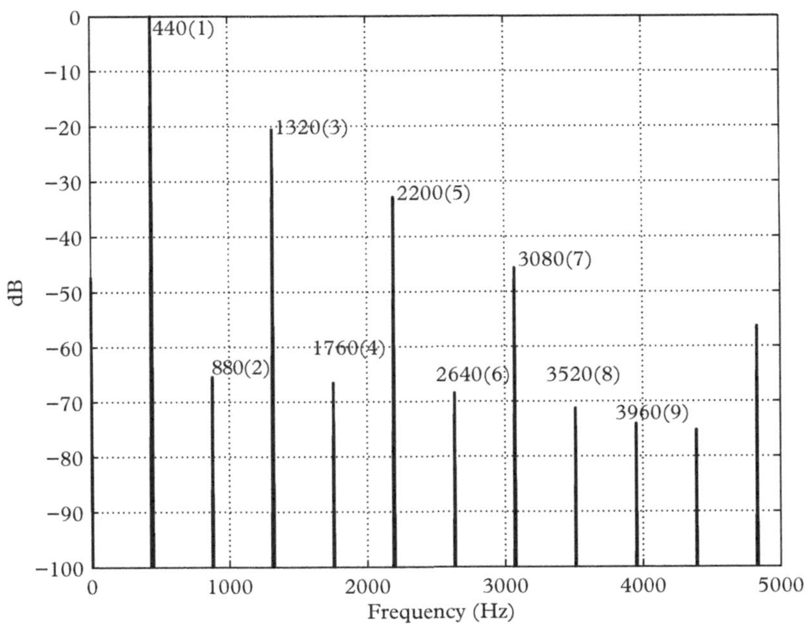

Figure 5.1.9 *Line spectrum of sum (over the string) of transverse vibrations*

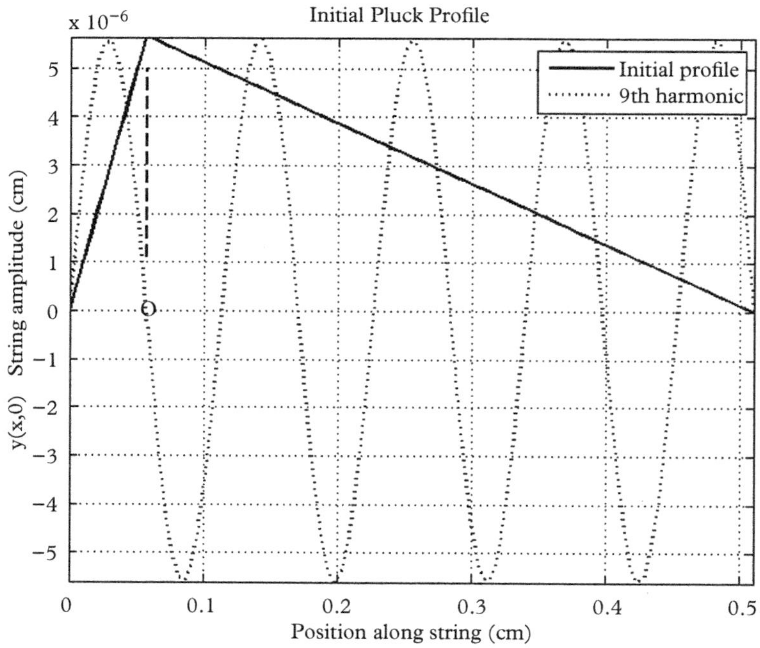

Figure 5.1.10 *Initial profile along with the ninth harmonic showing that the ninth harmonic has a zero at the peak of the initial profile*

The ninth harmonic is weak because, as Figure 5.1.10 shows, that component would have a wavelength such that each peak is cancelled by a valley. (Note that the first zero crossing in Figure 5.1.10, designated by the small circle and the vertical dashed line, coincides with the peak in the initial profile.)

To support this contention, Figure 5.1.11 shows the strengths of the harmonics required to fit the initial profile. (Sections 1.3 and 2.1 showed how a square wave is approximated by a sum of weighted harmonics. Also, you can look at the development of the Fourier series in Appendix 1.) Here we see that the ninth harmonic is missing. I am not going to show any math to support the following claim, which is that the string responds to the initial profile by treating each harmonic of that profile as an input. As time progresses, each harmonic is modified in amplitude. Since there is no ninth harmonic in the initial profile there is effectively none in the spectrum of the collected wave.

Thus, the effect of the wave propagating back and forth across the string is a minuscule back and forth movement of the surrounding air particles, which we have represented by $S(t)$ and that for someone with really sensitive ears would sound like 440 Hz.

Method Two—Sampling

We can theoretically collect the displacements at *just one point* along the string at every instant of time. For example, we collected the displacements at $L/9$, the spot where the initial pluck profile has its apex. The sampled sound wave, along with its line spectrum, is shown in Figure 5.1.12.

The reader should take a moment to think about the shape of the wave in Figure 5.1.12(a). The reason for its shape can be deduced by picking a spot along the position axis in Figure 5.1.4 or

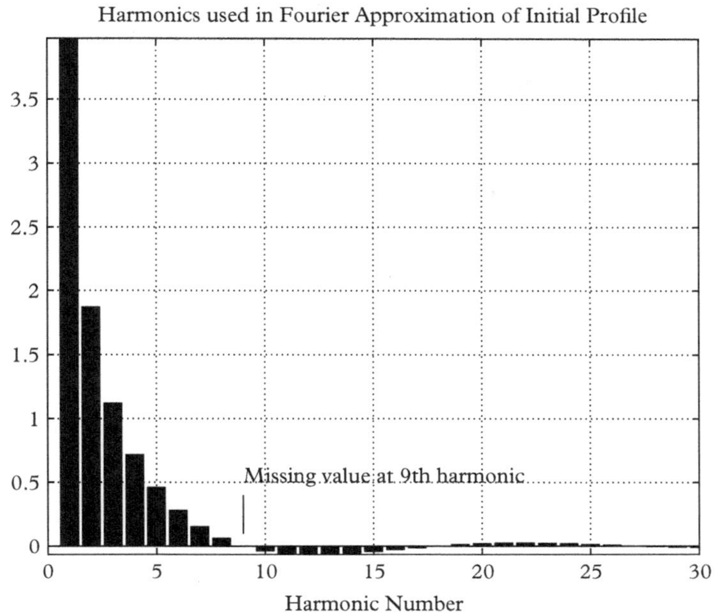

Figure 5.1.11 *Harmonics used in the Fourier series approximation of the initial profile showing the missing ninth harmonic*

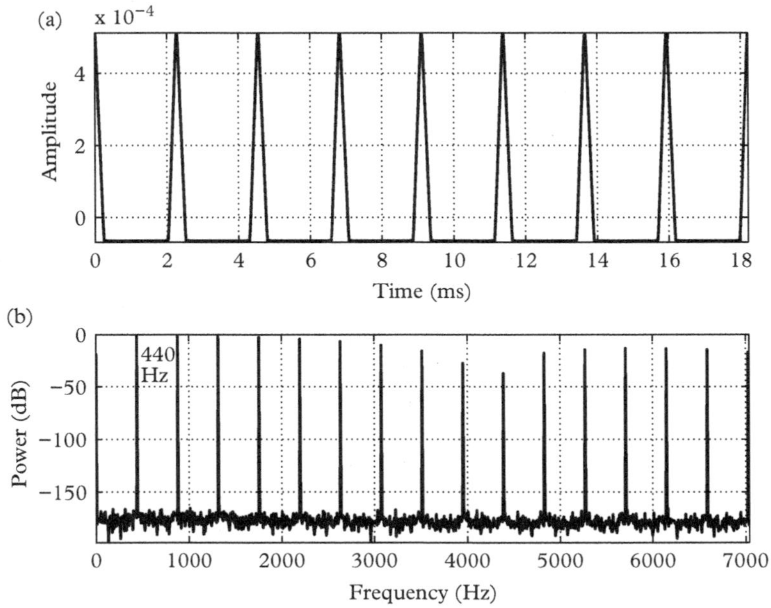

Figure 5.1.12 *(a) Samples at 5.7 cm along the wave, and (b) the line spectrum of the samples*

Figure 5.1.6; namely at 1/9th of the length, or 5.7 cm, and tracing the time behavior. Every 2.27 ms the wave passes by this spot and generates a pulse.

The line spectrum in Figure 5.1.12(b) shows both the odd and the even harmonics because there is no summing. Also, because of the sharpness of the wave, many harmonics occur in the spectrum. Compare this with Figure 5.1.9 where there is summing.

Method Three—Force on the Fixed Connection

Considering the force on the fixed connection at either end of the string due to the movement of the string would provide a third alternative to gathering the sound effect. The effect of the vibration of the string on the end connection would be a force proportional to the slope of the string at that point. This force could then be transmitted to a soundboard that would be rigidly attached to the fixed connection.

Why is the force of the vibrating string on the fixed connection proportional to the slope of the string with respect to position? Consider Figure 5.1.13, which shows the right-hand fixed connection (point D) with the string in three different orientations at different moments in time. The horizontal orientation B of the string has a rate of change with respect to position (or slope) of zero and exerts a horizontal force on the fixed connection proportional to the tension in the string, but there is no vertical force. Orientation A exerts a downward force on the fixed connection proportional to its slope, which is negative. Orientation C exerts an upward force on the fixed connection proportional to its slope, which is positive.

We have chosen the right-hand end connection where $x = L$ and have calculated the slope, which is simply the change in the transverse movement of the string y per change in position x at the right-hand end of the string. If the slope is high then the force is high, and vice versa.

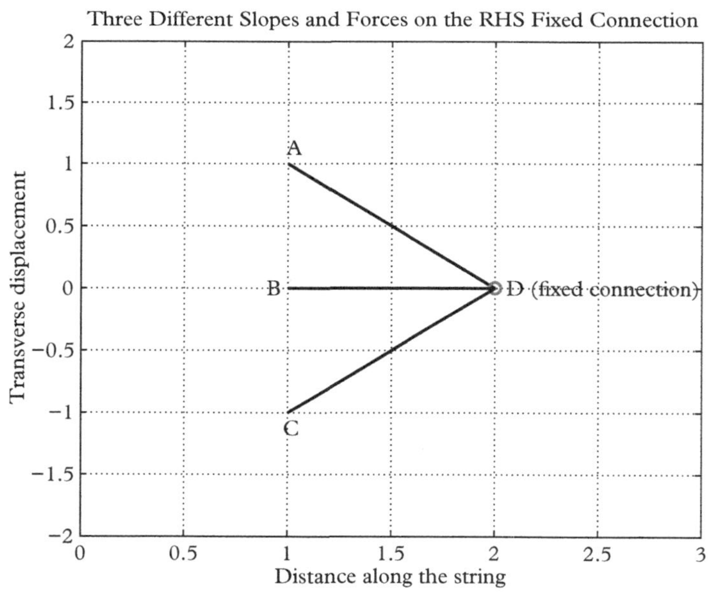

Figure 5.1.13 *Three different orientations of the string at the right-hand side*

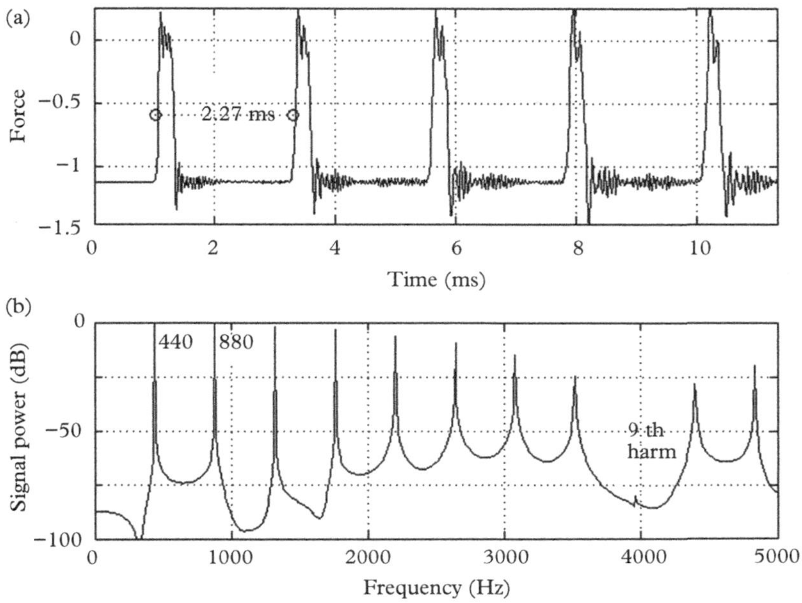

Figure 5.1.14 *(a) Time domain plot of the force on the right-hand side termination of string, and (b) the line spectrum of the force showing the missing ninth harmonic*

Figure 5.1.14(a) shows a plot of the force (actually the slope of the string) on the right-hand side fixed connection as a function of time, and Figure 5.1.14(b) shows the line spectrum of the force. Note the shape of the force in the time domain. A pulse results from the arrival of the traveling wave at the fixed end connections every 2.27 ms. The sharpness of the pulse is a result of the relatively sudden reflection and inversion of the wave at the fixed end. This kind of whiplash motion causes the appearance of the higher harmonics in the line spectrum. Note the absence of a peak in Figure 5.1.14(b) at the ninth harmonic due to the location of the pluck at 1/9th of the length of the string.

The tone from the force of a vibrating string on the bridge appears to be a result of the spike that occurs during the reflection and inversion of the traveling wave. This is contrary to the idea that a standing wave generates the sound of a vibrating string. We will deal with this in more detail when we compare flexible and stiff strings in Sections 5.8 and 5.9.

Newton's third law of motion is "When one body exerts a force on a second body, the second body simultaneously exerts a force equal in magnitude and opposite in direction to that of the first body." In this case the first body is the string and the second body is the fixed connection. Because this string does not have elasticity, the opposing force does not cause the shape of the string to change. Later on, when we look at stiff strings, that will not be the case.

Comments by Alfred Dolge in his book *Pianos and Their Makers*, published in 1911, are of interest. The following is a quote:

> The ear does not perceive sound through so-called tone waves, but because of the shock or jolt by which the sound is created. Consequently, Hansing [Siefried Hansing, author of *The*

Pianoforte and Its Acoustic Properties] looks upon the soundboard as a drum, upon which the vibrations of the strings, caused by the striking of the hammer, are delivered as shocks or jolts.

Before moving on to the effect of plucking the string at different locations, let us take a side trip and look at some actual strings.

5.2 Experiments on Single Piano Strings

A single-strand piano wire and a wrapped piano wire are compared when they are plucked and struck at various points along the wire. The tension is chosen such that the observed fundamental frequencies are about the same. The line spectrum, cumulative line spectrum, and autocorrelation are used in the comparison. Some of the results are consistent with expectations, while some are not.

Single-Strand Piano Wire—Plucked

A number 15½ piano wire (1 mm diameter and 79.1 in or 2.0 m in length) is stretched between two acoustic lag screws under a tension of 20 lbs (or 88.9 N) using weights from a weightlifting kit. The string density is assumed to be approximately 7800 kg m^{-3}. This means that the linear density is on the order of

$$\rho \pi D^2/4 = 0.0061 \text{ kg m}^{-1}.$$

A schematic of the experimental set-up is shown in Figure 5.2.1.

The time domain, line spectrum, and autocorrelation for *plucking* the string at one half of its length are shown in Figure 5.2.2.

One can see a beating-like variation in the amplitude (see points A and B) in Figure 5.2.2(a) that is probably the result of the transverse vibration plane varying from vertical to horizontal (out of the plane of the page in Figure 5.2.1). The line spectrum in Figure 5.2.2(b) shows a first partial at 29 Hz (labeled as 1). The higher partials follow in the expected order, with the odd harmonics

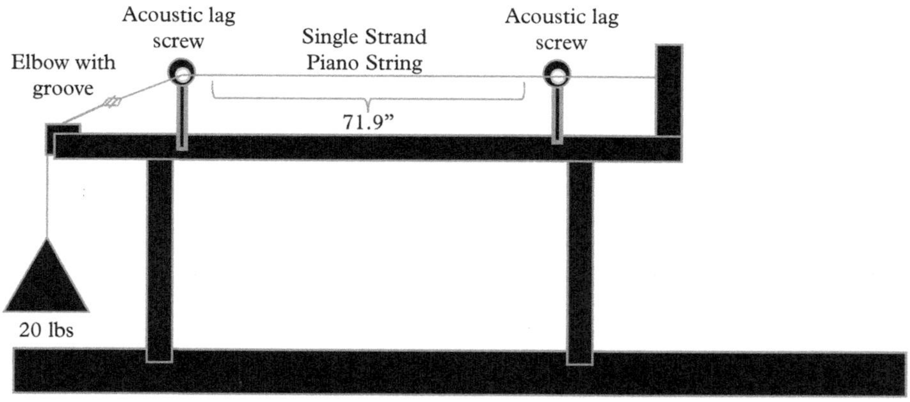

Figure 5.2.1 *Experimental setup for testing a single-strand piano string*

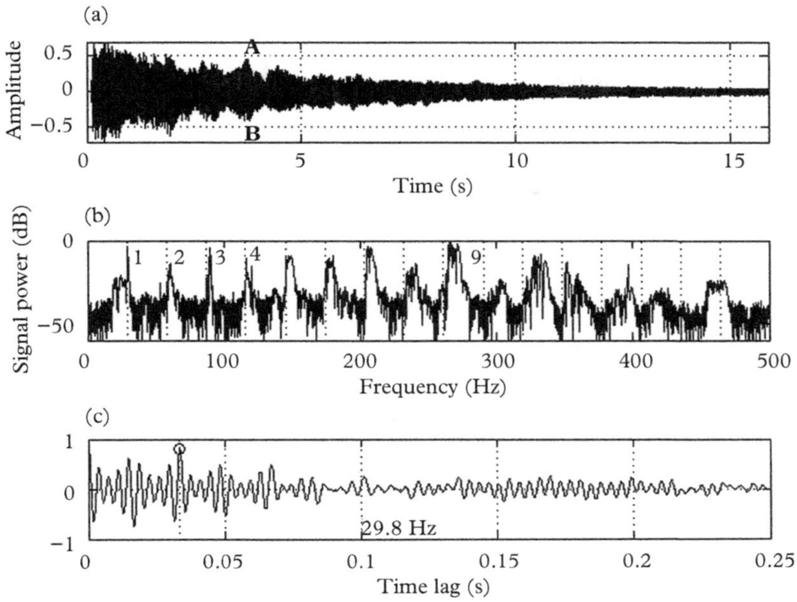

Figure 5.2.2 *Single-strand string plucked at one half of its length: (a) time domain, (b) line spectrum, and (c) autocorrelation*

slightly stronger than the even ones. Dotted vertical lines denote the locations of the harmonics. As the harmonic number increases one sees the location of the partial peak moving to the right of the associated harmonic frequency. This movement of the partial's frequency relative to that of the harmonic will be discussed in Chapter 12. Note that the fundamental component is strong, but that the ninth harmonic (or partial) is the strongest. The autocorrelation in Figure 5.2.2(c) suggests that the pitch is 29 Hz, which is consistent with the first partial.

Using Eq. 5.0.1 (for a perfectly flexible string) with the above-mentioned values of tension, diameter, density, and length, the calculated fundamental frequency is 30 Hz, which is pleasingly close to the observed value of the first partial:

$$f_{\text{nat}} = \sqrt{\frac{\tilde{T}}{4\tilde{\rho}\,L^2}} = \sqrt{\frac{88.96}{4 \times 0.0061 \times 2.01^2}} = 29.98\text{Hz}.$$

This agreement lends some support to the idea that a flexible string without elasticity can be a useable approximation to a real piano wire.

Single-Strand Piano Wire—Struck

The above experiment is repeated, except that the wire is *struck* (not plucked) at one half of its length with a striking stick that has muting felt at the striking point. The line spectrum and the autocorrelation are shown in Figure 5.2.3. The fundamental frequency and the estimated pitch match those of the plucked string. As with the plucked case, the even harmonics in Figure 5.2.3(a) are dominated by the odd ones, with the 7th, 9th, and 11th harmonics being the strongest. The

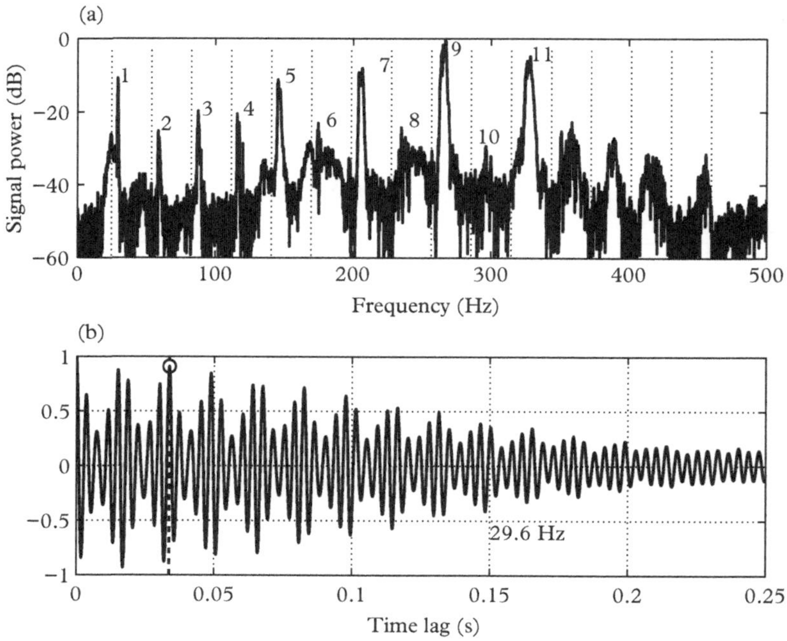

Figure 5.2.3 *Single-strand string struck at one half of its length: (a) line spectrum, and (b) autocorrelation*

dotted vertical lines denote the locations of the harmonics. As the harmonic number increases one sees the location of the partial peak moving to the right of the associated harmonic frequency. The autocorrelation plot in Figure 5.2.3(b) shows that there are several competing peaks for the struck string so the estimation of pitch is less clear than that for the plucked string.

Figure 5.2.4 compares the plucked and struck cases via the cumulative line spectra and the spectral centroids. In both cases the seventh and ninth harmonics dominate. Based on the shapes of the cumulative line spectra and the values of the spectral centroids, one might conclude that plucking generates more power in lower harmonics than does striking. In both the plucked and struck cases, the fundamental power is relatively weak.

Plucked Wrapped Piano Bass String

The above experiment is repeated with a 56 in (1.422 m) wrapped piano bass string subject to 250 lbs (1112 N) tension. The wrapped piano string is much denser and stiffer than the number 15½ wire used above, so the applied tension is much higher. Figure 5.2.5 shows the schematic of the device used in the experiment.

The string was *plucked* at 1/9th of its length. The time domain plot is similar to that in Figure 5.2.2 and is not shown. The line spectrum and autocorrelation are shown in Figure 5.2.6. The fundamental frequency of the plucked string is 23.6 Hz—labeled as 1 in Figure 5.2.6(a). The partials appear in the expected order and spacing except that the ninth harmonic at 211.5 Hz is missing (or greatly diminished), likely as a consequence of the location of the pluck. As with the plucked single-strand string, dotted vertical lines denote the locations of the harmonics. As the

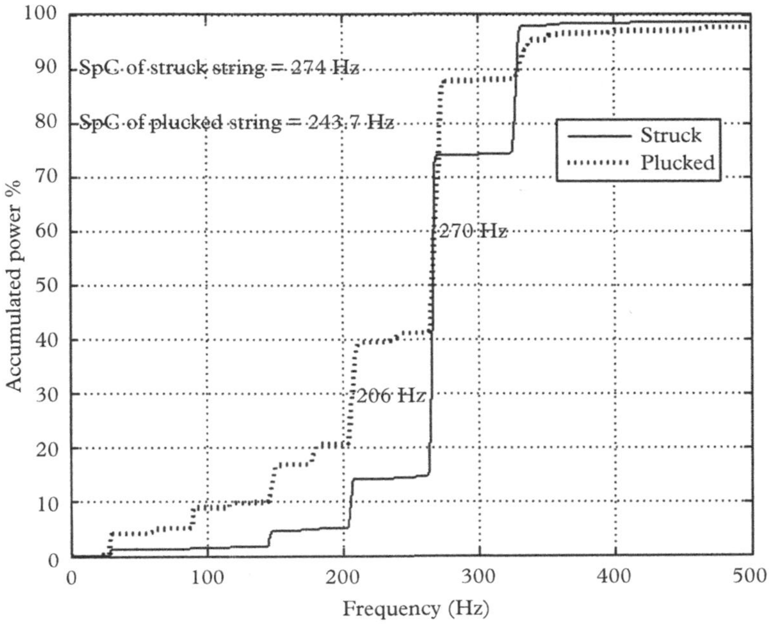

Figure 5.2.4 *Cumulative line spectra of single-strand string plucked and struck at one half of its length*

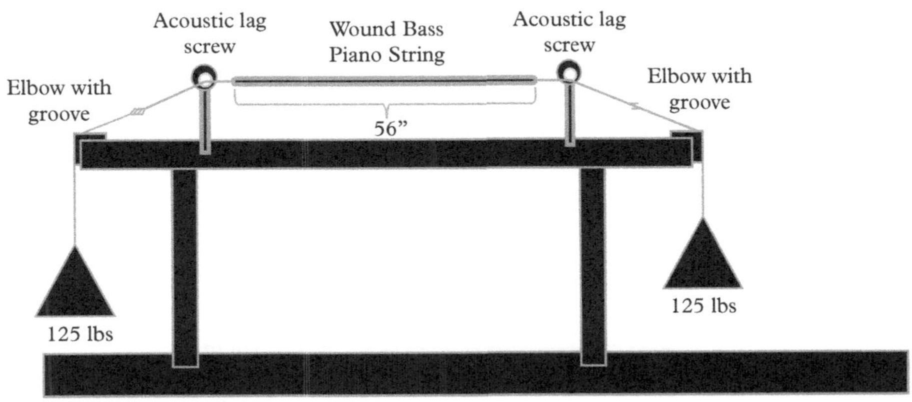

Figure 5.2.5 *Experimental setup for testing a wrapped piano string*

harmonic number increases one sees the location of the partial peak moving to the right of the associated harmonic frequency.

Regarding the autocorrelation in Figure 5.2.6(b), there is not a single prominent peak with a value over 0.45. The first major peak corresponds to a frequency of 23.6 Hz, which agrees with the line spectrum. However, the maximum peak corresponds to a frequency of 7.9 Hz. Hence, the idea of estimating the pitch of this wave from the autocorrelation is a bit strained.

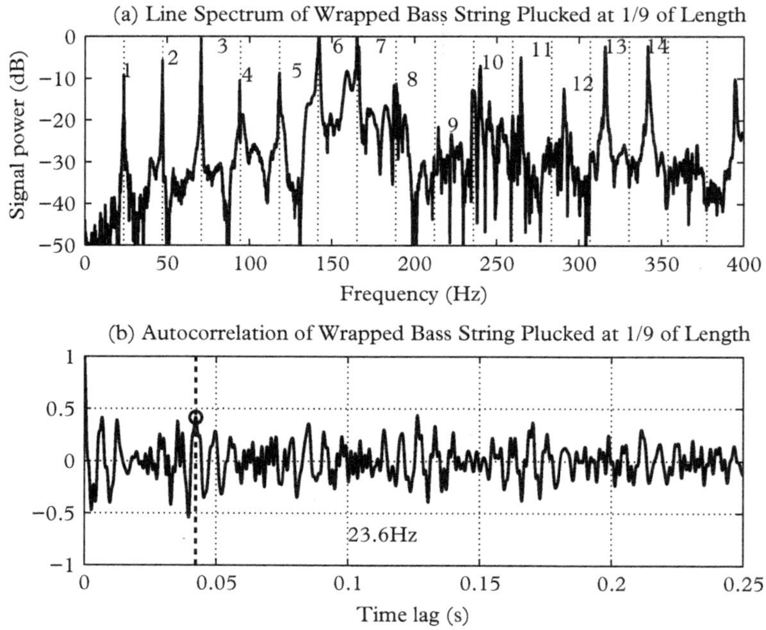

Figure 5.2.6 *A plucked wrapped bass string: (a) line spectrum, and (b) autocorrelation*

Struck Wrapped Piano Bass String

The string is *struck* at 1/9th of its length using a stick with a layer of piano felt on the tip. The fundamental frequency in Figure 5.2.7(a) is approximately 23.9 Hz, as it was for the plucked case. Again, note the diminished ninth partial. The autocorrelation in Figure 5.2.7(b) shows that there is one prominent peak associated with the fundamental of 23.9 Hz.

It does not make much sense to estimate the fundamental frequency using Eq. 5.0.1 because the winding on the string makes the idea of an overall density a bit fuzzy. Equation 5.0.1 can be turned around and solved for the effective linear density, as in

$$\tilde{\rho} = \frac{\tilde{T}}{f_{\text{nat}}^2 4L^2} = \frac{1112}{(23.5)^2 \times 4 \times (1.4224)^2} = 0.249\text{kg m}^{-1},$$

which is a couple of orders of magnitude greater than that for the single-strand piano wire. However, the wrapped wire is not only greater in diameter but significantly more stiff, so using Eq. 5.0.1 is quite a stretch.

Figure 5.2.8 shows the cumulative line spectra of the plucked and struck cases, which along with the spectral centroids suggest that the act of plucking the wrapped string excites the higher harmonics more than the act of striking. This is contrary to the results for the single-strand wire.

In both cases for the wrapped piano wire the strength of the fundamental is relatively weak, especially when compared to the performance of the single-strand wire where the fundamental is a little stronger but still weak compared to its other partials. We could, perhaps, attribute this

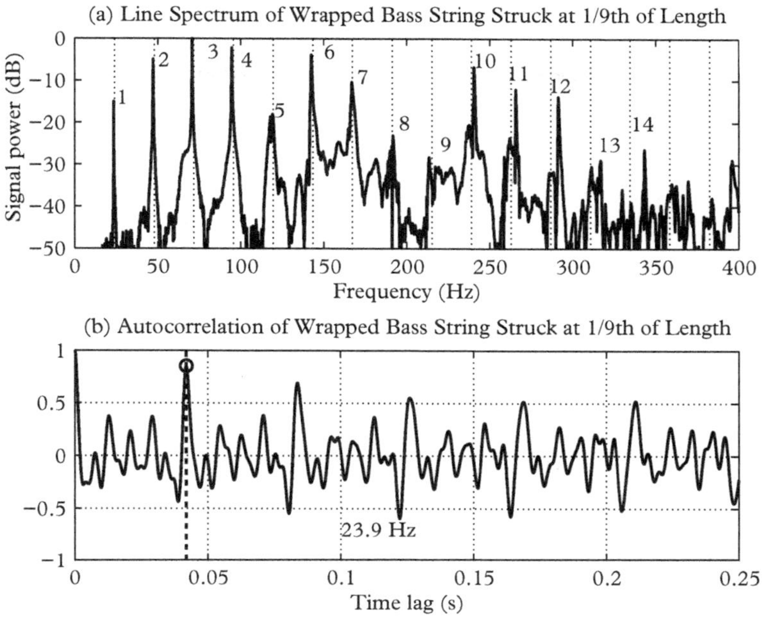

Figure 5.2.7 *A struck wrapped bass string: (a) line spectrum, and (b) autocorrelation*

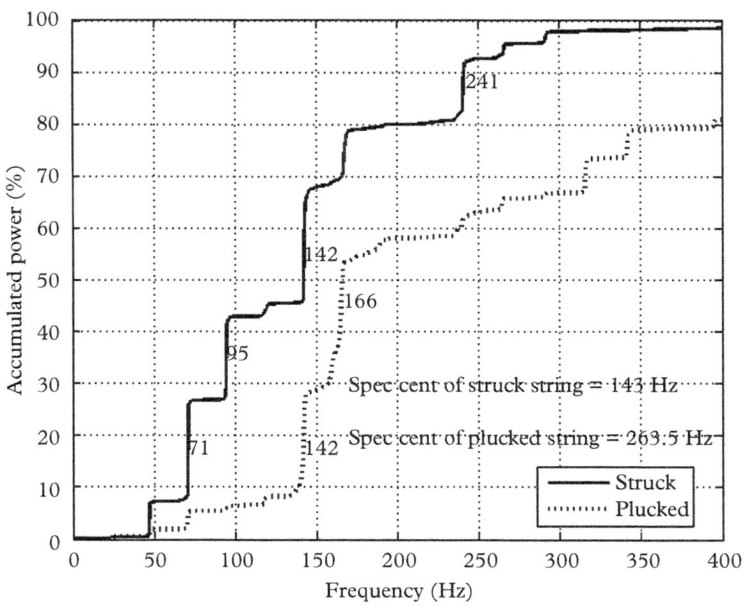

Figure 5.2.8 *Cumulative line spectra of plucked and struck wrapped bass strings along with the spectral centroids*

weakness in the fundamental for the wrapped wire (compared to the strength of the fundamental in the single-strand wire) to its stiffness and elasticity. This suggests that a stiff wire would have more resistance to the long-period vibration of the fundamental but more readily produce higher harmonics. Admittedly, this line of logic is a bit strained.

In Part 3 we will take a detailed look at all of the strings in a piano and find that the fundamental frequency of the wrapped strings for the lowest notes is, in fact, weak. We will also find that the partial frequencies tend to creep past the harmonic frequencies — a phenomenon called inharmonicity.

5.3 Two Stringed Musical Instruments

A Guitar

As an example of a plucked string, consider plucking a guitar string and generating the A3 note. Figure 5.3.1 shows the line spectrum, the cumulative line spectrum, and the autocorrelation of the guitar note. The fundamental is at 220 Hz and a second partial is at 441 Hz. Other partials appear in the line spectrum at 662 Hz, 884 Hz, and 1102 Hz, but, as the cumulative line spectrum shows, they are greatly diminished compared to the fundamental and the second harmonic. The autocorrelation produces several well-defined peaks. I chose the first one, which points to 220 Hz, to be the pitch since it is consistent with the line spectrum.

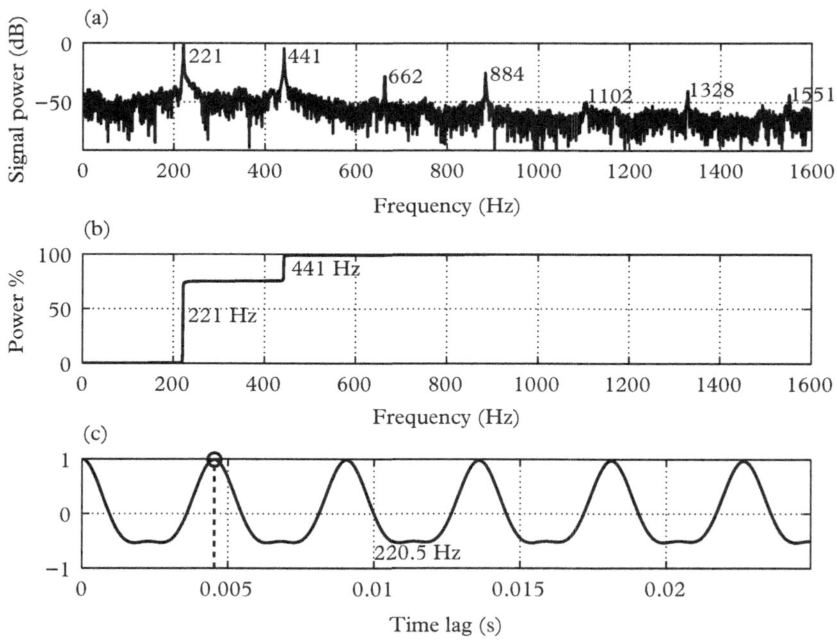

Figure 5.3.1 *Guitar A3: (a) line spectrum, (b) cumulative line spectrum, and (c) autocorrelation*

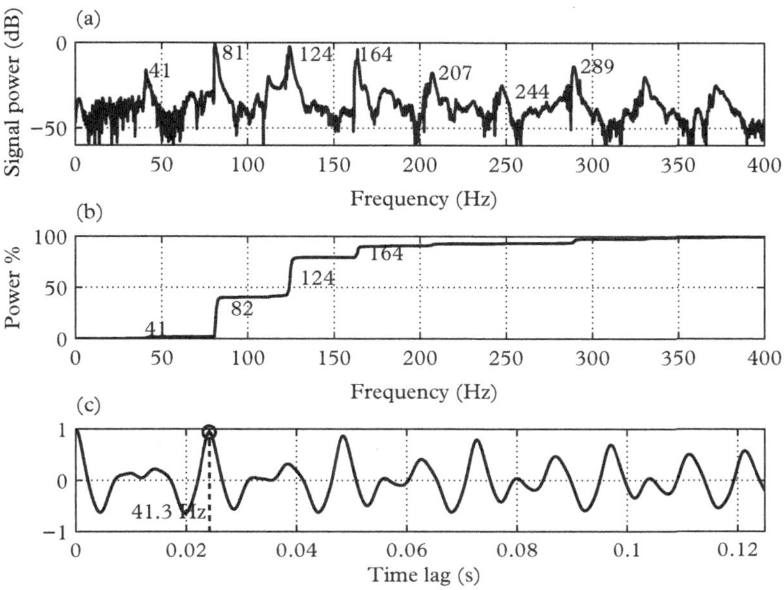

Figure 5.3.2 *Bass Violin E1: (a) line spectrum, (b) cumulative line spectrum, and (c) autocorrelation*

A Double Bass

As further examples of plucked strings, consider three notes from the double bass violin. Figure 5.3.2 shows the line spectrum and cumulative line spectrum of the E1 note on a double bass. The first peak in Figure 5.3.2(a) appears at 41 Hz, which corresponds to E1 (41.2 Hz). Even though it is weak compared to the second, third, and fourth partials, the spacing of the harmonics confirms that the E1 note has been plucked. Figure 5.3.2(b) shows that the second and third harmonics deliver the most signal strength while the fundamental contributes no more than 2% of the total signal. According to Figure 5.3.2(c), the pitch is 41 Hz.

Figure 5.3.3 shows the line and cumulative line spectra for the B2 note on the double bass violin. The fundamental at B2 (123.5 Hz) appears to have the most signal intensity (almost 75% of the signal strength).

Finally, Figure 5.3.4 shows the D4 note where the fundamental at 292.5 Hz (D4) represents approximately 90% of signal.

These nine graphs for the three bass string notes show that as the fundamental frequency of the note increases the fundamental harmonic takes on a higher percentage of the power. This is a characteristic we will see again in Part 3 when we look at pianos.

5.4 Plucking at the Center of the Idealized Flexible String

We return to the ideal vibrating string where the initial profile is now symmetrical and is shown in Figure 5.4.1(a). Note that the second harmonic is also plotted as a clue that the line spectrum of the collected wave will be missing the second harmonic. The line spectrum of the initial profile in Figure 5.4.1(b) shows that *all* of the *even* harmonics are missing. As mentioned above, the

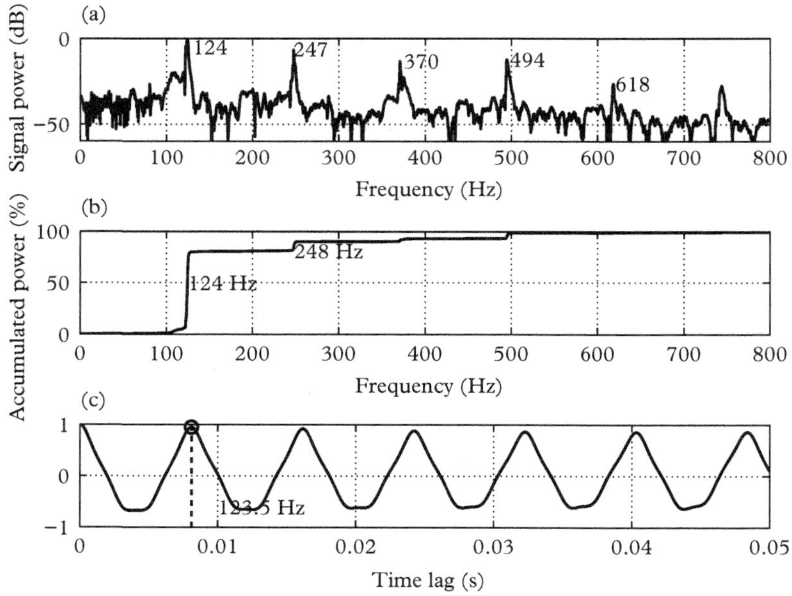

Figure 5.3.3 *Bass Violin B2: (a) line spectrum, (b) cumulative line spectrum, and (c) autocorrelation*

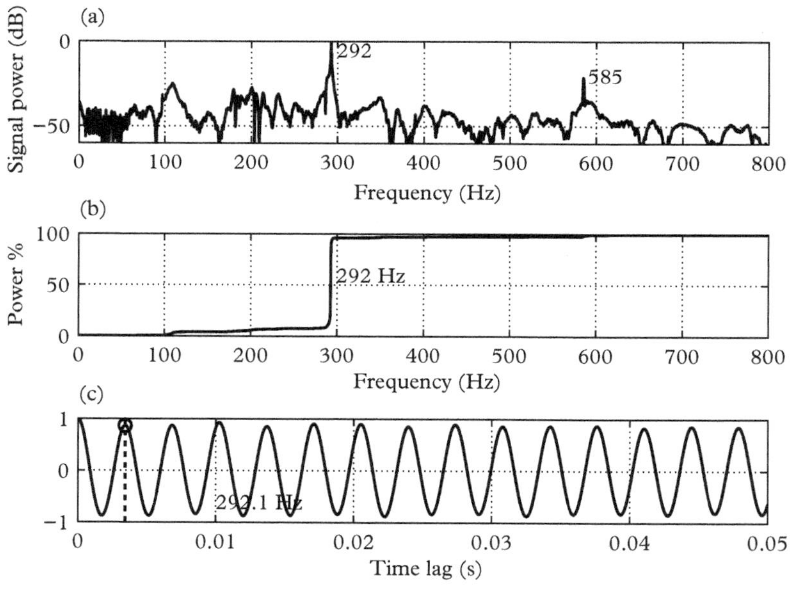

Figure 5.3.4 *Bass Violin D4: (a) line spectrum, (b) cumulative line spectrum, and (c) autocorrelation*

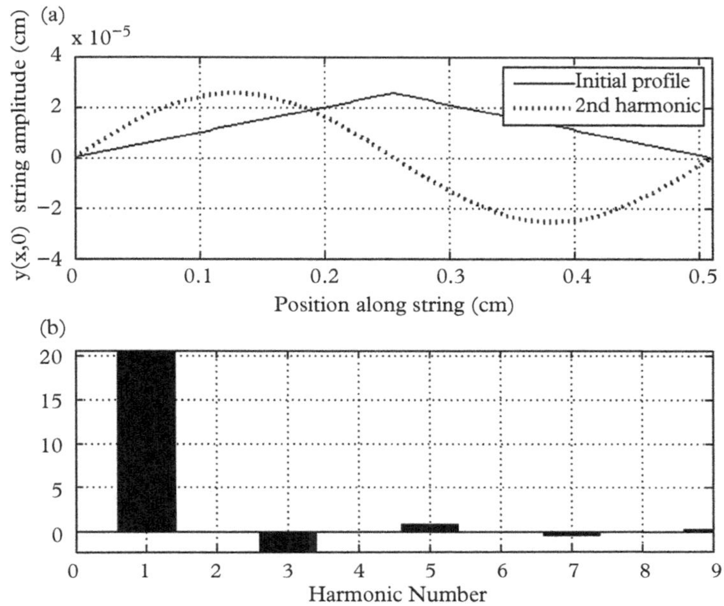

Figure 5.4.1 *Plucking at the center of a string: (a) initial profile and second harmonic, (b) harmonics used in the Fourier series approximation of the initial profile*

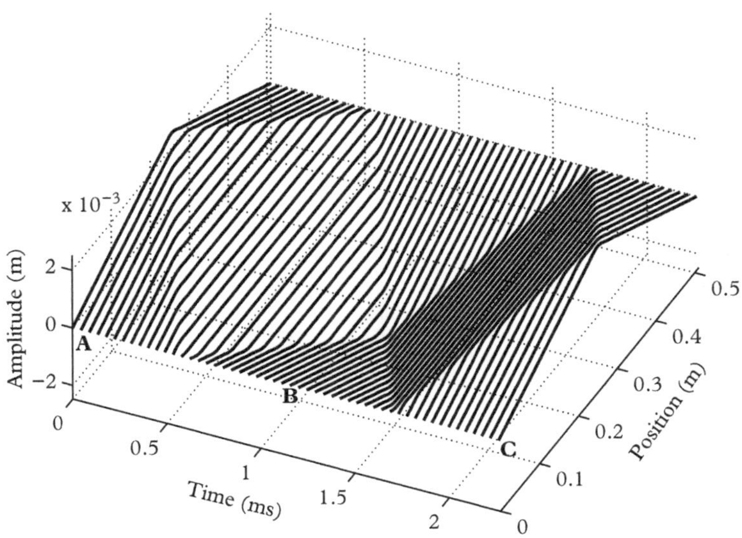

Figure 5.4.2 *Snapshots of the string response to plucking at the center*

harmonics of the initial profile are the inputs to the vibrating string, so the collected wave will exhibit effectively no even harmonics in its line spectrum. Note how dominant the amplitude of the first harmonic is.

Figure 5.4.2 shows a 3D plot of snapshots taken every 4.4 μs. The pluck is initiated at time A. At time B the wave has reached the lowest point and starts to return to the original shape at time C.

Because the pluck point was at the center of the string, the shape retains symmetry as time progresses.

The YouTube video referenced in the book's website (no. 5) shows the dynamic movement of the string for the two cases of a pluck at 1/9 and 1/2 of the string's length, plus a third case where the pluck point is slightly off center. The reader with access to Matlab can also see the movie by running the script in the file FiguresPart2_Chap5.m, which is available on the website.

5.5 A Symmetrical Initial Profile

Consider the initial profile shown in Figure 5.5.1 for a string 1.0 m long. We will use this shape later on and refer to it as Gaussian.

The response is shown in Figure 5.5.2, where the initial symmetrical "bump" at $t = 0\,\text{ms}$ (at point A) quickly splits into two smaller bumps moving away from each other. The two small bumps propagate to points B and C where they reach the fixed connections and are reflected (and inverted) and move back together, forming a single symmetrical trough at point D. Then the

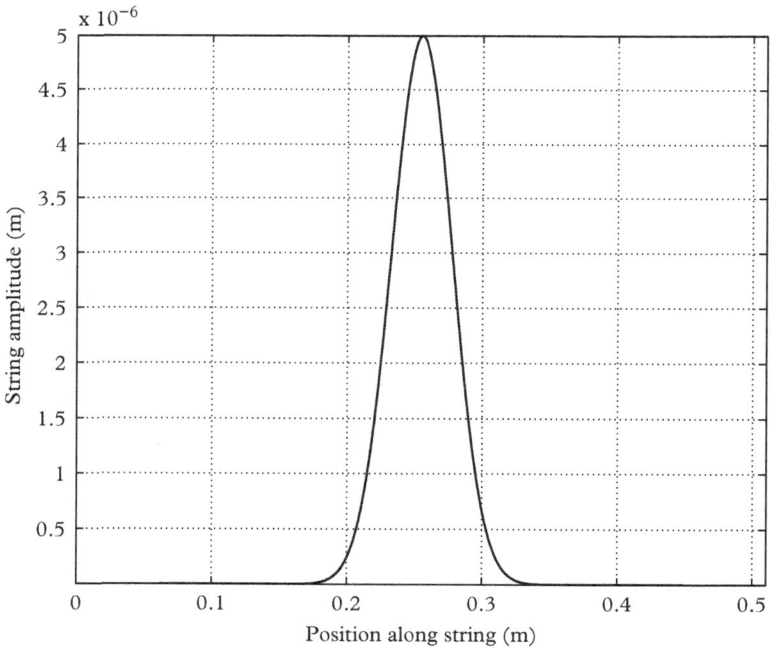

Figure 5.5.1 *A Gaussian initial profile*

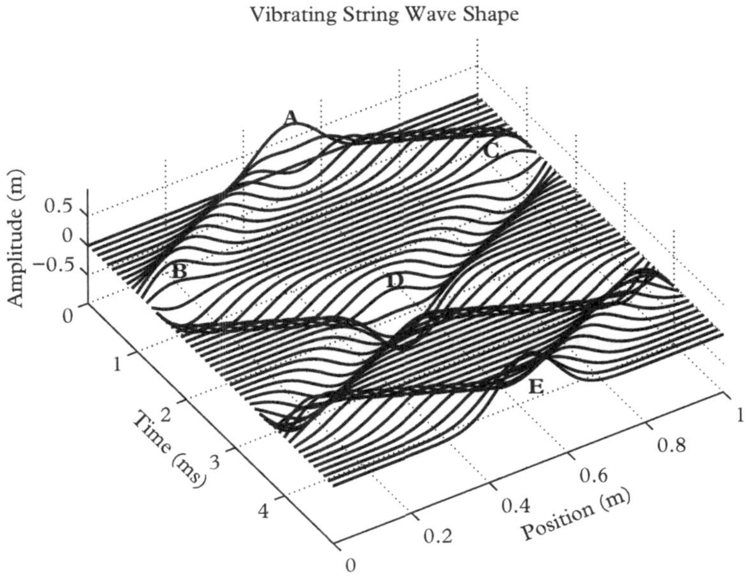

Figure 5.5.2 *Snapshots of the response of a flexible string to an initial Gaussian profile*

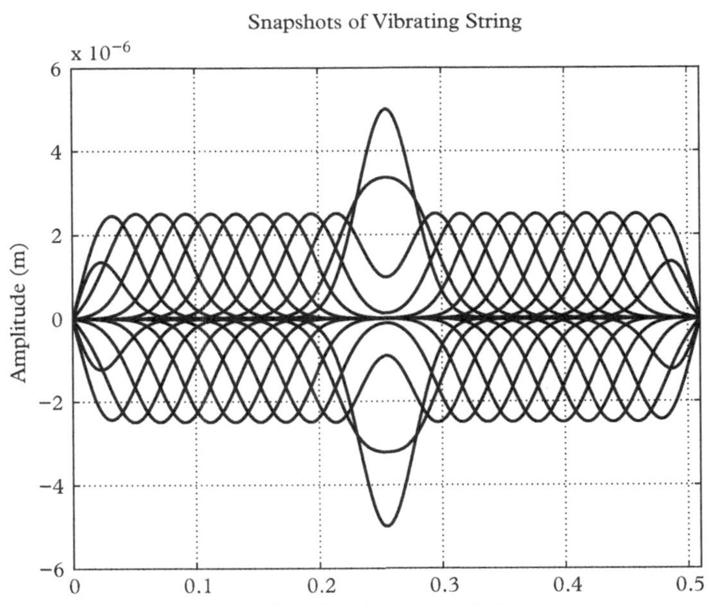

Figure 5.5.3 *Snapshots of the response of a flexible string to an initial Gaussian profile*

splitting repeats itself, completing the cycle at t = 5.0 ms at point E. You might remember that this splitting also occurred when the initial profile was a pluck at ⅙th of the string length; however, it was much less evident because one side was significantly shorter.

This behavior is to be compared to the symmetrical plucking case shown in Figure 5.4.2.

An alternative way of plotting the same data is shown in Figure 5.5.3. Here, the initial single peak breaks into two small peaks that retain their height until the cycle is completed when they coalesce into one higher peak. It will provide some insight when we look at the vibrating membrane. There is a dynamic presentation of this profile in the Matlab script in the file FiguresPart2_Chap5.m.

5.6 Adding a Soundboard at One End of the String

The diagram in Figure 5.6.1 is an attempt to illustrate the key parts of a piano soundboard. The fixed connection on the right end of the string is replaced with a non-rigid connection to a mass (representing the bridge and the soundboard) that can move in response to an applied force.

The string is attached at the tuning pin and the hitch pin. The tension can be changed by adjusting the tuning pin. The string is drawn over the bridge, which is rigidly attached to the soundboard, which, in turn, is rigidly mounted at both ends. When the string between the tuning pin and the bridge (the so-called speaking length) vibrates after it has been plucked or struck, the bridge and the soundboard also vibrate. The large surface area of the soundboard moves air and generates sound waves. The soundboard in a real piano is often a thin sheet of Sitka spruce with a surface area of approximately 1 m^2. A piano soundboard has a convex shape to add to the stiffness, and the non-speaking length of the string from the bridge to the hitch pin is often muted with felt. (Note that the board in Figure 5.6.1 purposely exhibits an exaggerated asymmetry, as do the soundboards in many pianos.)

For the sake of comparison, the idealized string studied in Section 5.1 had a length of 0.51 m and, if it had a diameter of 1 mm or 0.001 m, its cross-sectional area (lengthwise) would be 0.51 m × 0.001 m = 5.1×10^{-4} m^2. Therefore, one might guesstimate that the 1 m^2 of soundboard would displace a volume of air approximately 2000 times greater than that by the string by itself.

To study the dynamics of this mechanical system we need a simple model that has its basic features. A mass/dashpot/spring combination is the simplest ideal mechanical device that has some of these properties. The schematic of this arrangement is shown in Figure 5.6.2.

The string is, as before, attached at the left end. The other end is attached to a mass that can slide on a rod. The mass, representing the soundboard and bridge, is subject to a restoring force from springs and to a frictional force as it slides up and down the pole. This combination of a mass, springs, and friction is often called a mass/spring/dashpot model.

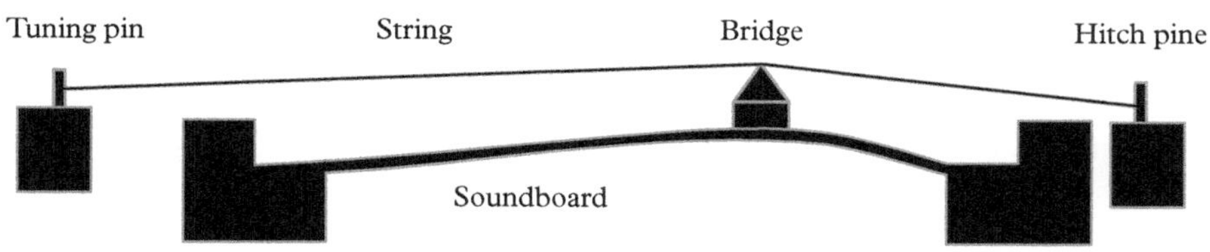

Figure 5.6.1 *Schematic drawing of a soundboard, string, and bridge*

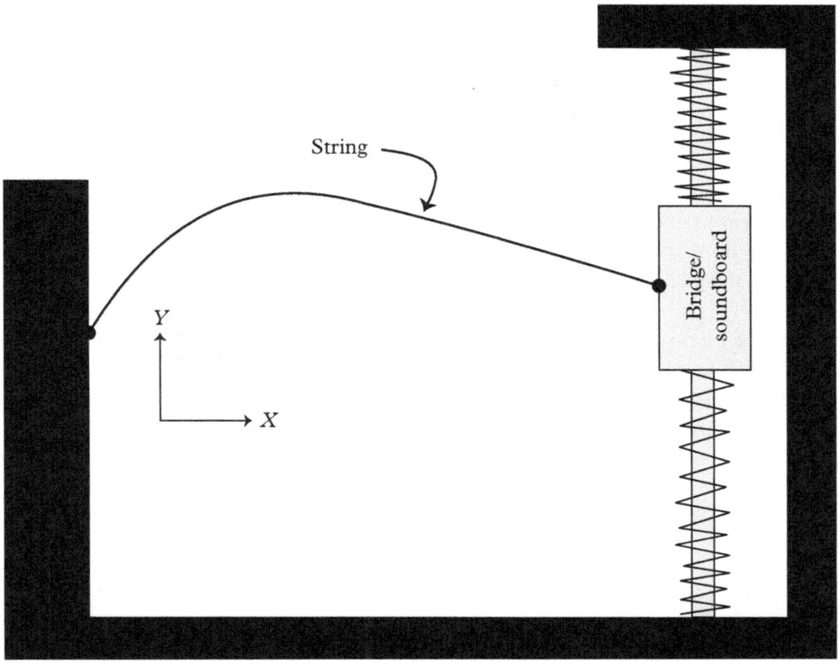

Figure 5.6.2 *Schematic drawing of a soundboard–string–bridge model based on a mass–spring–dashpot idealization*

The two schematics in Figures 5.6.1 and 5.6.2 have little in common physically, but in a dynamic sense the latter will behave approximately as the former. To gain insight, we will study the transient behavior of this idealized model. The bridge/soundboard in Figure 5.6.2 has a mass m, the spring has a spring constant k, and the dashpot has a coefficient of friction B. The movement of the string (under tension $\tilde{T}$) in the transverse direction causes a force to be applied to the mass. When the mass moves, the spring applies a restoring force in the opposite direction. Friction causes the movement to be damped.

The natural frequency of the mass/spring/dashpot system is given by

$$f_{\text{nat}} = \frac{1}{2\pi}\sqrt{\frac{k}{m}}.$$

Therefore, as the mass of the soundboard increases, its natural frequency decreases. Likewise, as the spring constant increases (a stiffer spring or a stiffer soundboard), the natural frequency increases. The natural frequency is not affected by friction; however, friction does affect the rate that oscillations fade away. Page 317 in the book by Arthur Benade, *Fundamentals of Musical Acoustics*, has an extensive qualitative discussion of this ideal model.

The soundboard moves in response to vibrations in the struck piano strings, and significant longitudinal sound waves in the surrounding air can be generated. The model in Figure 5.6.2 attempts to mimic this behavior.

Before moving on, it might be enlightening to present a quote from Del Fandrich (PianoTalk #2, February 1995):

> Soundboards are frequently, though incorrectly, called a piano's amplifier. In fact they do not amplify anything. In engineering terms, soundboards are transducers (a critical distinction). Amplifiers are devices that take small signals and make them larger; they add energy to the original signal. Transducers move energy from one system to another or change energy from one form to another, and this is what piano soundboards do. Energy is "transduced," or changed, from the mechanical wave energy in the string into sound energy in the air surrounding the piano. No energy is added or created in the soundboard.

Mr Fandrich is a piano designer and he often posts on the website "Piano World." The above quote was taken from another website, "PianoTalk" (he also contributed Chapter 21).

I will consider three cases. In each case the natural frequency of the string is 200 Hz, which is near the G3 note, and the string is plucked at 1/9th of its length. For the first case the natural frequency of the soundboard (mass/dashpot/spring) is 100 Hz; that is, lower than the natural frequency of the string. The movement of the string is nearly identical to that shown in Section 5.1 where both ends of the string had fixed connections. The displacement of the soundboard is significantly smaller than the string, and therefore not observable on the same scale as that for the string, but the effect is great because of the soundboard's surface area. Figure 5.6.3

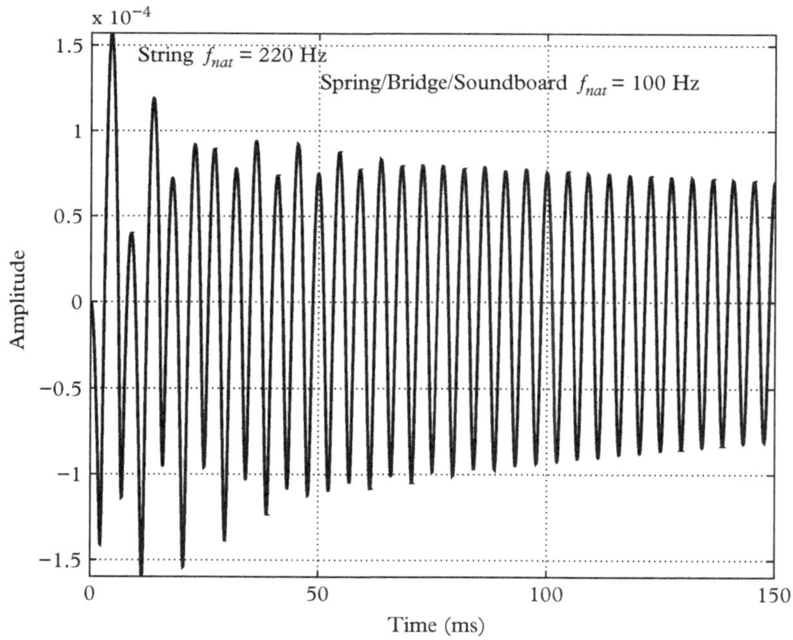

Figure 5.6.3 *Movement of right-hand side bridge/soundboard for a string natural frequency of 220 Hz and soundboard natural frequency of 100 Hz*

shows the displacement of the soundboard (on the right-hand side of Figure 5.6.2) in response to the string being plucked at ⅙th of its length.

The graph suggests that the amplitude of the displacement will gradually fade away due to friction.

The spectra and the autocorrelation for this wave are shown in Figure 5.6.4. The string's fundamental and its harmonics are present in Figure 5.6.4(a) along with a small peak at 100 Hz, which is the natural frequency of the soundboard. Note that all of the harmonics are present, not just the odd harmonics as was the case for the string with the rigid end points in the previous section. The cumulative line spectrum in Figure 5.6.4(b) shows that approximately 90% of the sonic power comes from the string's fundamental frequency. The autocorrelation in Figure 5.6.4(c) shows many peaks, but the second peak is slightly higher than the others and it points to a pitch of 100 Hz. The first peak has a time lag corresponding to 219 Hz, which is probably the better estimate of the pitch.

The second and third cases deal with soundboard natural frequencies of 300 Hz and 500 Hz. The following graphs summarize the results for the three cases. First, Figure 5.6.5 gives a comparison of the soundboard displacements for all three cases.

The wave shapes appear to change significantly as the soundboard natural frequency changes.

Second, the cumulative line spectra for all three cases are shown in Figure 5.6.6. For each case, the string's fundamental and harmonics are present. However, the distribution of the power is moved around depending on the natural frequency of the soundboard. As the soundboard's natural frequency increases, the strength of the string's fundamental decreases and power is moved to the higher harmonics. This observation is supported by the spectral centroids, which are denoted in Figure 5.6.6 by "SpC."

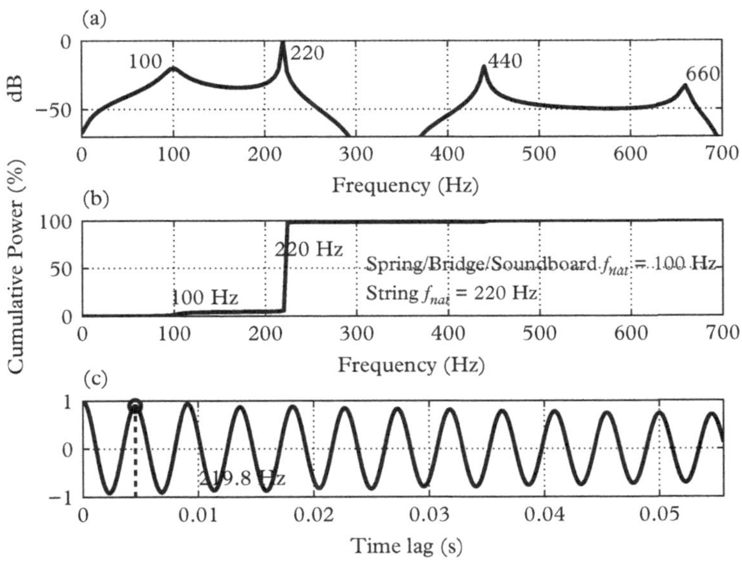

Figure 5.6.4 *Model soundboard with board natural frequency of 100 Hz and string natural frequency of 220 Hz: (a) line spectrum, (b) cumulative line spectrum, and (c) autocorrelation*

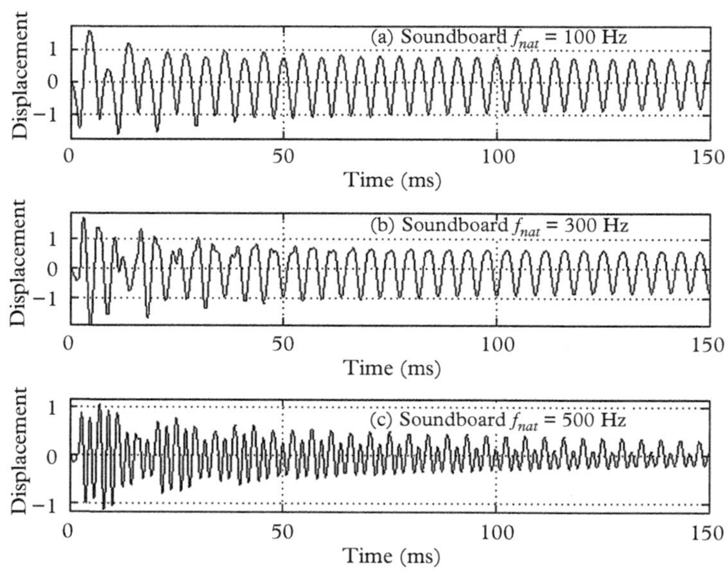

Figure 5.6.5 *Comparison of soundboard movement with string natural frequency of 220 Hz: (a) soundboard natural frequency of 100 Hz, (b) soundboard natural frequency of 300 Hz, and (c) soundboard natural frequency of 500 Hz*

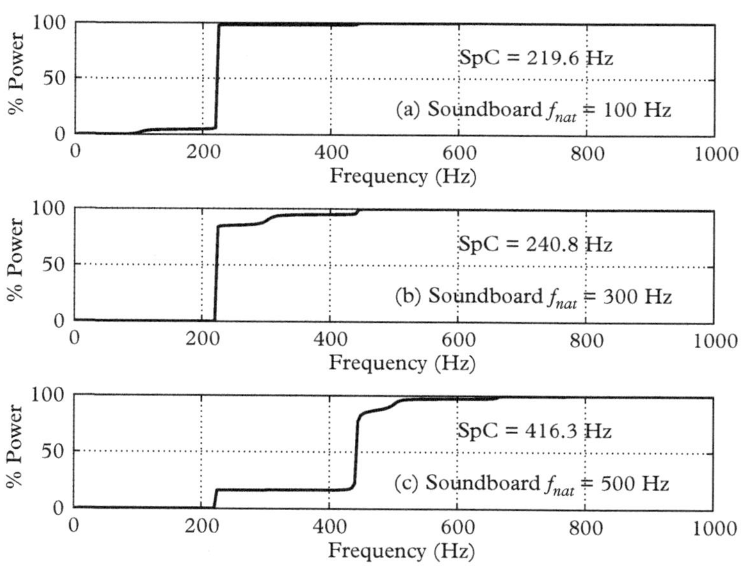

Figure 5.6.6 *Comparison of cumulative line spectra with string natural frequency of 220 Hz: (a) soundboard natural frequency of 100 Hz, (b) soundboard natural frequency of 300 Hz, and (c) soundboard natural frequency of 500 Hz*

As a reality check, remember that the "soundboard" in this simulation has only one natural frequency, while a real soundboard would have a natural frequency as well has harmonics, overtones, and modes since it is a two-dimensional plate. We will look at an idealized form of the two-dimensional plate later on in Chapter 8.

5.7 Forcing a Flexible String at One End

To the right-hand end of the string we will now attach an oscillator that can raise and lower the right end in a sinusoidal manner at a specified frequency while maintaining the tension. For this idealized frictionless string the forcing frequency cannot equal the natural frequency (or a multiple of same) of the string because an unstable resonance will result.

Forcing at 4.3 Times the Natural Frequency

Figure 5.7.1 shows a 3D plot of the string for the first 5 ms after the sine wave is applied to the right end of the string where the position equals 1.0 m (look at the back wall of the plot at point A). The string is 1.0 m long with a natural frequency of 200 Hz, which is near the G3 note (196 Hz fundamental frequency). The speed of sound through the string is 400 m s^{-1}. This means it will take

$$1.0\,\text{m}/\left(400\,\text{m}\,\text{s}^{-1}\right) = 2.5\,\text{ms}$$

for the wave to propagate from the back wall to the other end of the string at point C.

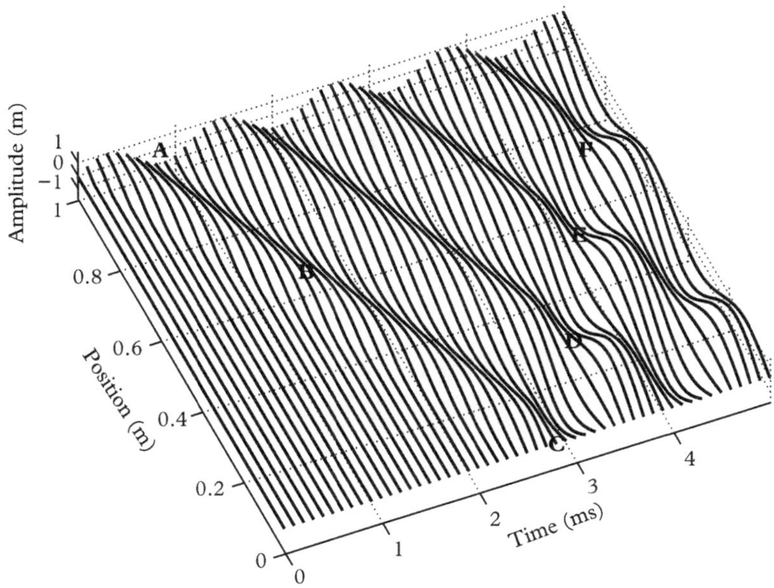

Figure 5.7.1 *Snapshots of the vibrating string in response to sinusoidal forcing where the forcing frequency is 4.3 times the natural frequency of the string*

The forcing frequency is 860 Hz. This means that every

$$1/(860\,\mathrm{Hz}) = 0.001163\,\mathrm{s} = 1.163\,\mathrm{ms}$$

the forcing signal will go through a cycle. Since the time of the simulation duration is 5 ms, there should be about

$$5\,\mathrm{ms}/(1.163\,\mathrm{ms}) = 4.3\,\mathrm{cycles}$$

graphed on the back wall of the plot.

As time progresses the wave moves from point A toward the other end of the string along the peak (see point B). The peak reaches point C at a time of 2.5 ms and is reflected and inverted such that a trough begins to propagate back toward the forced end of the string (the back wall designated by point A). The inversion develops its own depression that effectively moves along the trough designated by points C–F.

The front side of the graph, where the position is zero (or $x = 0$), is a rigid connection so there is no transverse displacement. At the back of the graph, where $x = 1$ m, one can see the 860 Hz sine wave in full bloom.

Forcing at 93% of the Natural Frequency

Figure 5.7.2 shows two views where the forcing frequency is 186 Hz (0.93 times the natural frequency). The speed of sound through the string is still 400 m s^{-1} so it still takes 2.5 ms for the

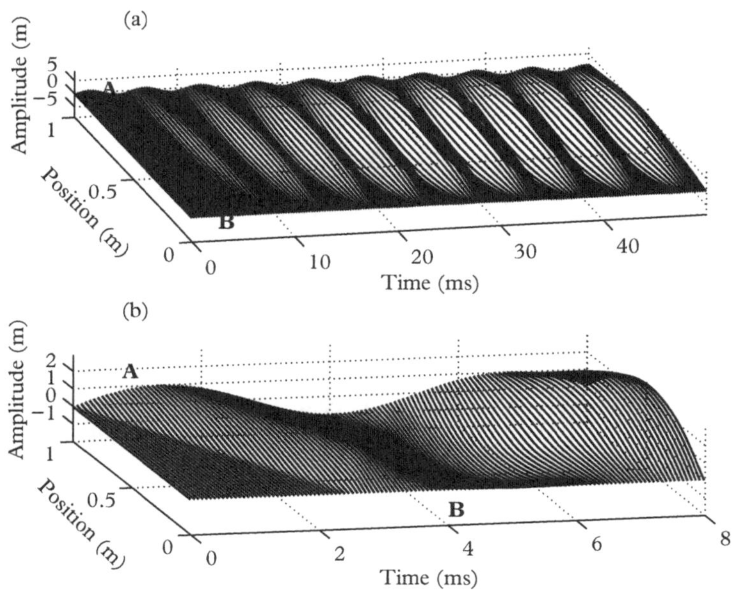

Figure 5.7.2 *Snapshots of the vibrating string in response to sinusoidal forcing where the forcing frequency is 93% of the natural frequency of the string*

wave starting at the back wall (point A) to reach the other end of the string at point B. However, because of the lower forcing frequency the duration of the simulation is 50 ms and there are almost 10 cycles of forcing at the back wall of the graph. If you look closely at Figure 5.7.2(a) you might be able to see that the amplitude is rising as time progresses because the string has not yet reached steady state. This is due to the closeness of the forcing frequency to the natural frequency.

Forcing with a Step Change

Figure 5.7.3 shows that the string has been snapped from 0.0 to 1.0 at $t = 0.0$ ms and at $x =$ 1 m (at point A). The step propagates to the other end of the string (point B) and is reflected *without inversion.* The reader should take a moment and compare Figure 5.7.3 with Figure 4.5.5 in Chapter 4 where basically the same partial differential equation was solved except for a different boundary condition at $x = 0$ (on the position axis). For Figure 5.7.3, the transverse displacement at $x = 0$ is forced to be zero because of a fixed connection at the non-forcing end (point B). In Figure 4.5.6 in Chapter 4, the slope of longitudinal displacement with respect to distance (not the displacement) was forced to be zero because the tube was open ended and exposed to the atmosphere. The simulation shown in Figure 5.7.3 would also apply to a tube closed at both ends with a step change disturbance at one end.

The figures presented in this section can be augmented with the movie-like presentation in the YouTube video referenced in the book's website under no. 25.

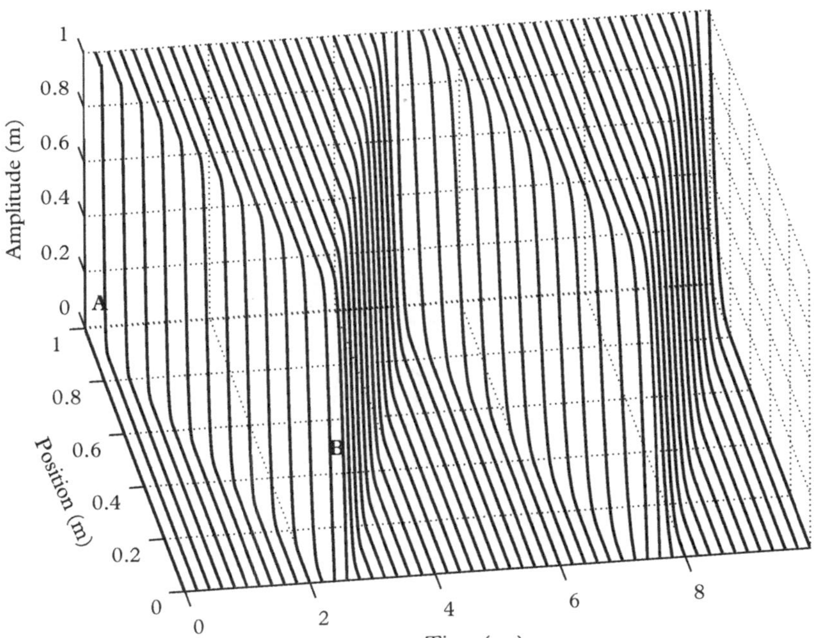

Figure 5.7.3 *Snapshots of the vibrating string in response to step forcing*

5.8 The Effect of a Hammer Striking the String

Morse's Approach

Guitar strings are plucked, but piano strings are struck by hammers. An analytical solution to the response of a string to a hammer strike appears on p. 103 in Section 10, Chapter 3 of the second edition of *Vibration and Sound* by Philip M. Morse, and it is the basis for several of the graphs in this section. Although some of the mathematical details are covered in Appendix 2, I should mention here that Philip Morse's solution comes from inserting an assumed functional shape for the applied force into the linear partial differential wave equation. This approach is to be compared with that of Nicholas J. Giordano, co-author of *Computational Physics*, who augments the wave equation with differential equations describing the dynamics of the hammer. A nonlinear expression relating the hammer force to the compression of the felt on the hammer head allows one to simulate the hammer force as well as the string response.

The first string to be studied is flexible, is fixed at both ends, has a length of 1 m and a natural frequency of 164.8 Hz (near the E3 note). We will use Morse's method. The picture of the force applied to the string by the striking hammer is shown in Figure 5.8.1. The hammer strikes the string at 1/9th of its length (at position 0.11 m). The force is applied over a period of approximately 1.0 ms (from $t = -0.5$ ms to $t = 0.5$ ms). The smooth shape of the hammer force as a function of time is an idealization. The alternative simulation approach designed by Professor Giordano shows an unevenness to the shape.

For the plucked string, the initial profile of the string was given in Figure 5.1.3. However, for this simulation, *before* the hammer strikes, the string displacement is zero all along its length. Figure 5.8.2 shows snapshots of the string during the early moments of the strike.

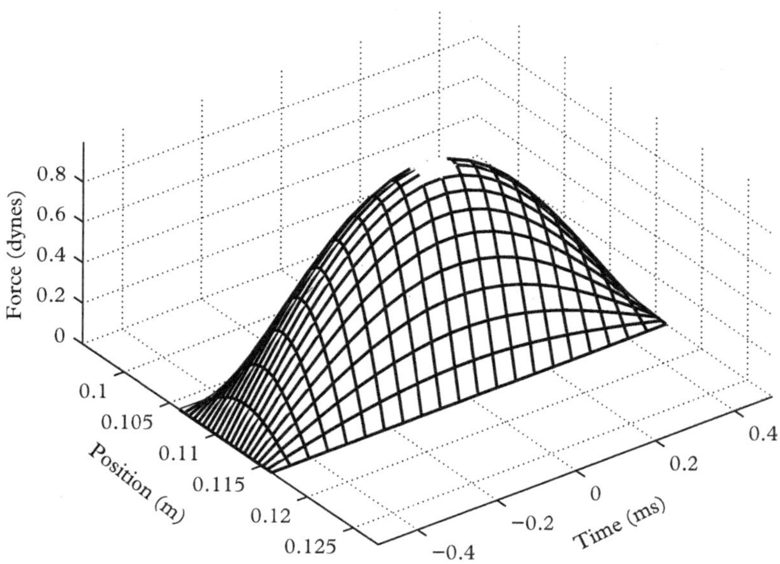

Figure 5.8.1 *Three-dimensional plot of the applied hammer force*

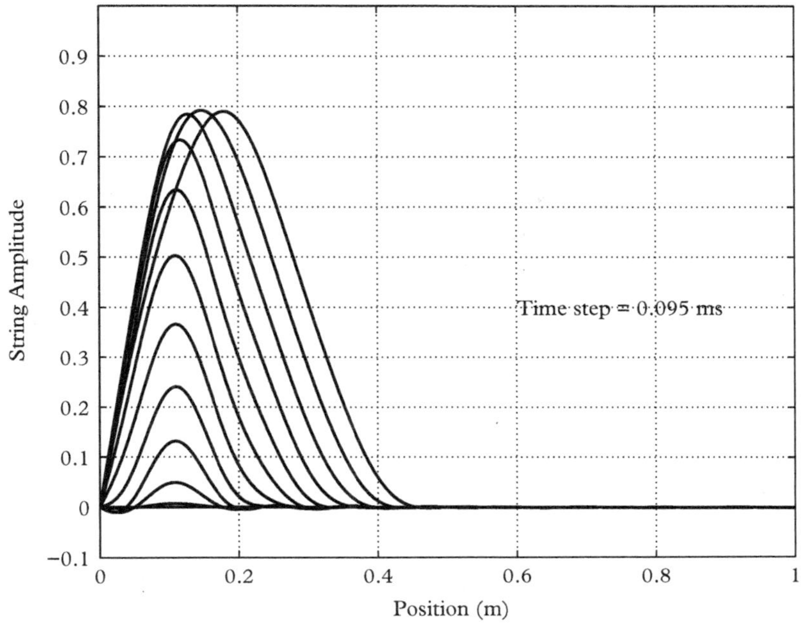

Figure 5.8.2 *Early snapshots of the vibrating string struck by a hammer: 0 to 4 ms*

Figure 5.8.3 shows snapshots of the string in three dimensions collected over the first 7.0 ms. The hammer strike takes place at point A. From there one can see the traveling wave increase in amplitude and move toward the other end of the string, where there is a reflection and an inversion at point B. The trough propagates to the other end of the string at point C.

Figures 5.8.4 shows the 3D surface created for 37 ms of simulation. As with Figure 5.8.3, point A denotes the beginning of the simulation. Compared to Figures 5.1.4 and 5.4.2 for the plucked string, one sees in Figures 5.8.3 and 5.8.4 that the propagation of the wave is similar but the shape is different. Although the pulse in the hammered string wave propagates back and forth along the string 165 times per second, the net effect is to set up standing waves. Note the flat "floor" in Figure 5.8.4, which is a consequence of the amount of time that the string lies flat on the initial axis while the pulse is moving back and forth (Exercise 2 at the end of this chapter deals with this in more detail). This is to be compared with the results of Section 5.7, where the string was sinusoidally disturbed at one end and also experienced standing waves. Furthermore, in Figure 5.1.4 one sees that the plucked string has no flat spots.

Figure 5.8.5(a) shows the force (actually the slope of the string) at the fixed end of the string. The line spectrum of the force is shown in Figure 5.8.5(b). Compare the two graphs in Figure 5.8.5 with those in Figure 5.1.4, which were for a plucked flexible string. Note that the ninth harmonic of the force in Figure 5.8.5(b) is weak because of where the hammer struck the string. Also, note that it appears that the force on the string is concentrated at those moments when the traveling pulse is reflected at the fixed ends.

The reader is encouraged view the YouTube video referenced in the book's website entitled "Response of a Flexible String to a Hammer Blow (no. 7)".

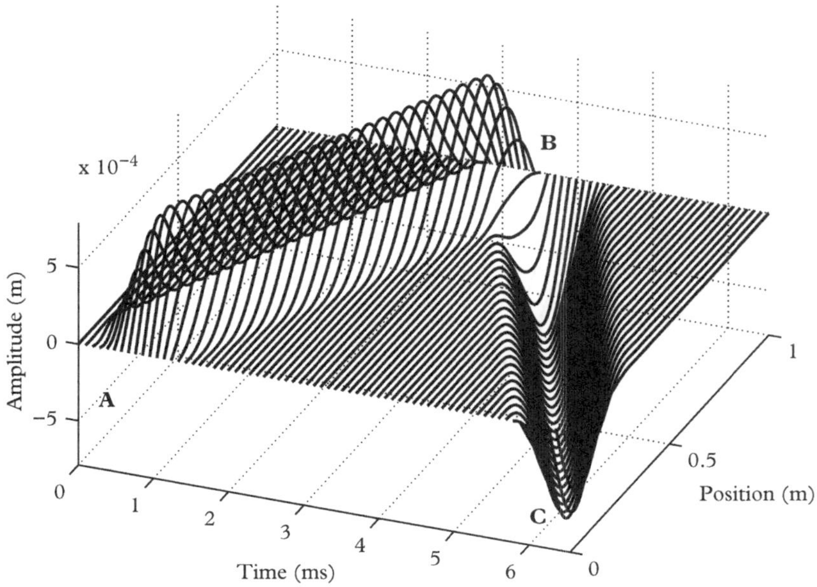

Figure 5.8.3 *Snapshots of a vibrating string in response to a hammer strike*

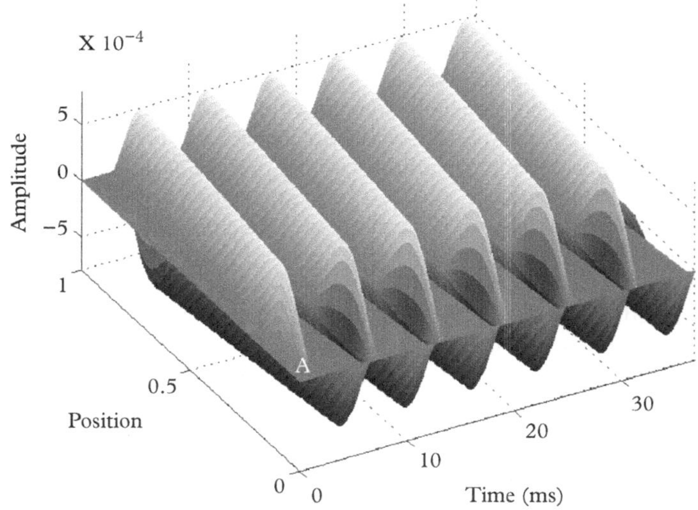

Figure 5.8.4 *Snapshots of a vibrating string in response to a hammer strike*

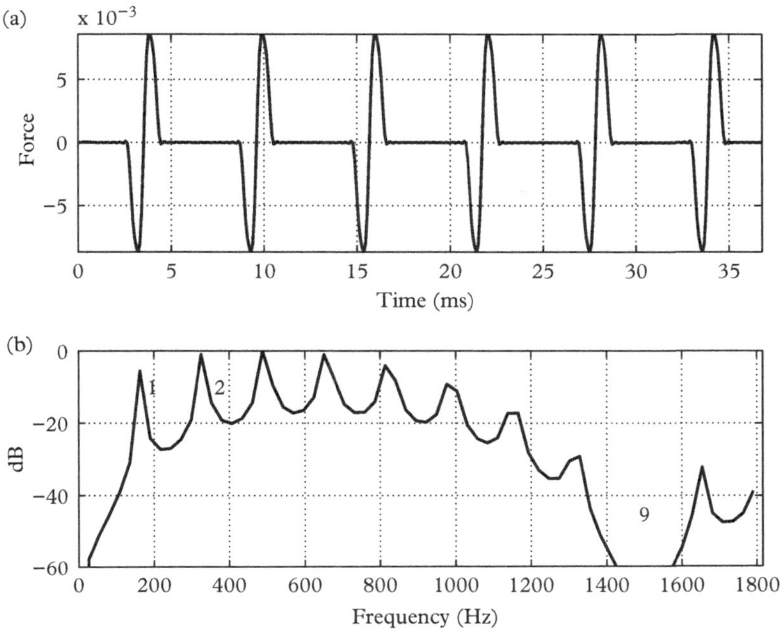

Figure 5.8.5 *Force on right-hand side termination point: (a) time domain plot, and (b) line spectrum*

Giordano's Approach

When the alternative approach presented in Nicholas Giordano's text is applied to the same string, one obtains Figure 5.8.6. This graph should be compared with Figure 5.8.3, which resulted from using Philip Morse's method. Note that the wave propagation speed is 329.6 m s^{-1}, so the time for the initial crest at point A to reflect off the other end at point B and return to the neighborhood of the hammer strike at point C is, using Eq. 5.0.6, approximately $2L/c = 6.1$ms.

The behavior of the felt compression, hammer position, and string displacement at the point of contact with the hammer is shown in Figure 5.8.7. The hammer force is proportional to the third power of the compression so it is not the smooth half cycle of a cosine wave that was used in Morse's method. Note the flat spots in the trace of the string displacement at the position of contact. This is a consequence of the time when the crest of the wave is moving along the string. Also, note that the amplitude of the string displacement appears to be decreasing with time.

If this last exercise is repeated for a string having a shorter length of 0.2 m and a higher natural frequency of 1046.5 Hz (parameters that are close to those for the C6 note on my Knabe), Figure 5.8.8 results. The wave propagation speed is higher at 418.6 m s^{-1} so it takes 0.48 ms for the initial peak at point A to reach the other end at point B.

The felt compression, the hammer position, and the string displacement at the point where the hammer strikes are shown in Figure 5.8.9. Note how the compression is now smooth, somewhat like the half-cosine shape that Morse used. The string displacement for the first crest is significantly higher than that for the subsequent crests, in contrast to those for the E3 simulation.

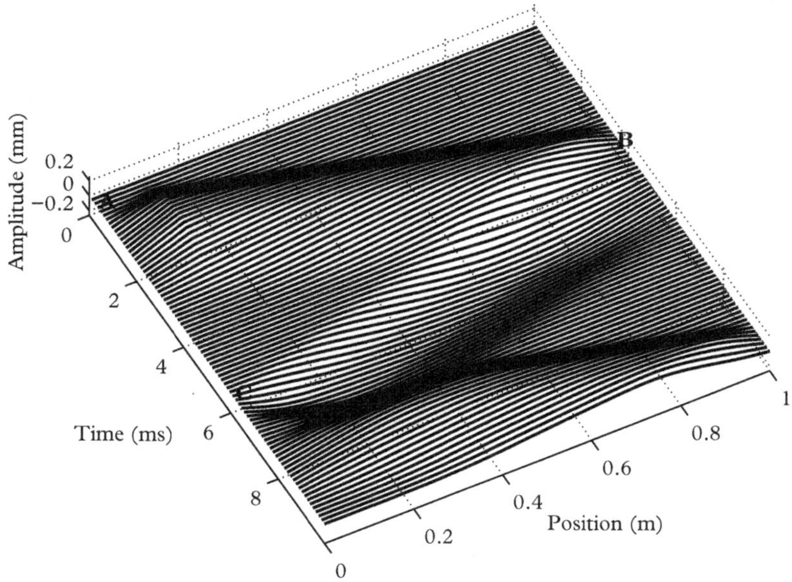

Figure 5.8.6 *Three-dimensional plot of vibrating string responding to a hammer strike using the Giordano method for a string 1.0 m long*

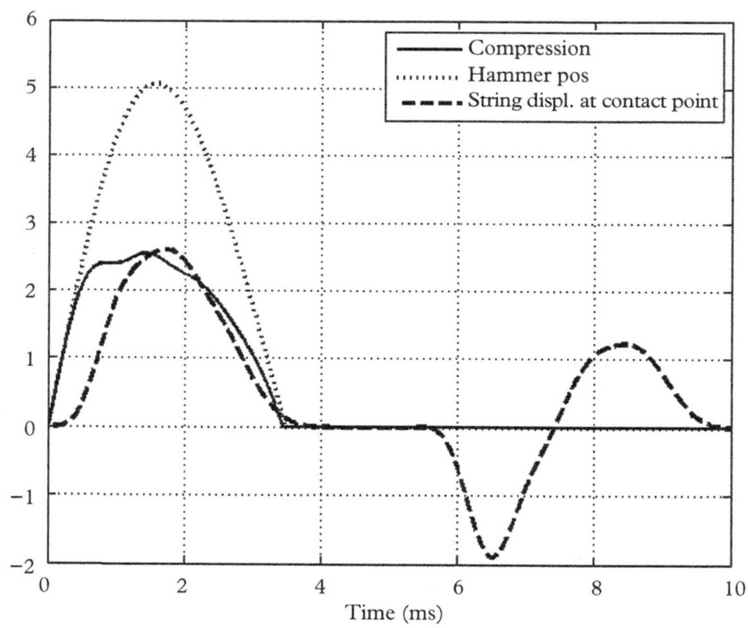

Figure 5.8.7 *Felt compression, hammer position, and string displacement at the point of contact with the hammer for a string 1.0 m long*

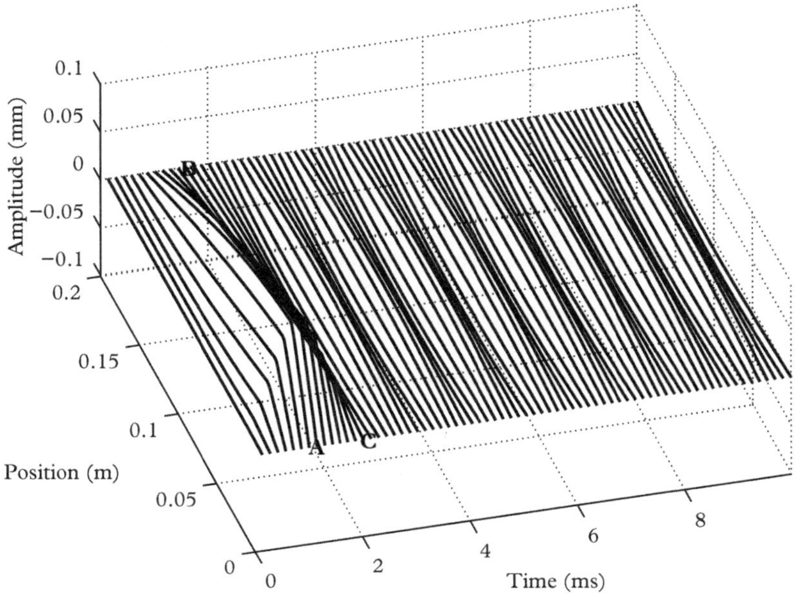

Figure 5.8.8 *Three-dimensional plot of vibrating string responding to a hammer strike using the Giordano method for a string 0.2 m long*

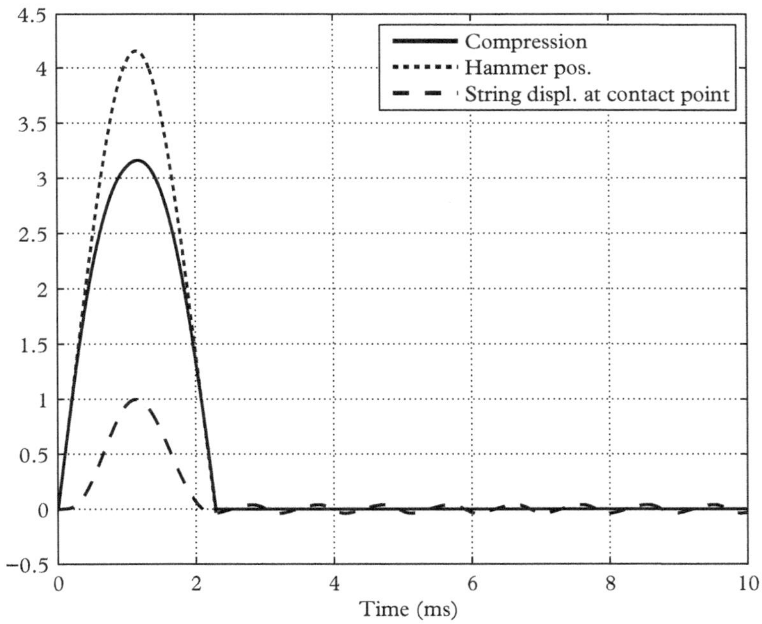

Figure 5.8.9 *Felt compression, hammer position, and string displacement at the point of contact with the hammer for a string 0.2 m long*

5.9 The Effect of String Stiffness in the Response to a Hammer Strike

The behavior of a stiff string is described by a fourth-order partial differential equation on which I shall not dwell here except to point out that it presents such mathematical challenges (at least to me) that I prefer to resort to numerical methods rather than analytical solutions (see Appendix 2 for a discussion of the mathematics). A stiff string is subject to tension but is also elastic, where elasticity is a characteristic measure of an object's resistance to stretching and its tendency to return to its original undeformed condition in the absence of tension (see the comments in Appendix 2 on elasticity). Elasticity will play a major role in the behavior of the string when the traveling wave generated by a hammer strike reaches a fixed end connection where it is reflected and inverted.

In this section we compare two one-meter-long strings with natural frequencies of 200 Hz. One string is the flexible string that we have been studying; the other has the same natural frequency but is stiff. Both strings are struck by a hammer at $^1/_{10}$th of their length and allowed to vibrate. Figure 5.9.1 shows how the shapes of the two strings evolve. The flexible string maintains its shape as the pulse generated by the strike propagates back and forth, left to right and right to left. The stiff string develops an increasing lead and it takes on shape changes with time such that by 34 ms it has developed three points of zero slope—two valleys and one peak.

Figure 5.9.2 is a composite of six snapshots taken between 28 and 30 ms after the hammer strike. The thin curve is the flexible string. The stiff string leads the flexible string and it exhibits ripples as the wave propagates back and forth. The flexible string, which is not subject to elasticity, maintains its shape indefinitely while the shape of the stiff string, which is subject to elasticity, continually evolves with time.

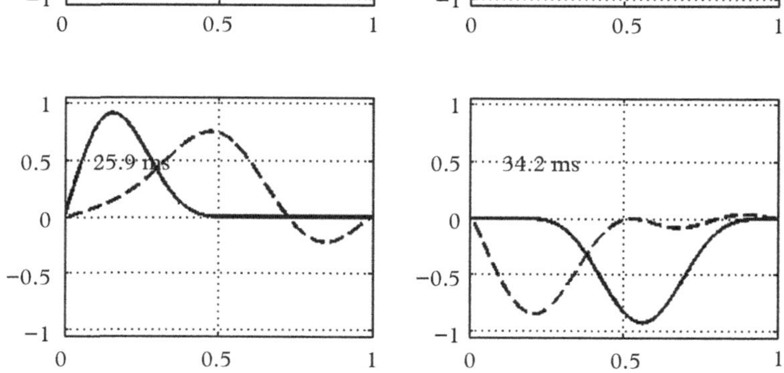

Figure 5.9.1 *Response of flexible and stiff strings to a hammer strike*

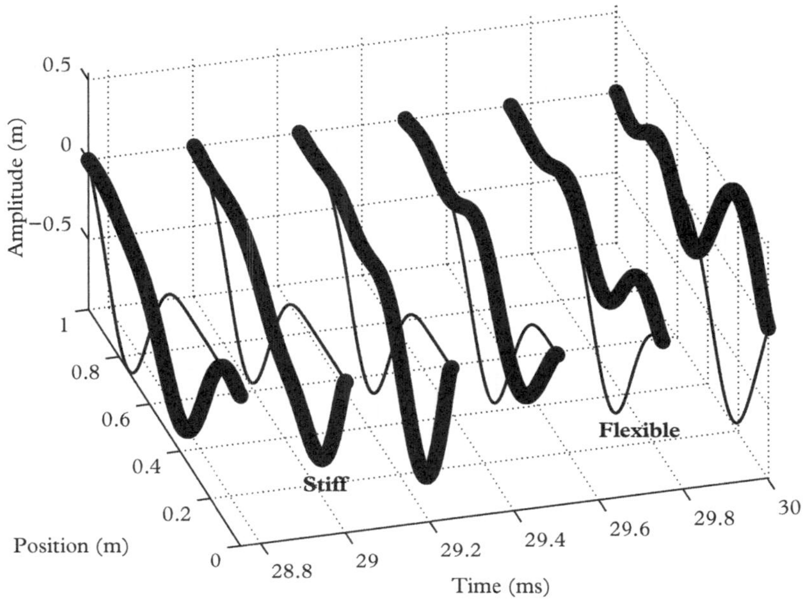

Figure 5.9.2 *Response of flexible and stiff strings to a hammer strike (the thin trace is for the flexible string and the thick trace is for the stiff string)*

In Section 5.1 we looked at the force generated at a fixed end connection and saw the appearances of separated pulses. In the next section we will pursue this.

I encourage you to look at the video referenced on the book's website for a study of two ideal wires and a real wire. It yields far more insight far more quickly.

5.10 The Force Generated on the Fixed End Connection by Struck Flexible and Stiff Strings

Two strings with a natural frequency of 200 Hz, one flexible and one stiff, are compared in response to a hammer strike at $^1/_{10}$th of their length. As shown in Section 5.9, the stiff string exhibits significantly more ripple and shape change than the flexible string. Therefore, the force on the fixed end connection will be significantly different for the stiff string when compared to that of the flexible string, and this is demonstrated in Figure 5.10.1. In Figure 5.10.1(a) we plot not the force but the *slope* of the string at the fixed end (which is proportional to the force), and one can see that the flexible string experiences a sharp and repeatable force pulse while the changes in the slope of the stiff string are considerably smaller in magnitude and accumulate more and more ripples with time.

Keep in mind that the force on the string is equal and opposite to that on the fixed connection. Therefore, because of elasticity, this abrupt application of force causes an acceleration at each reflection that propels the stiff string ahead of the flexible string. The time domain graphs in the previous section also show that the shape of the stiff string force changes with time.

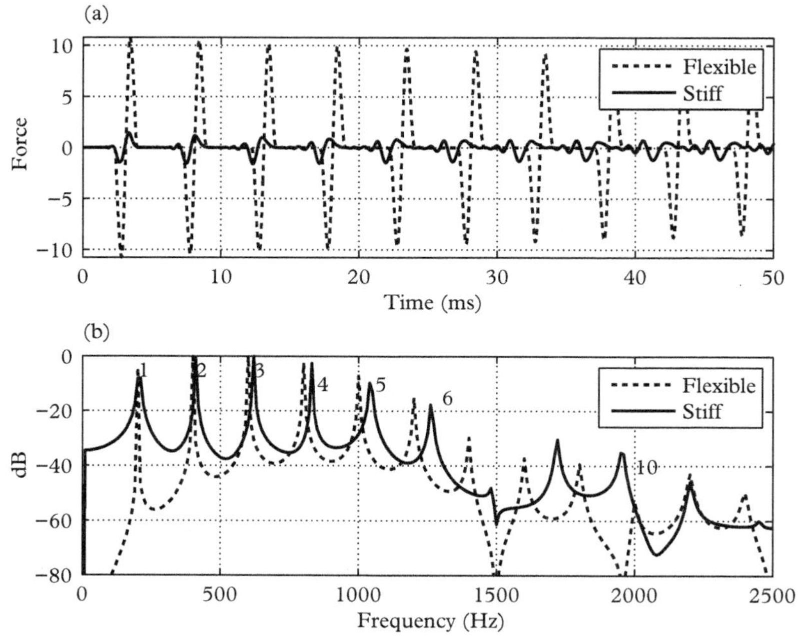

Figure 5.10.1 *Force on right-hand side termination point for flexible and stiff strings: (a) time domain plot, and (b) line spectrum (showing inharmonicity)*

The line spectrum of the forces, shown in Figure 5.10.1(b), shows that the locations of the stiff string partials exceed those of the flexible string by an amount that increases with frequency. The first three partials of the flexible and stiff strings are relatively close, but by the sixth partial there is a significant separation. By the tenth partial it is difficult to find the corresponding partial. In addition, because of where the strings were struck, the tenth partial is greatly diminished.

This empirical observation of the advancement of the partials is supported by mathematical analyses in the literature, which will not be dealt with here (it is touched upon in Appendix 2). When we look at pianos in Part 3, the impact of stiffness will be seen as one of the bases for something called inharmonicity (also called octave stretching).

Stiff strings resist bending and stretching while flexible strings have no resistance to either bending or stretching. As the fundamental frequency f increases, the associated wavelength λ, given by Eq. 5.0.3 for a flexible string, decreases (where c is the speed of propagation):

$$\lambda = \frac{c}{f}. \tag{5.0.3}$$

This "suggests" that the stiff string is being asked to make smaller, tighter bends as the frequency increases, especially at the points of reflection. This increased tightness is claimed to cause the harmonics of the flexible string to become non-integer multiples of the fundamental or partials of the stiff string. These partial frequencies are higher than the integer multiples of the fundamental and the effect is greater as the frequency increases (or as the wavelength decreases). Note that this explanation is presented as a "suggestion."

In the following sentences I will paraphrase some of the other attempts that I have come across to explain this phenomenon. I will mix them with some of my own.

Here is one version lifted from the Internet (so it has to be true): When the wavelength of a wave on a stretched flexible string is much greater than the thickness of the string, the wave velocity on the string is constant and the overtones or partials are at the harmonics. However, for high-frequency overtones with short wavelengths approaching the diameter of the string, the string behaves more like a thick metal bar: its mechanical resistance to bending becomes an additional force to the tension, which "raises the pitch" of the overtones. Assume that the string in the above simulation has a diameter of 1 mm or 0.001 m. At the fundamental frequency of 200 Hz the wavelength is

$$\lambda = \frac{c}{f} = \frac{448 \text{ m s}^{-1}}{200 \text{ s}^{-1}} = 2.2\text{m},$$

or about 2000 times greater than the diameter of the string. At the tenth harmonic at 2000 Hz, the wavelength would be around 0.2 m, which is still 200 times greater than the diameter of the string.

Another explanation starts with Eqs. 5.0.1 and 5.0.2, which give the fundamental frequency in terms of the string length for a flexible string:

$$f_{\text{nat}} = \frac{c}{2L} \qquad c = \sqrt{\frac{\tilde{T}}{\rho}}.$$

Stiffness, one might argue, causes the string to behave as though it were shorter than its actual physical length and causes the speed of wave propagation c to increase. (In the limit, as flexibility is overshadowed by stiffness, the string becomes a rod. This suggests that the speed of sound in a rod would be greater than that in a flexible string, which in fact it is.) Therefore, the effective denominator in the equation for f_{nat} gets smaller, the numerator gets larger, and the fundamental frequency increases. The higher partials increase even more because the wavelength λ is smaller and a tighter bend is needed.

The previous paragraphs present a couple of attempts to explain physically why a stiff string exhibits higher-pitched overtones (or partials). There are others in the book *Pianos Inside Out* by Mario Igrec.

An alternative discussion starts with a closer look at what happens when the pulse of the traveling wave reaches a fixed end connection and is reflected and inverted. (We have alluded to this in earlier discussions of the pulses in the slope of the string.) This is where the pulse in force on the fixed end connection is generated and where the nature of the resulting tone is defined. Figure 5.10.2 shows the shape of the flexible and stiff strings just before and after the *first* reflection and inversion at the fixed connection at $x = 1$ m. There is not much to differentiate the flexible from the stiff string. Point A denotes a snapshot at 2.4 ms after the hammer strike. Already one can see that the stiff string is slightly leading the flexible string. The extra shape changes experienced by the stiff string will be amplified and expanded upon each reflection, whereas the flexible string shape will remain the same after each reflection.

Figure 5.10.3 shows the same snapshots at the *third* reflection. Now the stiff string is beginning to exhibit some additional peaks and valleys compared to the flexible string, which is maintaining its shape. Point A now denotes a time of 7.2 ms after the hammer strike, while point B denotes a time of 9 ms after the hammer strike. The stiff string has covered more ground at point A and is therefore clearly in the lead.

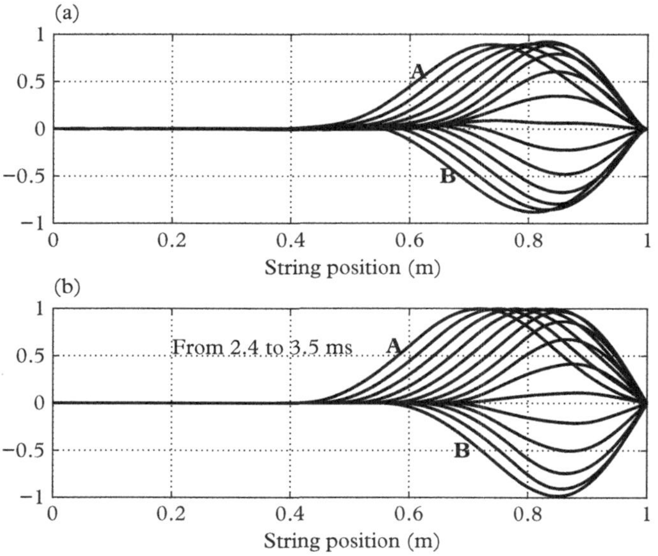

Figure 5.10.2 *Snapshots at the first reflection for (a) a stiff string, and (b) a flexible string*

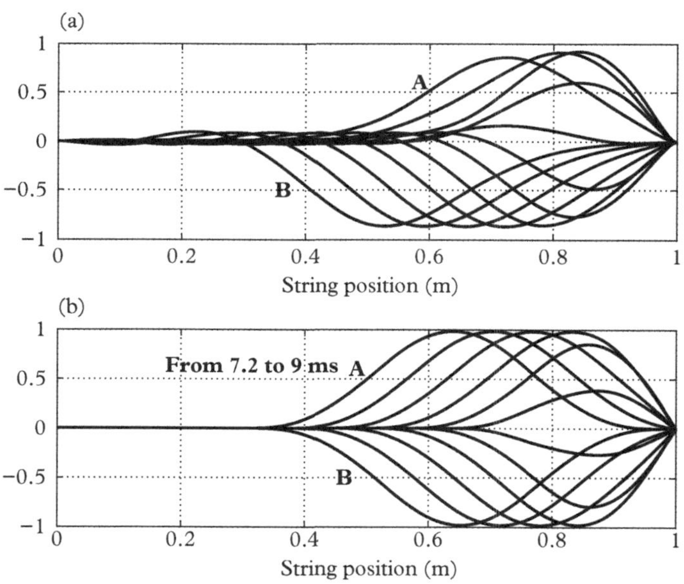

Figure 5.10.3 *Snapshots at the third reflection for (a) a stiff string, and (b) a flexible string*

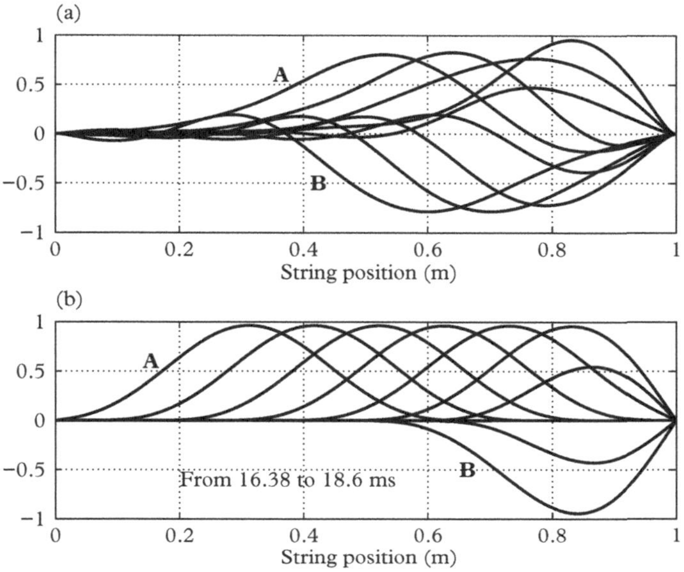

Figure 5.10.4 *Snapshots at the sixth reflection for (a) a stiff string, and (b) a flexible string*

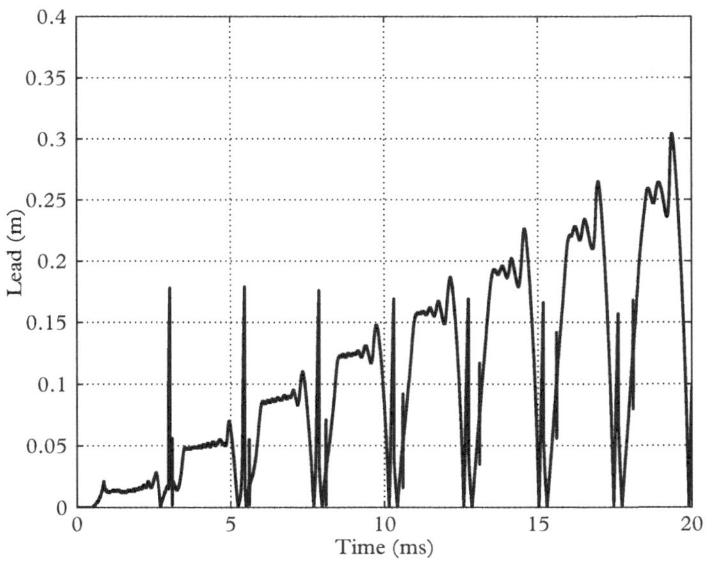

Figure 5.10.5 *The lead of the stiff string over the flexible string as a function of time*

Finally, Figure 5.10.4 shows the same snapshots at the *sixth* reflection. Now, point A denotes a time of 16.4 ms after the hammer strike. The difference between the two strings is even more pronounced. The flexible string continues to maintain its shape while the stiff string has developed three points of zero slope.

This difference is a consequence of elasticity. There is a "spring in the stiff string's walk" (as it were) due to elasticity. When the traveling pulse reaches the fixed connection and is reflected, elasticity causes the pulse to spring ahead of the flexible string thereby causing the stiff string to exhibit a higher speed of propagation, and therefore a higher frequency, which in turn leads to the inharmonicity that we will discuss again in Part 3 on the piano.

To test this idea that there is something special going on at the reflection points, one can find the position along the axis of the wave peak at each instant of time for the flexible and stiff strings. The difference between the stiff and flexible strings' wave peaks has units of meters and is a measure of the relative propagation speeds of the two strings. This difference is plotted in Figure 5.10.5. Here one sees that the lead of the stiff string grows in between reflections but experiences a step increase at the time of reflection every 2.5 ms.

5.11 Longitudinal Waves

In the previous section the flexible string model was modified and stiffness was added. Therefore, our idealized stiff string experienced both tension and elasticity. In the extreme case of complete stiffness and no flexibility, the string would evolve into a rod (or a "frozen string"). Longitudinal waves definitely exist in rods, so one might suspect the presence of longitudinal waves in the stiff string in addition to the transverse waves that we have been studying so far. These longitudinal waves, propagating along the axis of the string, would not be visible in the sense that the transverse waves are sometimes a blur to the eye. However, there might be back-and-forth movement along the direction of the string or wire as a consequence of localized variations in compression and extension.

Consider the time domain behavior of a C1 note on a Bösendorfer model 280 piano (more about this piano in Chapter 9) in Figure 5.11.1.

Now, take a closer look at the beginning of the wave in Figure 5.11.2.

Note the leading part of the wave from 0.1 s to 0.12 s, denoted by the horizontal dotted line with the circles at each end. This leading portion looks significantly different from the bulk of the wave starting at 0.12 s. If one posits that there are two events going on, then it appears as though one event arrives at the soundboard almost 0.02 s earlier.

The line spectrum of the whole wave is shown in Figure 5.11.3. There are regular peaks at multiples of the C1 fundamental, which is 32.7 Hz. But look at the strange appearance of a cluster of peaks in the neighborhood of 840 Hz, denoted by the horizontal dotted line with the circles at each end. Admittedly, these peaks are some 30 dB down, but they do stand out.

In Figure 5.11.4 we show just the leading edge, of duration 0.019 s, along with a fine grid line spectrum of that leading edge. The line spectrum in Figure 5.11.4(b) is centered at 1000 Hz with a range of 2000 Hz and a resolution of 2 Hz. This fine grid spectrum shows a dominant peak at approximately 898 Hz, which is consistent with the portion of the spectrum in Figure 5.11.3 for the whole wave. These results suggest that the peaks in the neighborhood of 900 Hz in Figure 5.11.3 come from the longitudinal component of the wave, which appears at the front of the wave in the time domain plot.

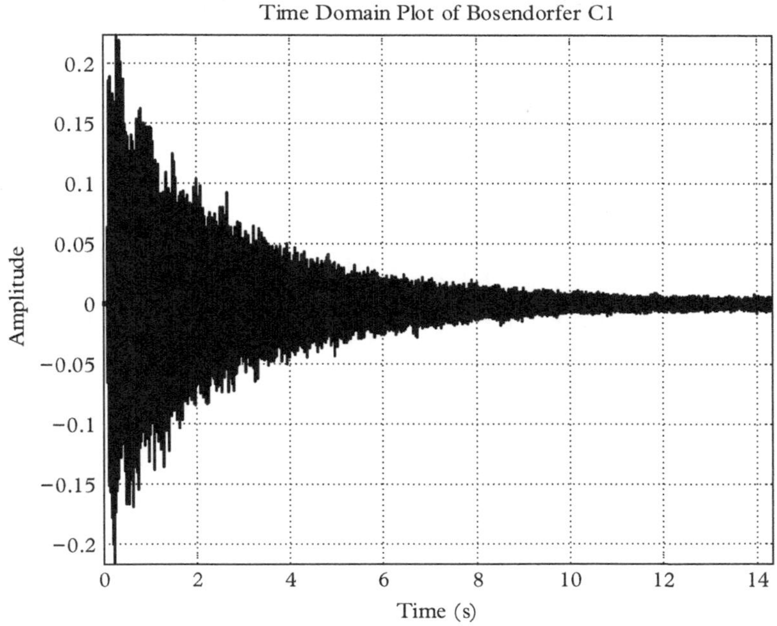

Figure 5.11.1 *Time domain plot of Bösendorfer C1 wave*

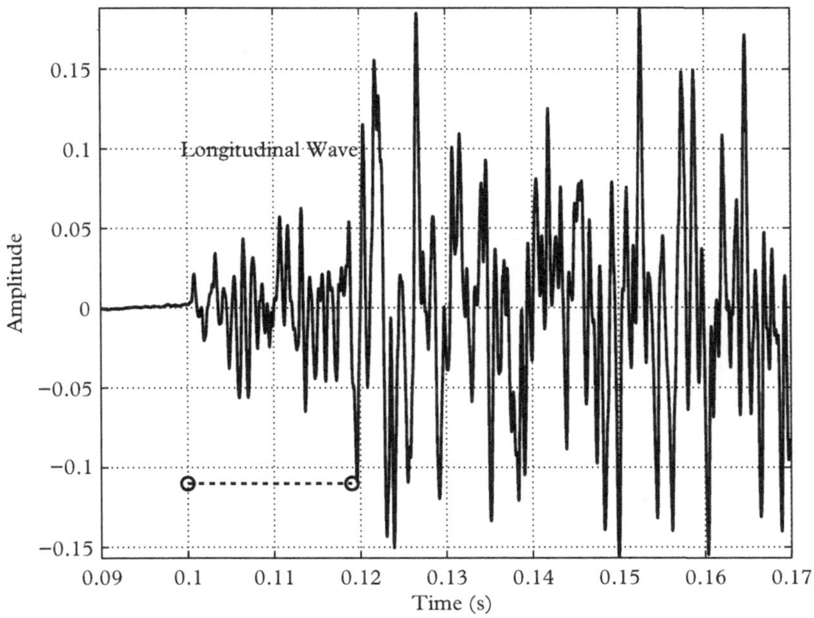

Figure 5.11.2 *A closer look at the early part of the Bösendorfer C1 wave*

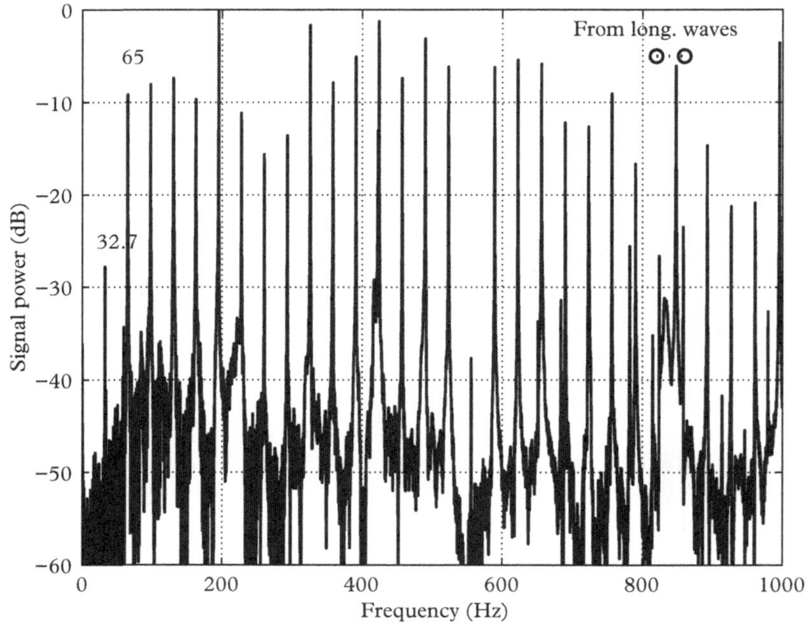

Figure 5.11.3 *Line spectrum of the Bösendorfer C1 wave*

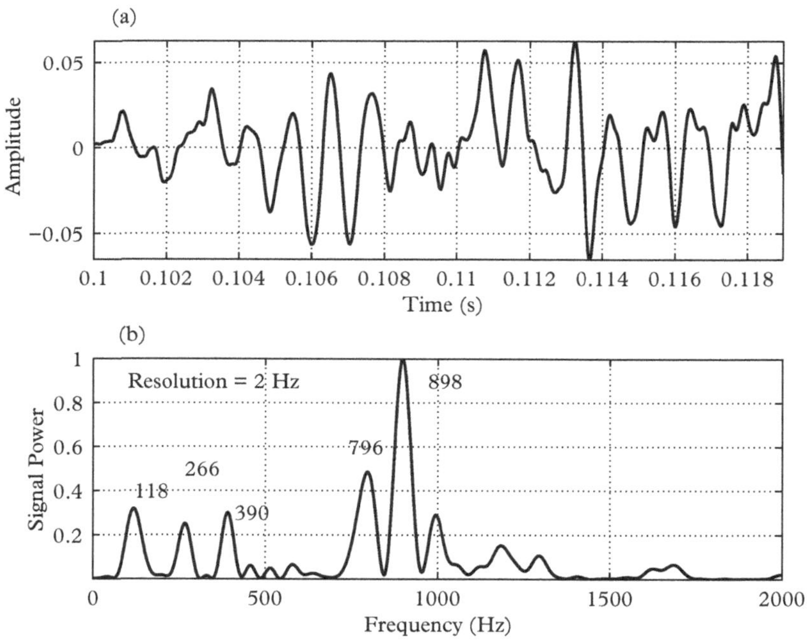

Figure 5.11.4 *Leading edge of Bösendorfer C1 wave: (a) time domain plot, and (b) fine grid line spectrum*

The longitudinal component shows up as a precursor because its speed is so much higher than the transverse wave. The propagation speed of a transverse wave in a perfectly flexible string (no elasticity) was shown in Section 5.0 to be about 400 m s^{-1} by means of Eq. 2.0.3:

$$c = \sqrt{\frac{\tilde{T}}{\tilde{\rho}}}, \tag{2.0.3}$$

where $\tilde{T}$ is the tension and $\tilde{\rho}$ is the linear density in g m^{-1}. The speed of a longitudinal wave in a rod (or a completely stiff steel wire or a "frozen string") with a Young's modulus of $192x10^6$ g m^{-1} ms^{-2}) and a density of $8x10^6$ g m^{-3} is

$$c = \sqrt{\frac{Y}{\rho}} = 4890\,\text{m s}^{-1}, \tag{5.11.1}$$

which is an order of magnitude faster. Equation 5.11.1 is discussed briefly in Appendix 2 (see Eq. A2.21).

As pointed out in Appendix 2, the same type of partial differential equation describes both the transverse displacements of a flexible string subject to tension only *and* the longitudinal displacements of a bar (which we will deal with in the next chapter) subject to elasticity only. The only difference is the dependence of the propagation speed c on the Young's modulus rather than the tension. Consequently, we can estimate the length of the C1 string using

$$L = \frac{c}{2f_1} = \frac{\sqrt{\frac{Y}{\rho}}}{2f_1} = \frac{4890}{2 \times 900} = 2.7\text{ m},$$

which is approximately the length of a Bösendorfer 280 C1 string.

5.12 Summary

We spent considerable time on the transverse waves generated by plucking and striking strings, both idealized and real. In general, the simulations suggest that a disturbance, like a strike, generates a pulse that propagates back and forth across the string. This concept seems to be best applied to long strings that have low natural or fundamental frequencies. As the string length decreases and therefore the fundamental frequency increases, the pulse becomes broader and less well defined.

The idea of disturbing a purely flexible string at one end was also considered briefly. The flexible string represents a model, which was shown to yield a reasonable representation of real single-stranded strings. We added stiffness to our model and studied its effect when compared to the perfectly flexible string. The impact of a vibrating string on its fixed ends was studied by looking at the force (actually the slope of the displacement with respect to position) of the string on the fixed end connection. On three separate occasions we found that the tone generated thereby depended heavily on the behavior of the traveling wave when it was reflected and inverted at the fixed end points. Furthermore, when the string is elastic the equal and opposite force of the end connection on the string produces an acceleration that causes the string to develop a lead over the flexible string traveling wave. This is manifested in a shifting of the partial frequencies associated with the vibrations set up at the end connection. We closed the chapter with a discussion of longitudinal waves in a vibrating string.

The next chapter will deal with the vibrating bar, which plays an important part in musical instruments such as the vibraphone and marimba. It also serves as a stepping stone to the study of vibrating membranes (drums) and vibrating plates (piano soundboards).

5.13 Exercises

1. Set the variable iPluck to 0 (pluck at ⅙th of the length) and run the script in the file VibStringPrepFor3DBook.m. Experiment with different string lengths. Set iPluck to 2 (pluck at ⅖th of the length) and repeat the run. How does the plucking point affect the results?
2. Find the code in the file HammerStrike3TextBook.m where Figure 5.8.4 is generated. Run this script, and when Figure 5.8.4 is constructed use the Matlab "Rotate 3D" button to rotate the graph and see where the "flat spots" are located.
3. Using the script in the file VibStrgHammerGio3Mod.m, try to repeat the experiments in Section 5.8 using the Giordano method with string lengths that are longer and shorter than those in the two examples. Make sure that you pick the correct natural frequencies for the different lengths. How and why are the results different?
4. Find the Matlab script in the file ImplicitVibStrgForceOneEndBook.m, and look in the file FiguresPart2_Chap5.m to find where this script is used to generate Figure 5.7.1. Make a modification to force the string at a frequency that is just greater than the natural frequency. How does the result compare with the other forcing choices?
5. Why is the force on the bridge in Figure 5.10.1 so much higher for the flexible string as compared to the stiff one? Follow the shape of the flexible string as it is reflected. Does it have sharper corners than the stiff string? How would these sharper corners affect the slope of the string as it is reflected?

6

The Vibrating Bar

Part 2 started with a totally flexible string under tension. Stiffness was then added and the resulting model string was subject to tension *and* elasticity. That model approximately simulated the activity of a piano string. Now, the tension is removed and we have a bar that is not under tension but is purely elastic. Such a bar is a component in marimbas and xylophones.

We will start by reviewing the values of some basic parameters associated with the vibrating bar. Using those parameters, a simulation will be carried out. An analysis of a toy xylophone will follow.

6.1 Basic Parameters of the Vibrating Bar

The sketch in Figure 6.1.1 is designed to simulate the transverse (not longitudinal, as in Section 5.11 of the previous chapter) displacements of a vibraphone or marimba bar. Both ends of the bar are free and the bar is supported at two points.

In the following paragraphs several formulas will be used to calculate parameters of the bar to be simulated. These formulas can be obtained from books on acoustics like those by Morse, Kinsler and Frey, or Olsen. If the formulas appear confusing, try to focus on the numbers that I will extract from these formulas. If that is confusing, just trudge onward until we get to some graphs resulting from a simulation.

First, the speed of sound c in the bar is given by (same equation as in Section 5.11)

$$c_{\text{bar}} = \sqrt{\frac{Y}{\rho}}, \tag{6.1.1}$$

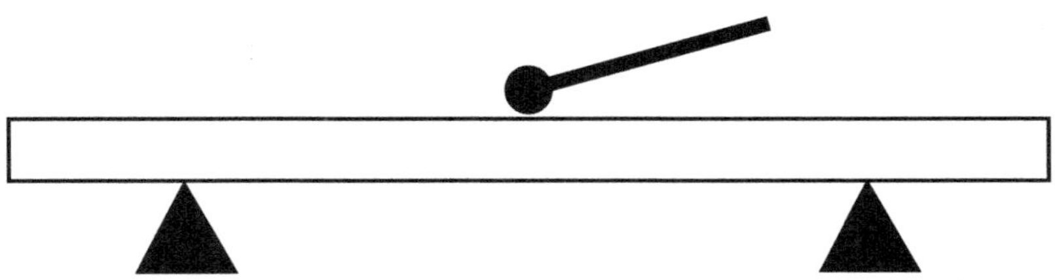

Figure 6.1.1 *Schematic of transversely vibrating bar showing the support points, the mallet, and the strike point*

Spectral Analysis of Musical Sounds with Emphasis on the Piano. First Edition.
David M. Koenig, with comments by Delwin D. Fandrich.
Published in 2015 by Oxford University Press.

where Y is Young's modulus (a characteristic measure of elasticity) and ρ is the bulk density. As the Young's modulus Y increases or as the material becomes more stiff and inelastic, the speed of sound c_{bar} increases. For example, the density of aluminum is 2700 kg m^{-3} and the Young's modulus is 7.1×10^{10} N m^{-2}, so Eq. 6.1.1 gives the speed of sound in aluminum as 5128 m s^{-1}. A material like Lucite has a density of 1200 kg m^{-3} and a Young's modulus of 0.4×10^{10} N m^{-2}. The speed of sound in Lucite is 1826 m s^{-1}.

The issue is complicated by the fact that this speed of sound c_{bar} refers to longitudinal waves propagating within the bar in a direction parallel to the bar's axis. For our vibrating string of the last chapter we were interested in the speed of transverse traveling waves moving in a direction parallel to the string's axis but vibrating transversely (perpendicularly to the string's axis).

If you were to strike a long bar at one end at its tip, sound waves would propagate along the bar's length (or longitudinally) at a speed dictated by Eq. 6.1.1. Figure 6.1.2 shows a schematic of the tip being struck.

Although longitudinal waves in a bar are important, we are interested in the propagation of transverse waves along a bar so we have to keep this difference in mind.

The length of the bar L used for our simulations is 0.4064 m. The only other parameter needed for the calculations is something called the radius of gyration κ given by

$$\kappa = \frac{\delta}{\sqrt{12}} = 0.0013 \text{ m}, \tag{6.1.2}$$

where δ is the thickness of the bar (in this case 0.0043 m). The radius of gyration is used to describe how the cross-sectional area in a beam or bar is distributed around its axis. We will *not* need to gain familiarity with this parameter to proceed.

Kinsler shows that the fundamental or natural transverse frequency of a bar is given by Eq. 6.1.3:

$$f_{1,\text{bar}} = \frac{3.01112^2 \pi \kappa}{8L^2} \sqrt{\frac{Y}{\rho}}, \tag{6.1.3}$$

which for our bar is 138.9 Hz (near C3#). Equation 6.1.3 can be compared to Eq. 2.0.2 for the flexible string:

$$f_{1,\text{string}} = \frac{1}{2L} \sqrt{\frac{\tilde{T}}{\tilde{\rho}}}, \tag{2.0.2}$$

where Young's modulus Y, a measure of elasticity, is replaced by the tension $\tilde{T}$ and the dependence on the length L is different. Equation 6.1.3 shows us that stiffer bars have a higher natural frequency.

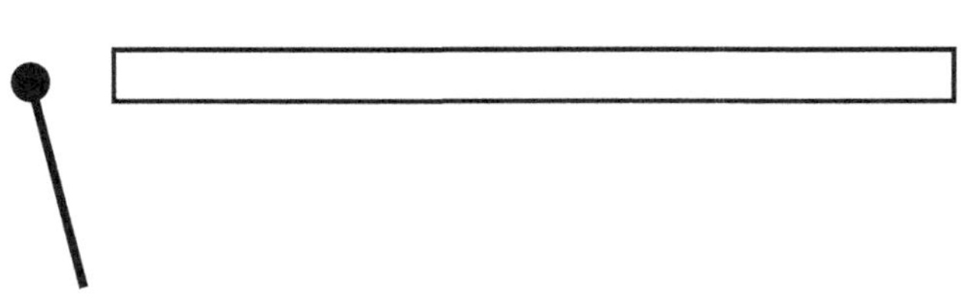

Figure 6.1.2 *Schematic of longitudinally vibrating bar showing the mallet and strike point*

So, after this barrage of formulas we have accomplished little more than the ability to estimate the natural frequency of a transversely vibrating bar. We will have reason to refer to these formulas later when we carry out some simulations and want to check our results.

The propagation velocity v of a *transverse* wave disturbance in our test bar is given by

$$v = \sqrt{2\pi f_{1,\mathrm{bar}}\, c_{\mathrm{bar}} \kappa} = 74.9 \text{ m s}^{-1}, \tag{6.1.4}$$

where we have already determined numbers for $f_{1,\mathrm{bar}}$, c_{bar}, and κ. For the case of the flexible vibrating string the speed of the propagating wave c_{string} was given by

$$c_{\mathrm{string}} = \sqrt{\frac{\tilde{T}}{\tilde{\rho}}}, \tag{6.1.5}$$

where $\tilde{T}$ is the tension in the string. The propagation velocity for the bar v in Eq. 6.1.4 is dependent on $f_{1,\mathrm{bar}}$, the natural frequency of the bar, and on c_{bar}, the speed of sound longitudinally through the bar. Consequently, in bars, transverse waves of different frequencies propagate at different speeds, unlike the flexible string.

The fundamental wavelength of the transverse wave propagating along the bar is given by

$$\lambda_1 = \frac{v}{f_{1,\mathrm{bar}}} = \frac{74.9}{138.9} = 0.539\,\mathrm{m}. \tag{6.1.6}$$

In summary, we have introduced a gaggle of new variables: κ, Y, $f_{1,\mathrm{bar}}$, v, and c_{bar}, and have estimated numerical values for them. So, now that we have a few numbers under our belt and an awareness that the bar is going to behave significantly differently than the string, let us proceed to crank out some simulations and try to get some insight into the dynamics of a vibrating bar.

6.2 A Simulation of a Vibrating Bar

Assume that a mallet strikes the bar in the center such that the bar takes an initial shape as illustrated in Figure 6.2.1. This initial shape is the opposite of what might happen if the bar were struck from the top but we will use it because it is comparable to the initial Gaussian-shaped bump used for the string in Figure 5.5.1. Note the vertical dotted lines representing the support points for the bar. These support points were chosen as the "nodal points" of the bar's fundamental frequency, which are presented in the books by Morse, Kinsler and Frey, and Olsen without proof (they were derived first in Lord Rayleigh's *Theory of Sound*, published in 1887).

The amplitude in Figure 6.2.1 is scaled so that the initial amplitude profile lies between 0.0 and 1.0.

A sequence of snapshots of the response over a duration of 0.576 ms is given in Figure 6.2.2, where the dashed vertical lines again show the points of support. This graph can be compared to that in Figure 5.5.3 for the flexible string. Because of the absence of fixed end connections, there is no reflection or inversion as there was for the string.

A three-dimensional plot shows another view of snapshots over a period of 1.44 ms in Figure 6.2.3. The location of *one* of the support points, which are stationary, is represented by the line of small circles.

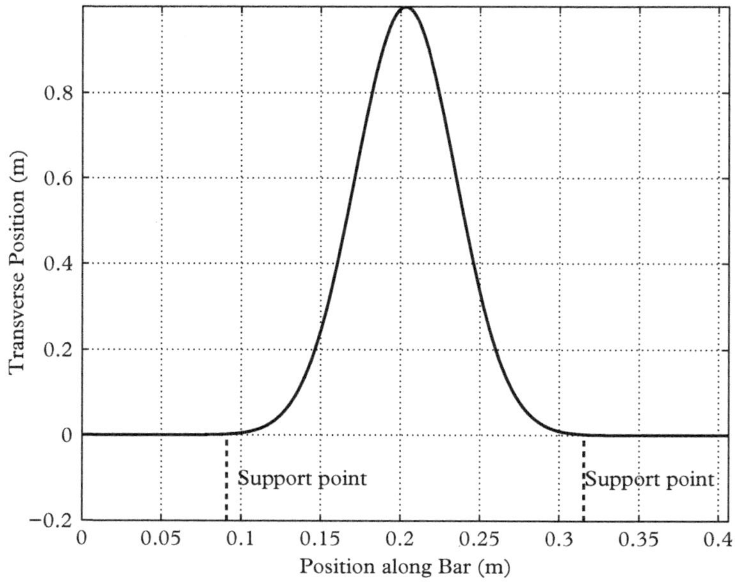

Figure 6.2.1 *Initial profile of vibrating bar*

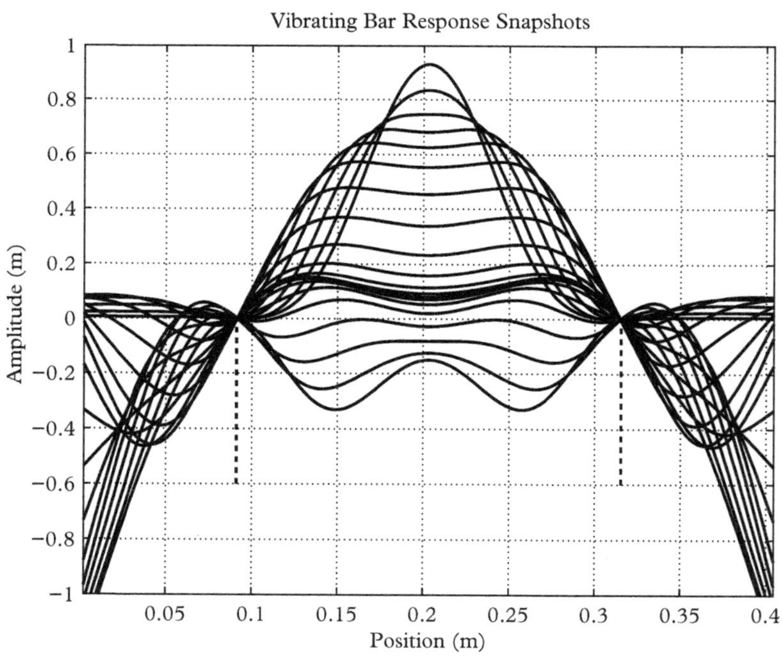

Figure 6.2.2 *Response of vibrating bar to initial condition shown in Figure 6.2.1*

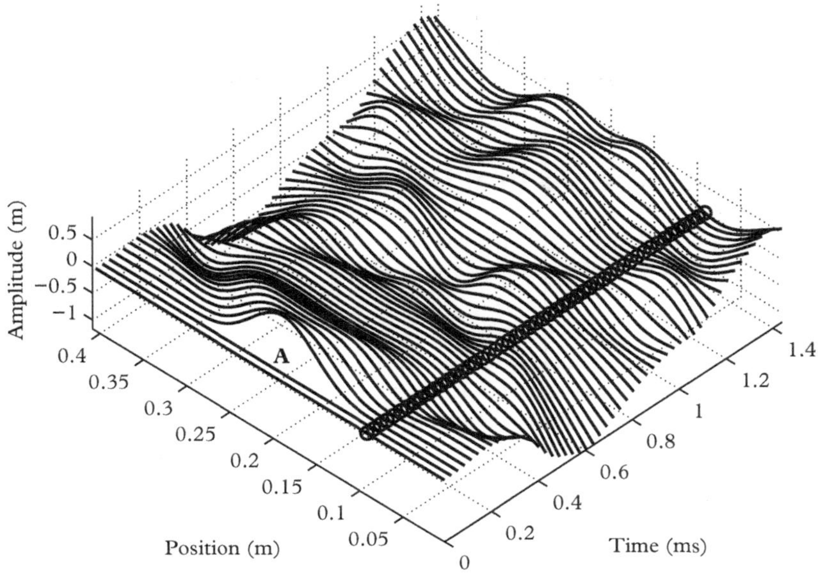

Figure 6.2.3 *Snapshots of vibrating bar response to initial condition at point A*

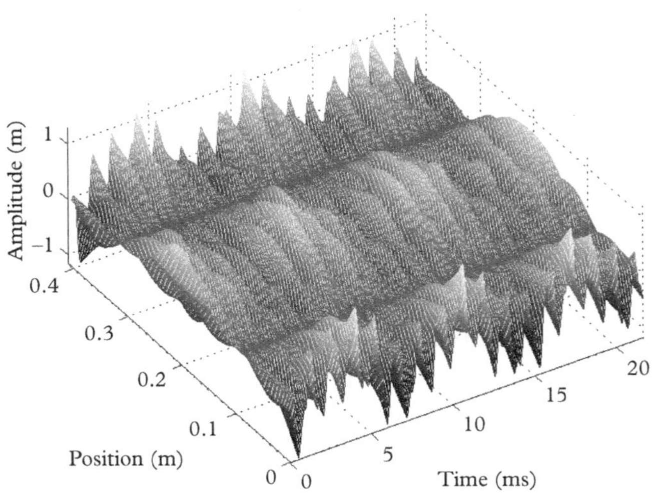

Figure 6.2.4 *Response of vibrating bar to initial condition shown in Figure 6.2.1*

The initial profile is located at point A. As time progresses the initial profile morphs into a variety of shapes.

The results of a longer simulation are shown in Figure 6.2.4, where one can see the two troughs associated with the support points and a repeating structure.

6.3 Averaging the Vertical Displacements

To get a feel for the sound that might emanate from the bar, we averaged the transverse displacements over the bar at every instant. A small portion of these average transverse displacements is shown in Figure 6.3.1.

The trace in Figure 6.3.1 looks periodic with a period of about 1.26 ms, which corresponds to 793 Hz. The line spectrum, cumulative line spectrum, and autocorrelation of this record having a duration of 720 ms are given in Figure 6.3.2.

There is a small spike at 140 Hz in the line spectrum, Figure 6.3.2(a), which is near the fundamental frequency of the bar (138.9 Hz, which is near C3#). The cumulative line spectrum, Figure 6.3.2(b), indicates that only the components at 750 Hz and 3223 Hz have significant strength. The pitch of the bar according to the autocorrelation is 789 Hz (this estimate is supported by the fine grid autocorrelation). This pitch estimate does not agree closely with the estimated fundamental frequency.

There are no harmonics of 140 Hz, or of 750 Hz. Instead there are components that we will call "partials" or "overtones" at 750 Hz, 1825 Hz, 3223 Hz, 4677 Hz, 5925 Hz, 6847 Hz, 7446 Hz,

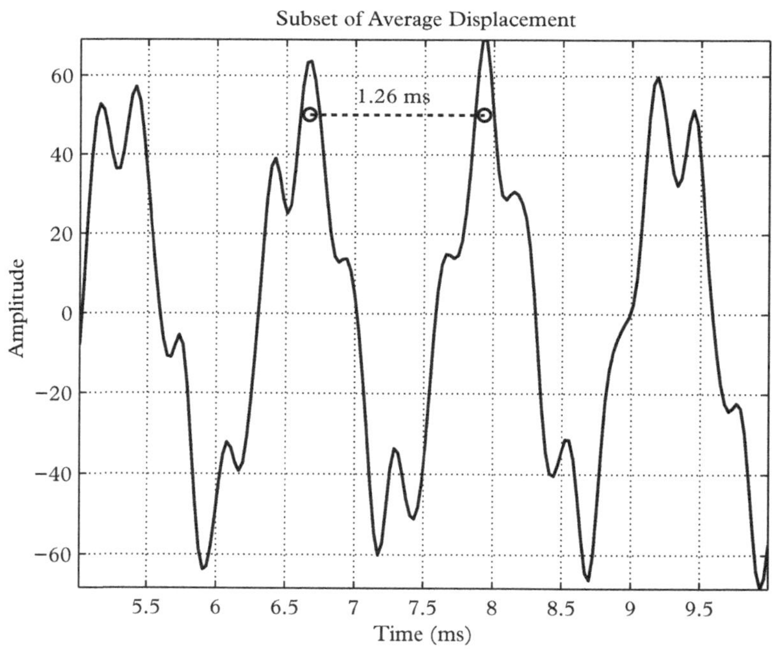

Figure 6.3.1 *Average displacement of bar showing the fundamental period of approximately 1.26 ms*

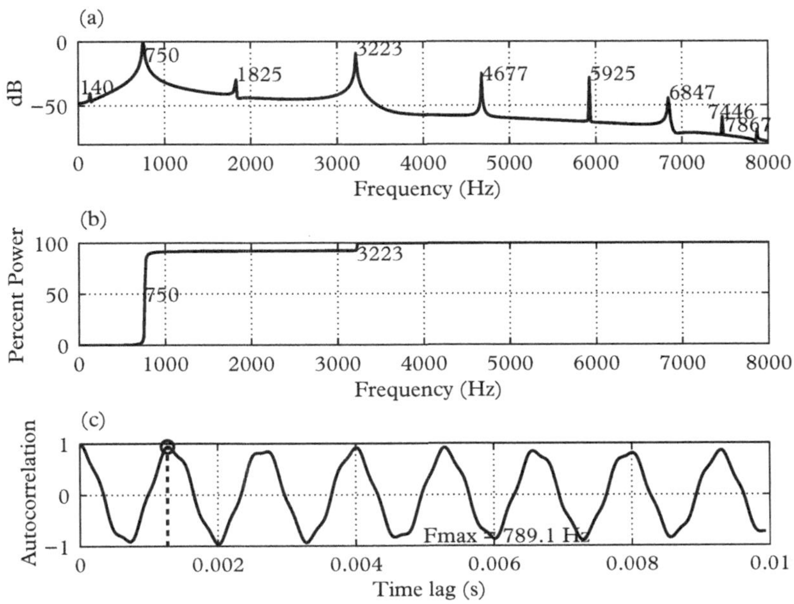

Figure 6.3.2 *Average displacement: (a) line spectrum, (b) cumulative line spectrum, and (c) autocorrelation*

and 7867 Hz, which are plotted in Figure 6.3.3 where the partials are seen to be much larger than the theoretical harmonics.

According to the reference books on sound and acoustics (Morse, *Vibration and Sound*; Kinsler and Frey, *Fundamentals of Acoustics*, and Olson, *Music, Physics and Engineering*), the vibrating bar has a theoretical fundamental frequency f_1 but the overtones or partials, $f_2, f_3, f_4, \ldots$ are not integer multiples of the fundamental. Instead, the partials or overtone frequencies are given approximately by

$$f_2 = \frac{5^2}{3.0112^2} f_1 = 2.757 f_1$$

$$f_3 = \frac{7^2}{3.0112^2} f_1 = 5.4 f_1$$

$$f_4 = \frac{9^2}{3.0112^2} f_1 = 8.933 f_1$$

$$f_5 = \frac{11^2}{3.0112^2} f_1 = 13.3 f_1$$

$$f_6 = \frac{13^2}{3.0112^2} f_1 = 18.6 f_1$$

$$f_7 = \frac{15^2}{3.0112^2} f_1 = 24.8 f_1$$

etc.

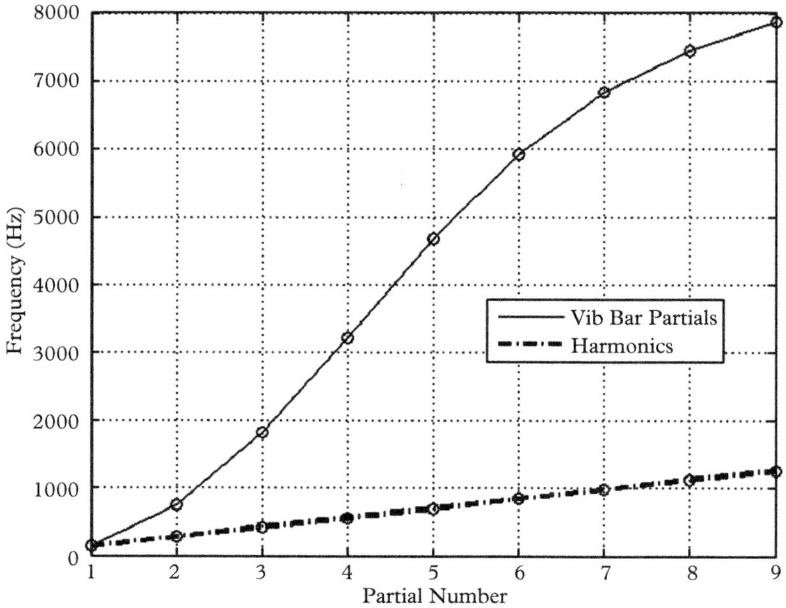

Figure 6.3.3 *Vibrating bar partial frequencies along with those for multiples of the fundamental frequency*

Therefore, the fundamental and overtone frequencies have a strict relationship—it is just not quite as simple as that for the vibrating string.

The first observed primary spike in Figure 6.3.4 at 750 Hz turns out to be the following factor of the natural frequency:

$$\frac{750}{140} = 5.35,$$

which is close to the third partial. The second primary spike at 3223 Hz turns out to be the following factor of the natural frequency:

$$\frac{3223}{138.9} = 23.2,$$

which is close to the seventh partial. The other partials are 30 dB down (or more) and do not seem to have an impact, as evidenced by the cumulative line spectrum.

The sound of the collected signal is available in the file VibBar4.mp3. If you listen, it will not sound even remotely close to C3# but it does sound reasonably close to G5, which has a frequency of 784 Hz, which is close to the estimated pitch using the autocorrelation.

These results suggest that the performance of the vibrating bar is significantly different from that of the vibrating string.

6.4 An Example: A Toy Xylophone

An example of a vibrating bar comes from my grandson's eight-bar (C, D, ..., C) xylophone. Although a toy, it has the same construction as that presented in Figure 6.1.1 The line and cumulative line spectra derived from striking the third bar (as in C D E) is given in Figures 6.4.1.

The first peak in Figure 6.4.1 (a) resides at 1300 Hz, which is reasonably close to E6 (1318 Hz). The other peaks at 3376 Hz, 5434 Hz, 6886 Hz, 8708 Hz, and 10,036 Hz are not multiples of the first peak, so there is no apparent integer-based harmonic structure. However, the ratios of the partials to the fundamental (in parentheses) bear a striking resemblance to those mentioned on p. 51 of Arthur Benade's book:

$$\frac{3376}{1300} = 2.59 \;\; (2.68 \;\; \text{second partial})$$
$$\frac{5434}{1300} = 4.18$$
$$\frac{6886}{1300} = 5.29 \, (5.25 \;\; \text{fourth partial}) \, .$$

Determining the pitch from the autocorrelation in Figure 6.4.1 (c) is a bit muddy. The first peak denoted by the small circle corresponds to a frequency of 6300 Hz, which is beyond the range of the piano. The second peak, located at about a time lag of 1 ms, corresponds to a frequency of 1075 Hz. The absolute strongest peak resides at a time lag of about 3.1 ms and corresponds to a frequency of 320 Hz, which is near the fundamental frequency for E4. To my ears the sound is

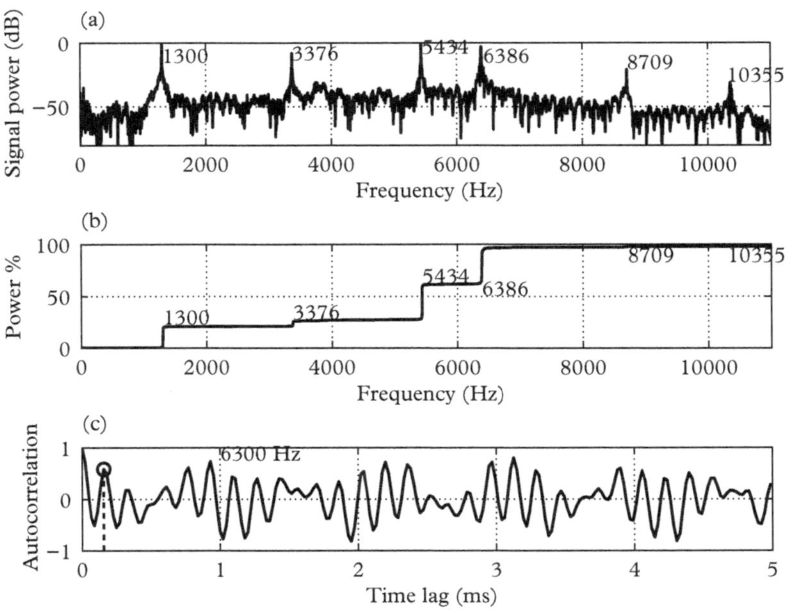

Figure 6.4.1 *Toy xylophone: (a) line spectrum, (b) cumulative line spectrum, and (c) autocorrelation*

closest to E6 (1318 Hz), which actually corresponds to a lesser highest peak at a time lag of about 0.76 ms. The reader is invited to listen to the sound of the toy xylophone in the file ToyXylo.mp3.

6.5 Summary

This short chapter attempted to illustrate some of the differences between the vibrating string, both purely flexible and partly stiff, and the vibrating bar, which is totally stiff. Using simulation, we found that the harmonic structure is different, as is the nature of the traveling transverse waves. An examination of a toy xylophone supported the idea that the harmonic structure is indeed different and it also showed that one has to carefully interpret the results of the autocorrelation.

6.6 Exercises

Download the Matlab® script file VibBarExplicit.m. This script simulates the movement of a bar that is clamped at one end and free at the other. Vary the parameters and compare the time domain behavior and the spectra with the bar that is free at both ends used in this chapter. Is the spacing of the partials for the clamped bar similar to that of the free bar? What tone would you expect to hear? Try to match the tone with a note on a piano. Which peak in the line spectrum does the sound correspond to? Did the autocorrelation give a pitch estimate that makes sense? By the way, to find the peak when the autocorrelation is displayed, you have to click twice: once on the left of where the peak is and once on the right.

7 The Vibrating Membrane

The circular flexible membrane can be considered an idealization of a drum head. We will look at the response of the membrane surface to a variety of initial conditions and compare it to the bar and string.

7.1 Basic Parameters of the Vibrating Membrane

We start with a symmetrical disturbance at the center. A disturbance located away from the center of the membrane will also be studied. Like the vibrating string, the membrane is subject only to tension and has no stiffness. It may interest the engineer-reader that our idealized membrane is actually described by the same partial differential equation as for the vibrating string except that it is defined on a two-dimensional circle rather than a one-dimensional line.

7.2 Response to a Center Displacement

The drum head presented in Figure 7.2.1 can be idealized by a flexible membrane, much like a guitar string can be approximated by a flexible vibrating string. Figure 7.2.1 shows the Gaussian-shaped initial condition of such a membrane (radius 0.5 m) that has been displaced from quiescent conditions at the center.

For the purposes of simulation, the circular drum head has been partitioned into 20 sectors, and hence is not a perfect circle. To really get a feel for how the drum head moves in response to this initial displacement one should view the YouTube video referenced in the book's website entitled "Circular Membrane Dynamics" (no. 18). In lieu of that, consider the eight snapshots in Figures 7.2.2 and 7.2.3.

The sequence in Figures 7.2.2 and 7.2.3 shows that the initially displaced surface evolves into an outward-moving circular wave with a crest that appears to be as high as the initial peak. At about 7 ms it reaches the boundary and is reflected into an inward-moving circular wave with a trough. This back and forth movement of the circular wave continues indefinitely. Compare it to the movement of the initial bulge for the vibrating string in Figures 5.5.2 and 5.5.3, where a central peak evolves into two outward moving waves. A similar discussion is presented on p. 186 in Morse's book.

At each instant, the average displacement of the surface is computed and plotted in Figure 7.2.4, where there are periodic components. The overall up and down movement of the membrane surface is represented by a component with an approximate period of 52 ms. This might be considered the fundamental movement. There are other possible periodic components,

Spectral Analysis of Musical Sounds with Emphasis on the Piano. First Edition.
David M. Koenig, with comments by Delwin D. Fandrich.
Published in 2015 by Oxford University Press.

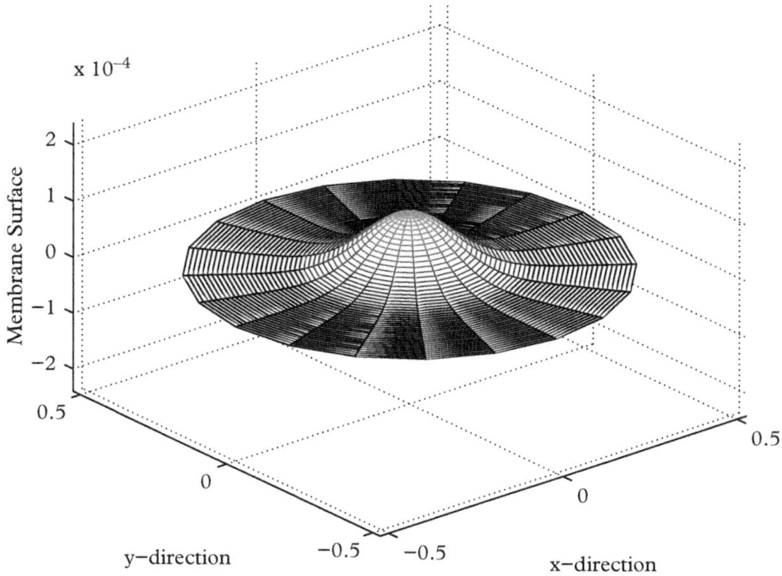

Figure 7.2.1 *Initial condition for vibrating membrane*

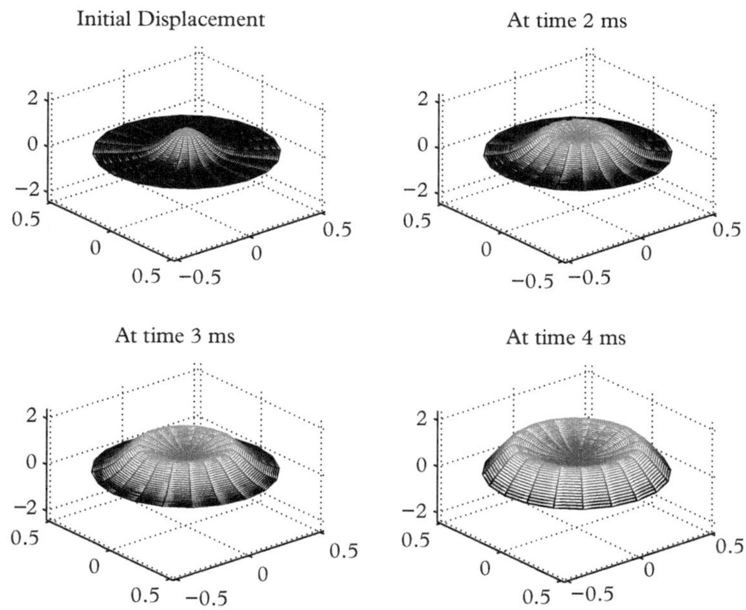

Figure 7.2.2 *Snapshots of vibrating membrane at 2 ms, 3 ms, and 4 ms*

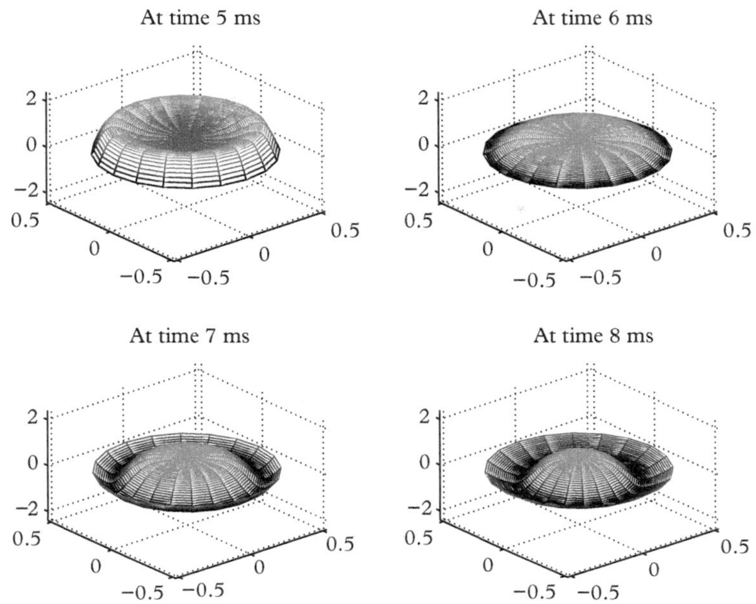

Figure 7.2.3 *Snapshots of vibrating membrane at 5 ms, 6 ms, 7 ms, and 8 ms*

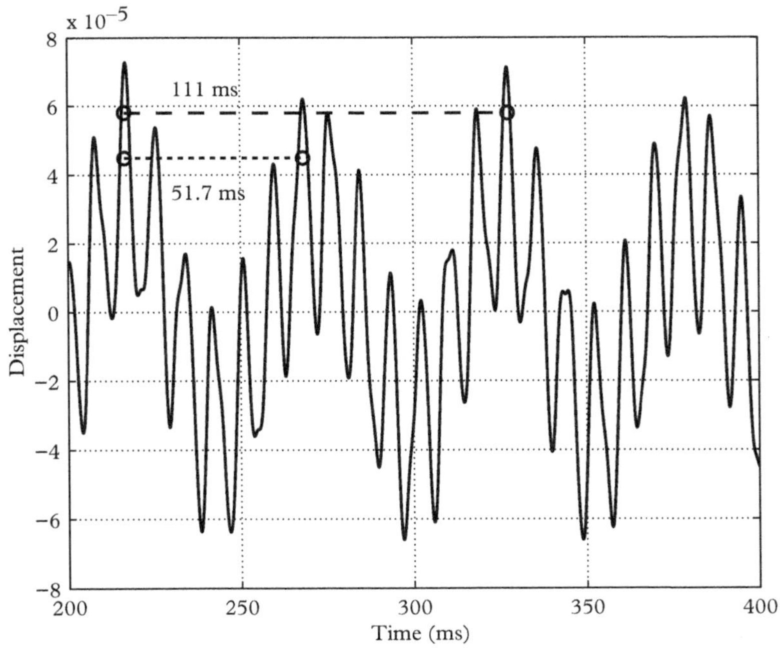

Figure 7.2.4 *Average vibrating membrane disturbance at each instant of time showing two periods*

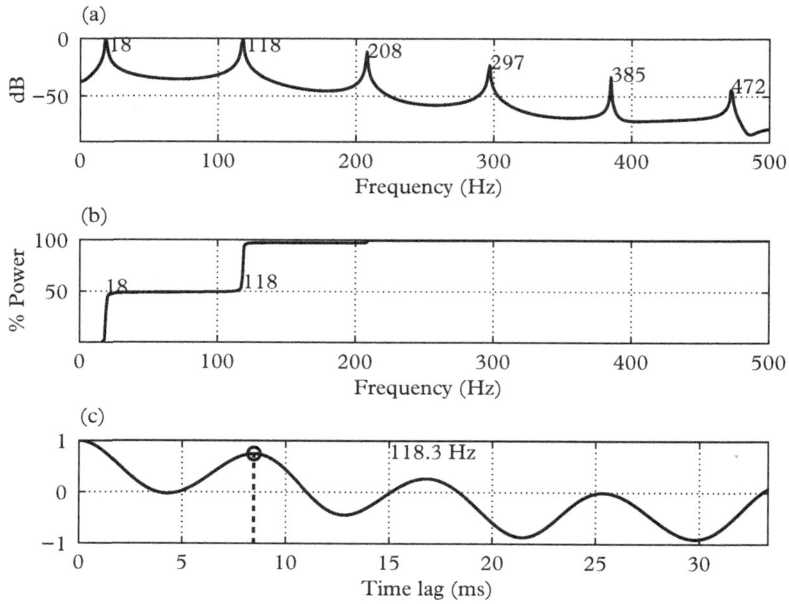

Figure 7.2.5 *Average vibrating membrane disturbance: (a) line spectrum, (b) cumulative line spectrum, and (c) autocorrelation*

one of which is denoted by the dotted line labeled 111 ms. The line spectrum and cumulative line spectrum confirm this and are shown in Figure 7.2.5.

The first line spectrum peak in Figure 7.2.5(a), which *might* be considered a fundamental harmonic, occurs at 18 Hz and corresponds to the 52 ms period observed in Figure 7.2.4. There are also peaks at 118 Hz, 208 Hz, 297 Hz, 385 Hz, and 472 Hz, but there is no integer-based harmonic structure as there was with the vibrating string. The differences between the partials are 99 Hz, 89.5 Hz, 89.5 Hz, 88.5 Hz, and 87.5 Hz (to the nearest 0.5 Hz). This suggests that the drum might have a sound reminiscent of something with a fundamental of about 90 Hz. However, the autocorrelation in Figure 7.2.5(c) suggests that the pitch is 118 Hz, which is close to A2#.

The sound of this membrane responding to this initial displacement is in the file Membrane.mp3, and to my ear it sounds closest to A3#.

An Alternative Visualization of the Membrane Dynamics by Slicing Across a Diameter

Since the initial displacement is symmetrical, one could ignore the angular dependence of the surface displacement and instead concentrate on just the radial dynamics. In effect, we would be looking at a slice across the diameter of the circular membrane. Figure 7.2.6 shows 5 ms of the response to the initial Gaussian-shaped displacement.

This shows how the initial Gaussian displacement splits into two segments and actually grows in peak amplitude.

Another view of the same data is given in Figure 7.2.7, where the growth in the peaks is shown more dramatically.

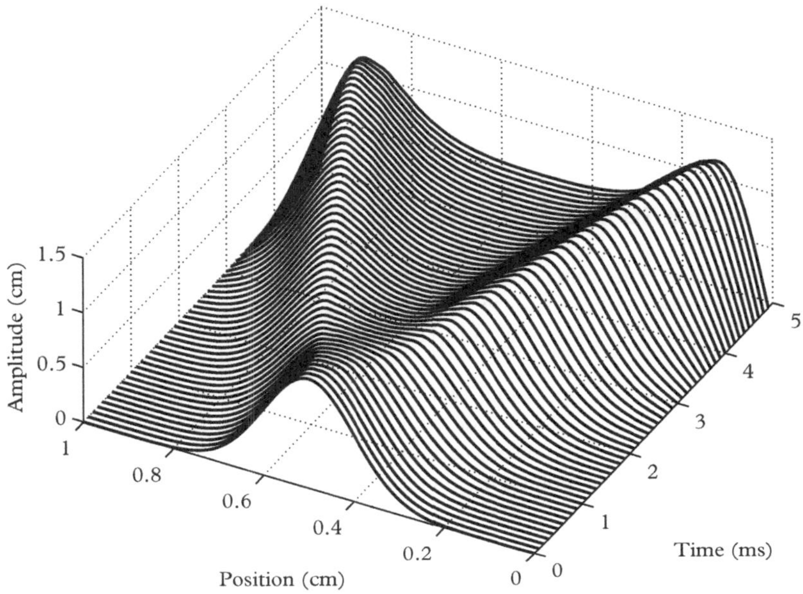

Figure 7.2.6 *Snapshots of slices through the center of a vibrating membrane in response to an initial Gaussian condition*

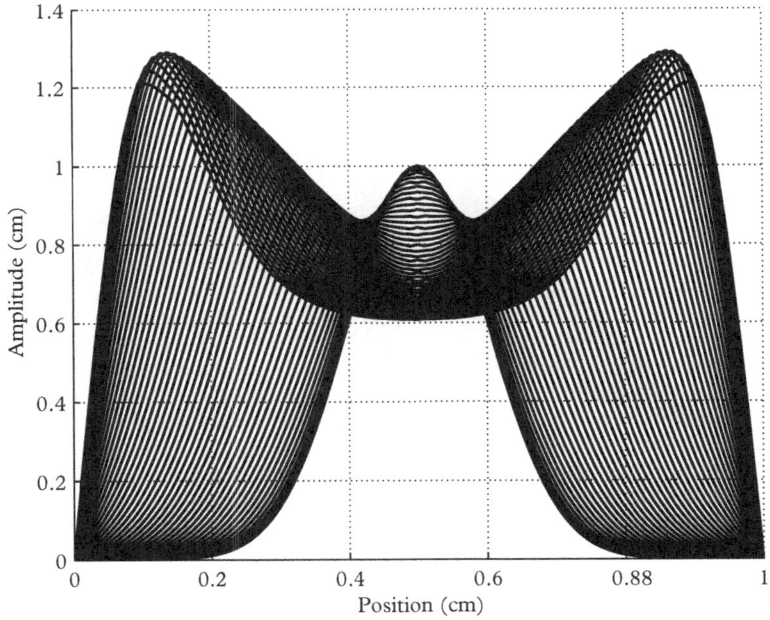

Figure 7.2.7 *Snapshots of slices through the center of a vibrating membrane in response to an initial Gaussian condition (alternative view)*

Figures 7.2.6 and 7.2.7 show how the surface of the membrane actually grows in height in response to the initial Gaussian displacement. The reader should take a look at Figures 5.5.2 and 5.5.3 for the case of the vibrating string and Figures 6.2.2–6.2.5 for the vibrating bar. Remember that the presentations for the membrane are from slices taken through the center, whereas those for the vibrating string and bar are without modification.

7.3 Response to a Cone-Shaped Initial Condition

Consider the case where the initial displacement has a symmetrical cone shape, or a triangular shape. Figure 7.3.1 shows the response when a slice is taken through the center point. Note the ripples in the tails of the response compared to the smoothness when the Gaussian initial condition is used.

The initial cone or triangle splits into two segments but it does not gain in amplitude. This is to be compared with the two cases of the vibrating string where the initial displacements were Gaussian and triangular. The responses are substantially different. This conical initial condition might be the result of somehow pulling the membrane up at the center and then letting go, and is probably more realistic.

7.4 Response to an Off-Center Displacement

If the initial displacement is off center (but still of Gaussian shape) then the visual response of the drum head is different, as shown in Figures 7.4.1 and 7.4.2.

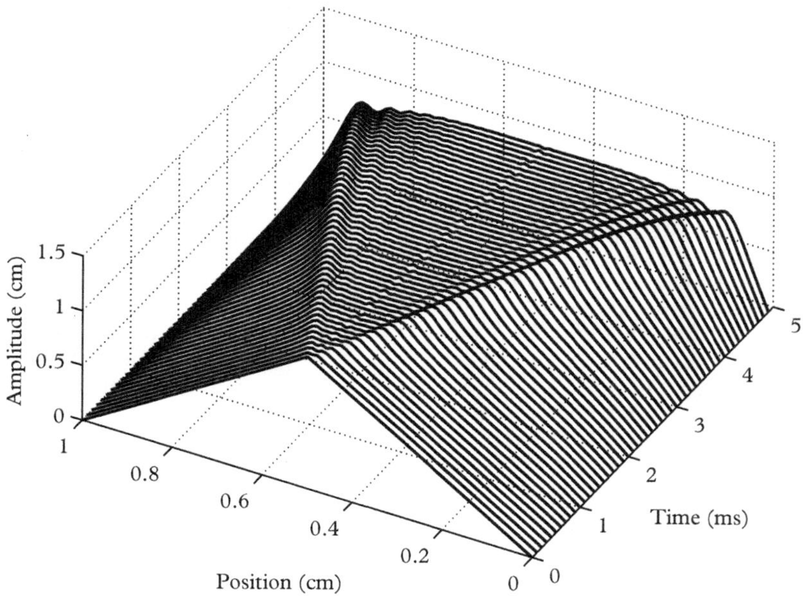

Figure 7.3.1 *Snapshots of slices through the center of a vibrating membrane in response to an initial cone-shaped condition*

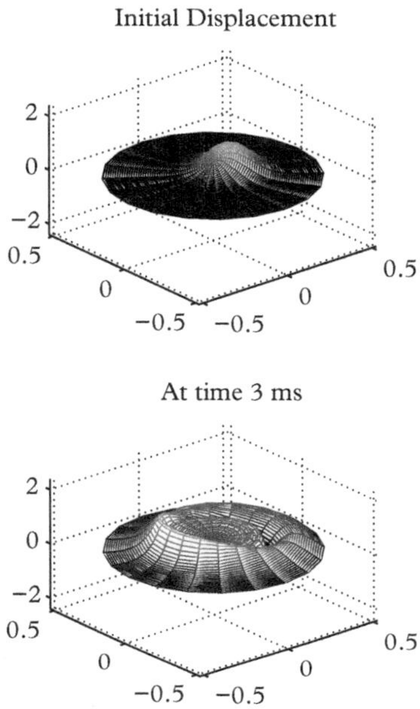

Figure 7.4.1 *Snapshots of vibrating membrane at 2 ms, 3 ms, and 4 ms with off-center initial condition*

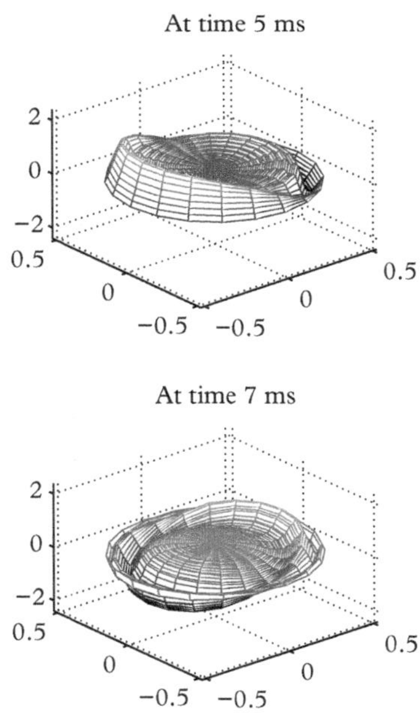

Figure 7.4.2 *Snapshots of vibrating membrane at 5 ms, 6 ms, 7 ms, and 8 ms with off-center initial condition*

Figure 7.4.1 initially shows one main wave. One side of the wave moves toward the rim and has a larger amplitude. The other side of the wave moves toward and through the drum center and has a smaller amplitude.

The average overall displacement at each instant of time has a shape similar to that for the center displacement case in Figure 7.2.4. The line spectrum and cumulative line spectrum of the average displacement are shown in Figure 7.4.3.

The line spectrum in Figure 7.4.3(a) shows peaks at the same locations as for the center displacement case, but with the splitting at each peak that was mentioned earlier. That is, there are doublets at 118 Hz, 208 Hz, 297 Hz, 385 Hz, and 472 Hz. The amount of the separation is on the order of 6 Hz. This suggests that the off-center initial displacement generates a secondary set of waves that have partials that are, on average, about 6 Hz different. The autocorrelation in Figure 7.4.3(c) for this offset initial condition is essentially the same as that in Figure 7.2.5(c), except that the peak location is slightly displaced such that the associated pitch frequency is 120 Hz instead of 118 Hz.

Figure 7.4.4 shows a close-up of the splitting at 208 Hz using the fine grid line spectrum.

A mathematical analysis of the membrane dynamics involves solving a partial differential equation with the three independent variables being the time, the radial direction from the center, and the angular direction around the center. This analysis would show that the radial movement of the surface is described by Bessel functions. The application of the boundary condition on the rim, which constrains the surface to have zero displacement, generates "eigenvalues" that in turn yield associated "eigenfrequencies." Although interesting and elegant, this approach is beyond

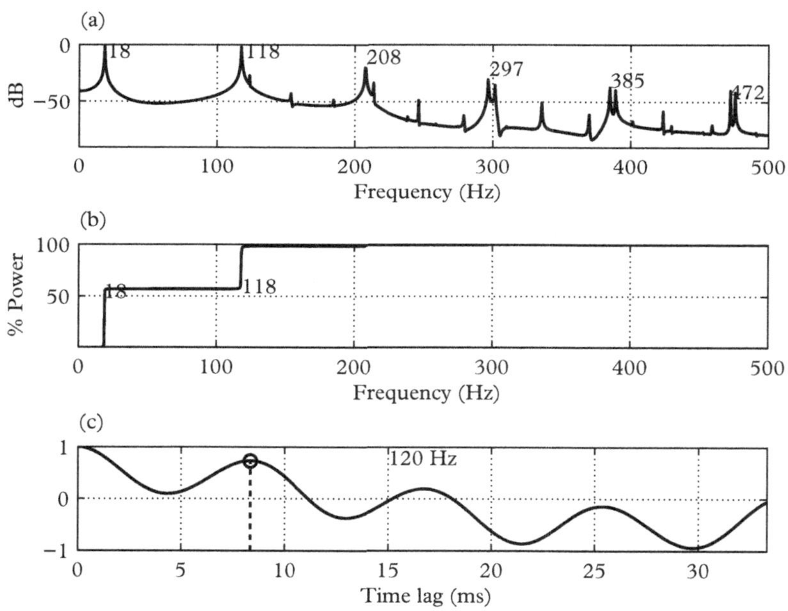

Figure 7.4.3 *Average vibrating membrane disturbance (with off-center initial disturbance): (a) line spectrum, (b) cumulative line spectrum, and (c) autocorrelation*

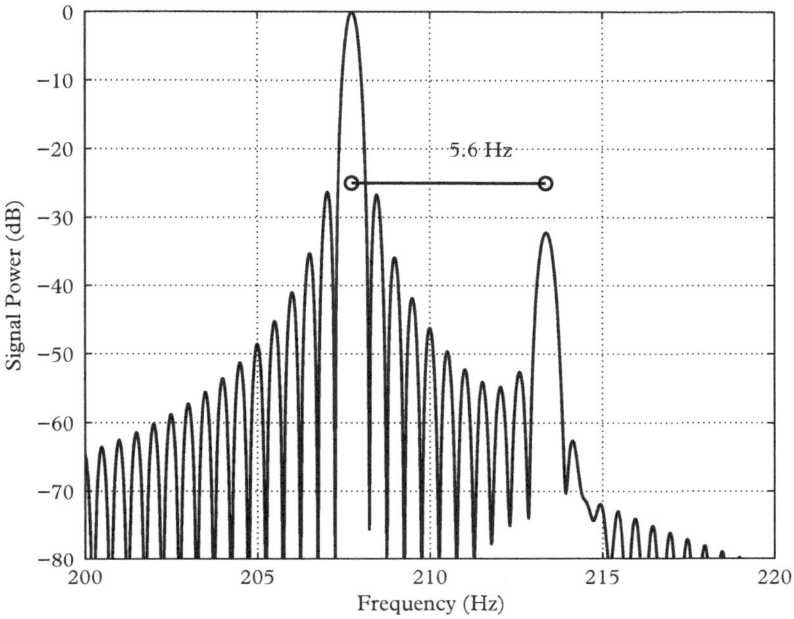

Figure 7.4.4 *Fine grid line spectrum of average displacement with off-center initial condition disturbance*

the scope of this book (see Morse, Kinsler and Frey, and Appendix 2). However, it shows that there are different "modes" of surface movement that are excited by the initial displacement. This sophisticated method offers no physical explanation for the observed phenomena. There are several Internet sites that show movies of these different modes.

We chose to solve the partial differential equation numerically and observe the development of the different modes empirically using the line spectrum. For the case of the centrally located initial displacement, symmetrical modes are excited. For the off-center initial displacement, a different type of mode was also excited such that a variety of waves moved back and forth across the drum head surface. Some of the frequencies of these additional asymmetrical modes are close to those of the symmetrical modes, hence the line splitting in the spectra.

7.5 Summary

This chapter has perhaps provided some insight into how the two-dimensional cousin of the vibrating string behaves. Like the string, the membrane has no stiffness, is subject only to tension, and has fixed end points (actually a fixed circular boundary). On the other hand, the vibrating bar was stiff, not subject to tension, and lay on two support points, but did not have fixed end connections. Unlike the string, neither the membrane nor the bar have the harmonic structure of a fundamental frequency along with harmonics that are integer multiples of the fundamental frequency. The next subject, the plate, is also a relative of the membrane and the bar: it has physical dimensions like the membrane and a physical characteristic, elasticity, like the bar. Along with the vibrating string, the plate is an integral component of the piano, which is the subject of Part 3.

7.6 Exercise

Using the Matlab® script for the chapter 7 figures (in the file FiguresPart2_Chap7.m), find the code that generates Figure 7.3.1 and modify it so that a plot like that in Figure 7.2.7 can be generated. How do they compare and why are they different?

8

The Vibrating Plate

The circular plate is a logical progression from the circular membrane studied in the last chapter. Both Gaussian and cone initial conditions will be used in the simulation of its dynamics. A square plate will also be analyzed. The responses of both the circular and square plate exhibit symmetry. In contrast, the asymmetrical response of a triangular-shaped plate is also numerically simulated. The analysis of a vibrating string struck by a hammer in connection with a square plate and a bridge is compared with a similar analysis of a square plate not activated by the hammer but by an isolated pulse to the bridge. Finally, an experiment is performed on my Knabe grand by giving the bridge a thump and collecting the response via a transducer mounted on the soundboard.

8.1 Basic Parameters of the Vibrating Plate

The soundboard of a guitar and the soundboard of a piano, although dimensionally quite different, vibrate in response to vibrating strings attached by means of a bridge. In Chapter 5, soundboard dynamics were simulated by a mass/spring/dashpot model with a resonant frequency. We also studied the force of a vibrating string on a fixed end point as an indicator of bridge/soundboard dynamics. In this chapter a circular, a square, and a triangular soundboard will be studied. The dynamics of soundboards (or plates) are dictated by stiffness whereas those of the membrane depend on tension. The ideal plate made of Sitka spruce studied here is not under tension and is clamped rigidly on the boundary.

8.2 Circular Plate—Gaussian Initial Condition

In this section a circular plate will be briefly studied. As with the membrane, we will initially specify radial symmetry and mostly look only at a *slice across a diameter*. Figure 8.2.1 shows the response of the plate with an initial Gaussian displacement represented by point A. At each instant of time, Figure 8.2.1 shows a slice of the plate made across a diameter.

The initial Gaussian displacement is dispersed into waves that move outward and are reflected from the boundary. When compared to the responses of the flexible string in Figure 5.5.2 and the membrane in Figure 7.2.6, the plate shows yet a different set of dynamics.

The line spectrum, the cumulative line spectrum, and the autocorrelation of the average displacement over the whole plate at every instant of time are given in Figure 8.2.2.

The fundamental frequency in Figure 8.2.2(a) is 30 Hz. There are peaks at 110 Hz, 231 Hz, 593 Hz, 831 Hz, and so on, shown in the cumulative line spectrum in Figure 8.2.2(b).

Spectral Analysis of Musical Sounds with Emphasis on the Piano. First Edition.
David M. Koenig, with comments by Delwin D. Fandrich.
Published in 2015 by Oxford University Press.

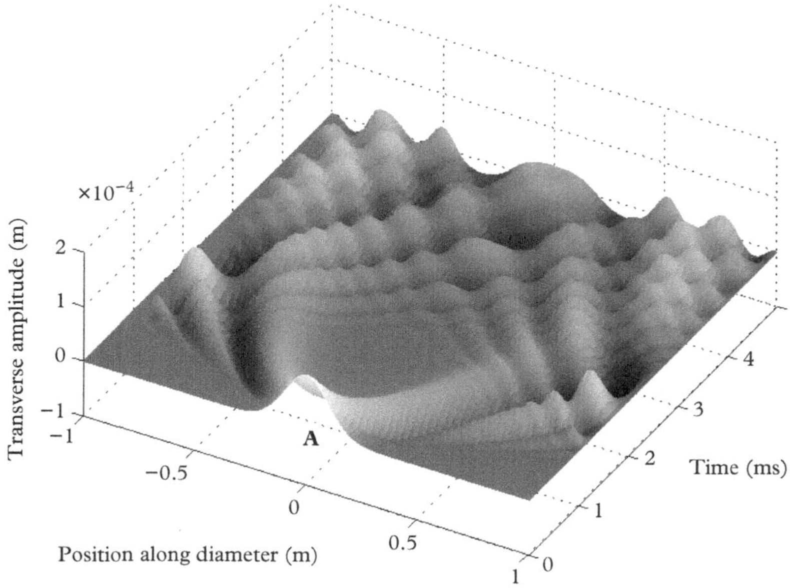

Figure 8.2.1 *Three-dimensional plot of slices over a diameter of a circular plate responding to an initial Gaussian condition*

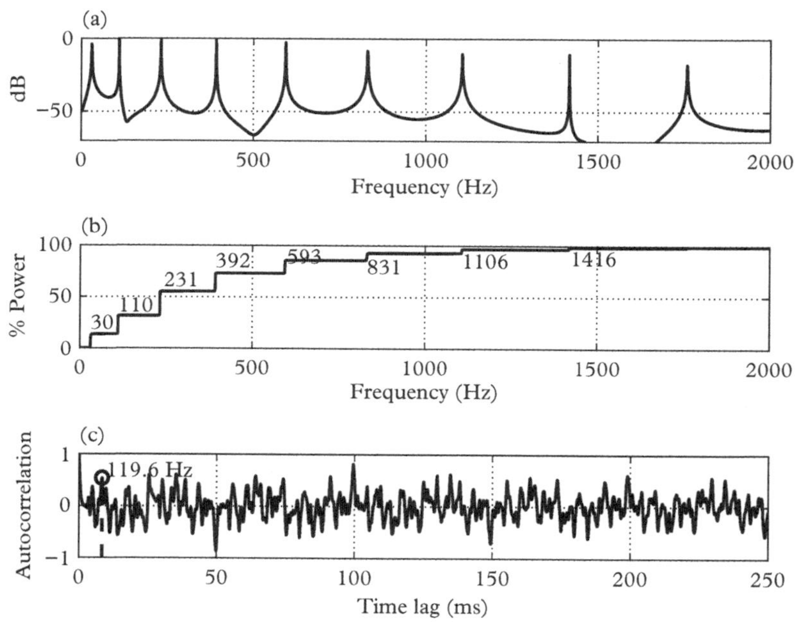

Figure 8.2.2 *Average displacement of the plate: (a) line spectrum, (b) cumulative line spectrum, and (c) autocorrelation*

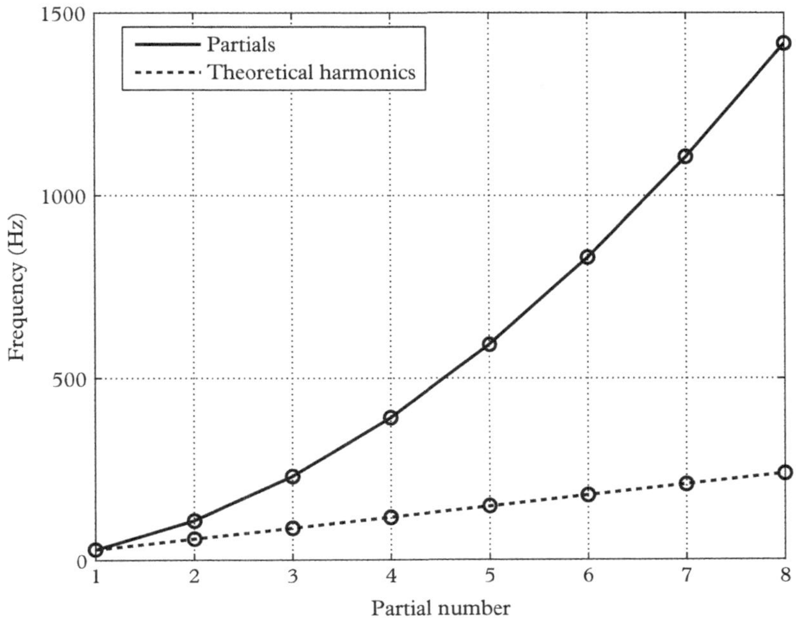

Figure 8.2.3 *Partials for the vibrating plate*

The autocorrelation in Figure 8.2.2(c) indicates that the pitch, based on the first peak, is 120 Hz, although it may not be significant because of the plethora of similar-sized peaks. The spacing between the overtones or partials increases with frequency, as is demonstrated in Figure 8.2.3 where the harmonics based on the first partial are also plotted.

A similar analysis for the case of an asymmetrical initial displacement would generate splitting in the partials as was the case for the membrane.

8.3 Circular Plate—Cone Initial Condition

Figures 8.3.1 and 8.3.2 show some snapshots of the plate surface in response to an initial cone-shaped displacement.

Using a perspective similar to that taken to construct Figure 8.2.1, i.e., taking a slice across a diameter, the response to the initial cone displacement is shown in Figure 8.3.3. Just like the response to the initial Gaussian displacement, outward waves are generated and reflected at the boundary.

Compare Figure 8.3.3 to Figure 7.3.1 for the membrane. The stiffness of the plate causes a significantly different response. For the case of the initial cone displacement, the triangular-shaped membrane wave maintains its shape as it moves outward. However, the triangular-shaped plate wave loses its initial shape and disperses into smaller waves as it moves outward.

A longer view of the response is shown in Figure 8.3.4, where the snapshots are taken over a 40 ms period. Note that the overall trend of the center point is downward and then, after about

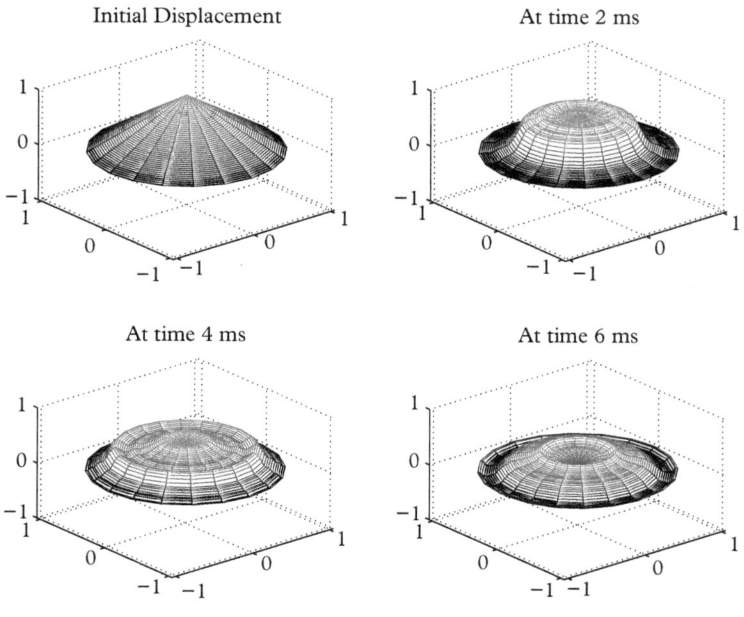

Figure 8.3.1 *Snapshots of vibrating plate at 2 ms, 4 ms, and 6 ms*

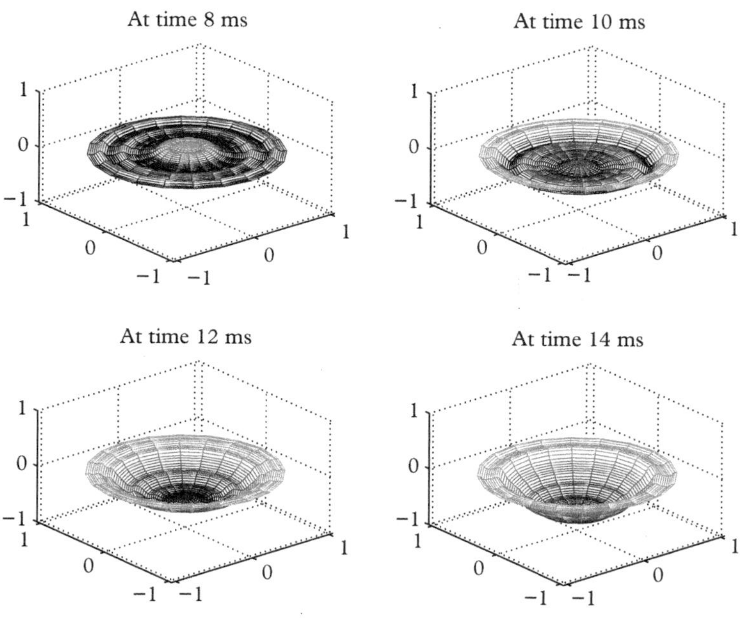

Figure 8.3.2 *Snapshots of vibrating plate at 8 ms, 10 ms, 12 ms, and 14 ms*

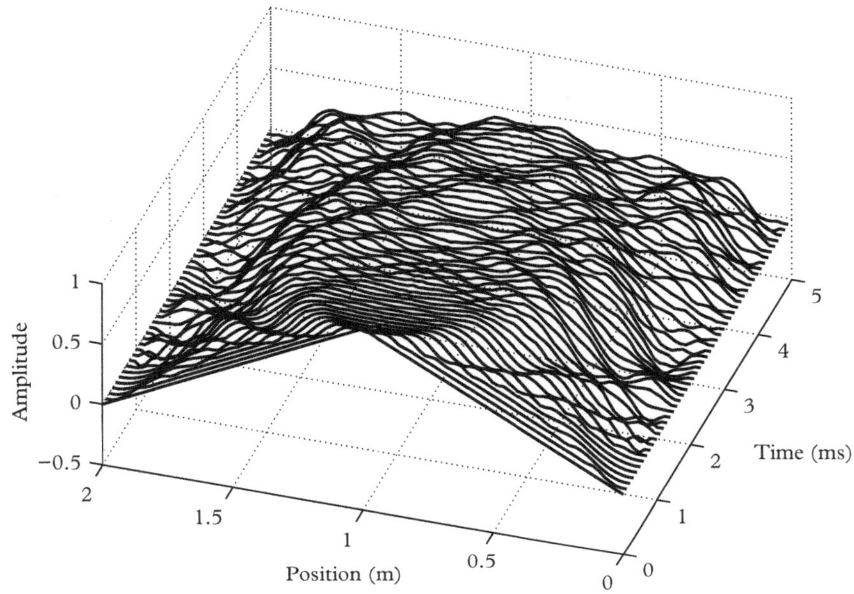

Figure 8.3.3 *Snapshots of vibrating plate via slices across a diameter for 5 ms*

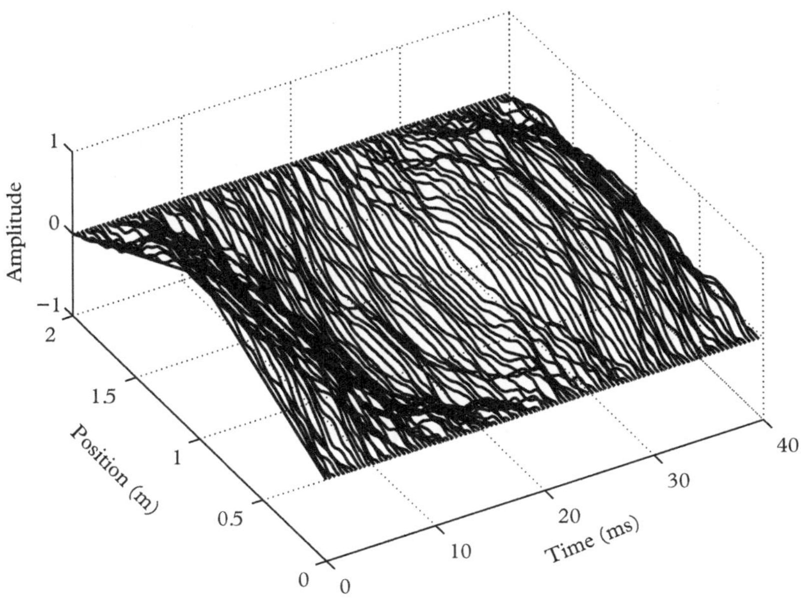

Figure 8.3.4 *Snapshots of vibrating plate via slices across a diameter for 40 ms*

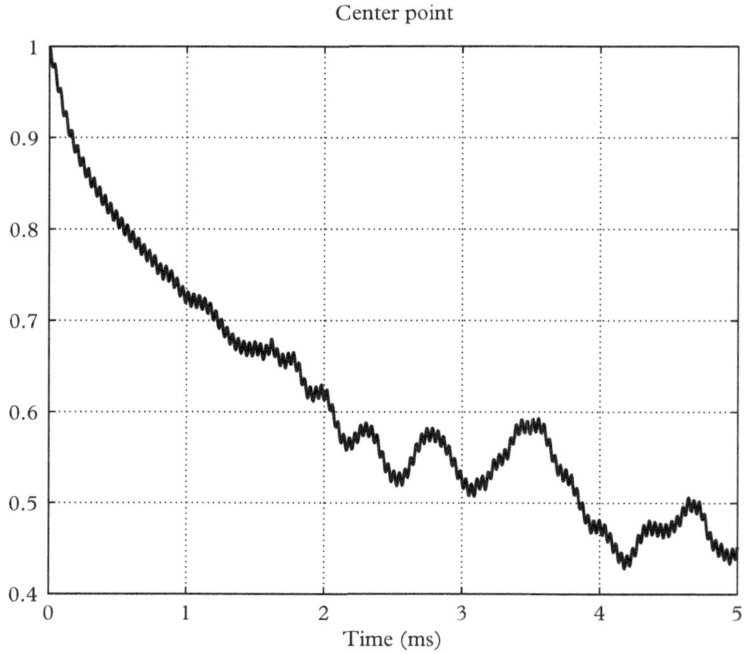

Figure 8.3.5 *Behavior of plate center position*

15 ms, upward. During this descent and ascent there are numerous faster ripples resulting from waves being reflected at the boundary. Also, note that although the surface in Figure 8.3.4 appears "wiggly" and unruly, it is actually symmetric about a center line where the position is 1.0 m.

You can see that there are at least two things going on. First, one sees the overall slower descent and ascent of the center point. Second, the faster waves going from the center to the edge and back are causing a "bobbing" of the center point. This bobbing of the plate center is borne out in Figure 8.3.4 and in Figure 8.3.5, where the center of the plate is shown as a function of time.

A movie of the circular plate responding to this conical initial condition is referenced on the book's website (no. 19).

8.4 Square Plate—Gaussian Initial Condition

The circular dimensions of the plate were chosen for comparison with the flexible membrane, which ideally simulates a drum head. However, we are more interested in rectangular or square plates because they simulate a piano soundboard in an idealistic manner. Figures 8.4.1 and 8.4.2 show the square board's dynamics, in response to an initial Gaussian displacement in the center of the square, at 5 ms intervals.

The initial displacement is symmetrical around the center of the plate and the disturbance waves move outward in a circular fashion until they hit the non-circular edges and are reflected.

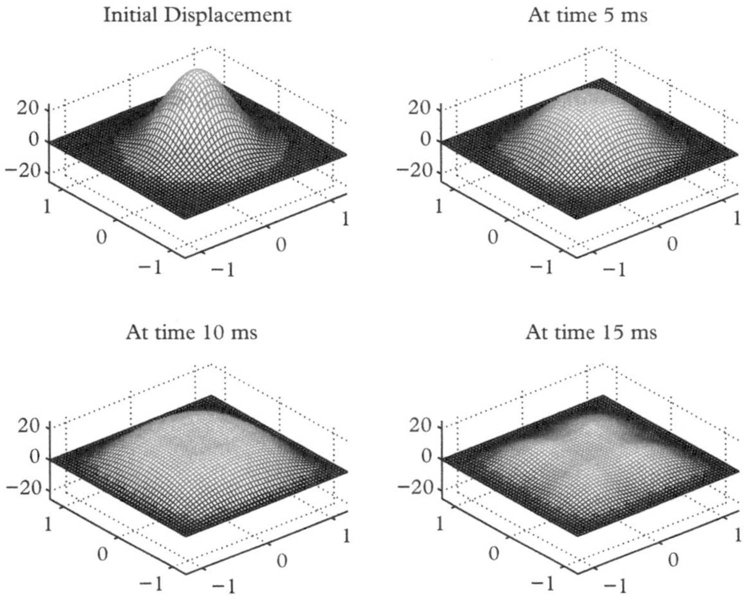

Figure 8.4.1 *Snapshots of vibrating plate at 5 ms, 10 ms, and 15 ms*

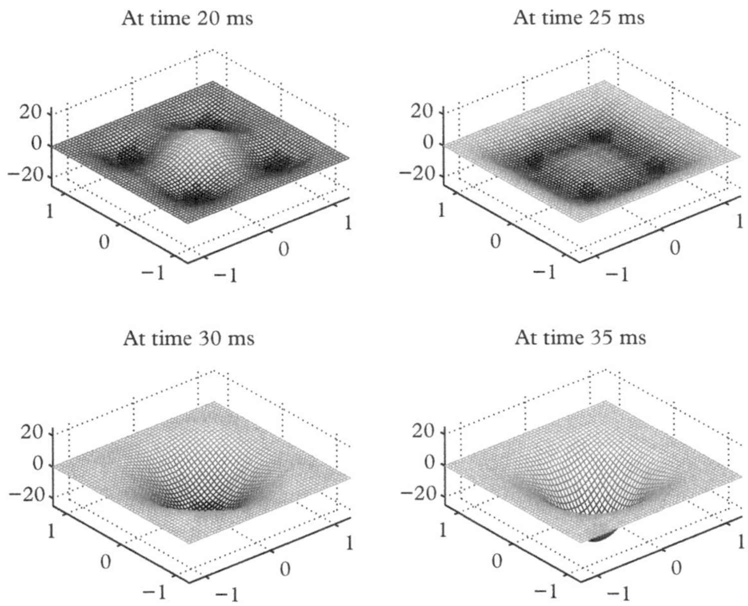

Figure 8.4.2 *Snapshots of vibrating plate at 10 ms, 25 ms, 30 ms, and 35 ms*

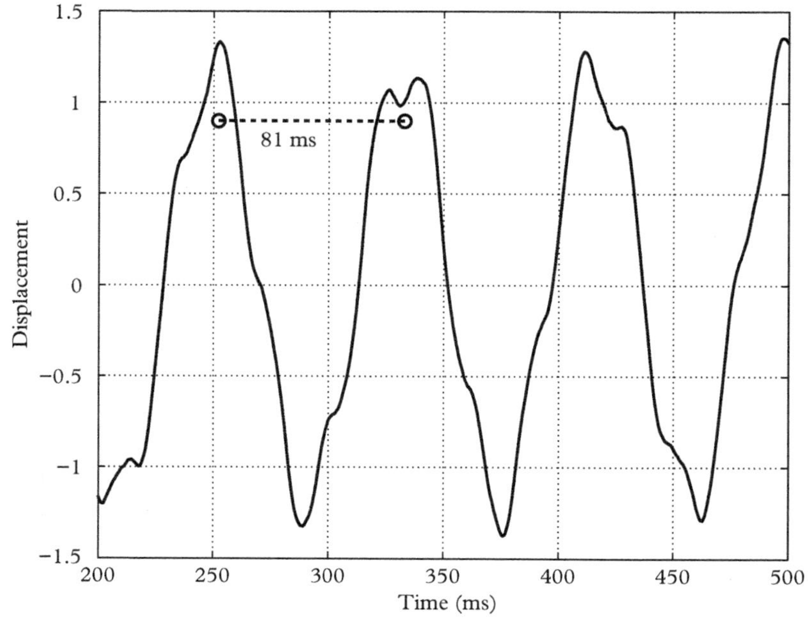

Figure 8.4.3 *Average displacement of vibrating plate showing a period of 81 ms*

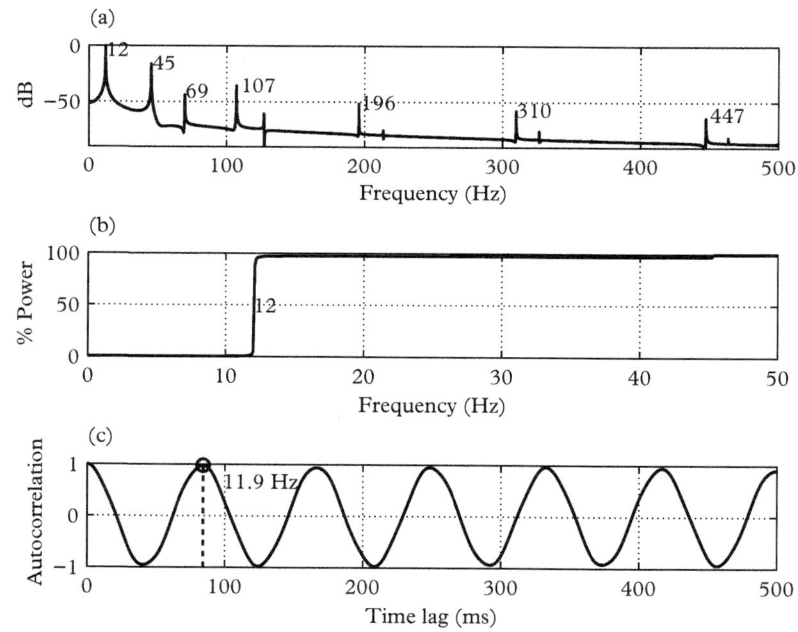

Figure 8.4.4 *Average displacement of vibrating plate: (a) line spectrum, (b) cumulative line spectrum, and (c) autocorrelation*

At this point a four-quadrant displacement results, which then morphs back into a single symmetrical bump. This pattern repeats indefinitely. Once again a video is more enlightening—please take a look at the YouTube video referenced in the book's website (no. 20).

The average displacement of the board is shown in Figure 8.4.3, where there is an apparent periodicity of 81 ms.

The line spectrum, cumulative line spectrum, and autocorrelation of the average displacement is shown in Figure 8.4.4, confirming that a 12 Hz component is dominant.

There is a system of overtones in the line spectrum in Figure 8.4.4(a) with the spacing between peaks increasing with frequency. There is also some splitting of the peaks, probably due to the different modes excited by the reflection of the initially symmetrical displacement by the square sides. Compare these graphs with those for the circular plate where the initial disturbance is symmetrical and consistent with the shape of the plate. The spectrum for the circular plate shows no splitting because symmetry is maintained by the reflections.

8.5 Triangular Plate—Gaussian Initial Condition

A right-triangular plate with sides of 2.4 m and 3.6 m will understandably exhibit a nonsymmetrical series of displacements, as shown in Figure 8.5.1. Note the dark lines outlining the boundaries of the triangular plate surface.

The line spectrum, cumulative line spectrum, and autocorrelation of the average displacement is shown in Figure 8.5.2. Unlike the symmetrical plates in previous sections, the triangular plate shows strength at a main or fundamental frequency at 21.3 Hz along with weaker strength at a host of irregularly spaced frequencies, probably the consequence of the asymmetrical shape

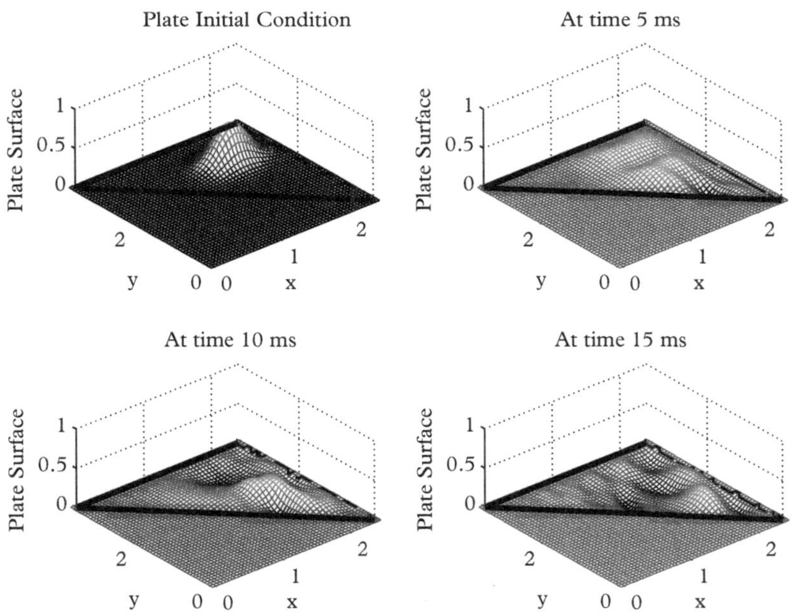

Figure 8.5.1 *Snapshots of triangular vibrating plate at 5 ms, 10 ms, and 15 ms*

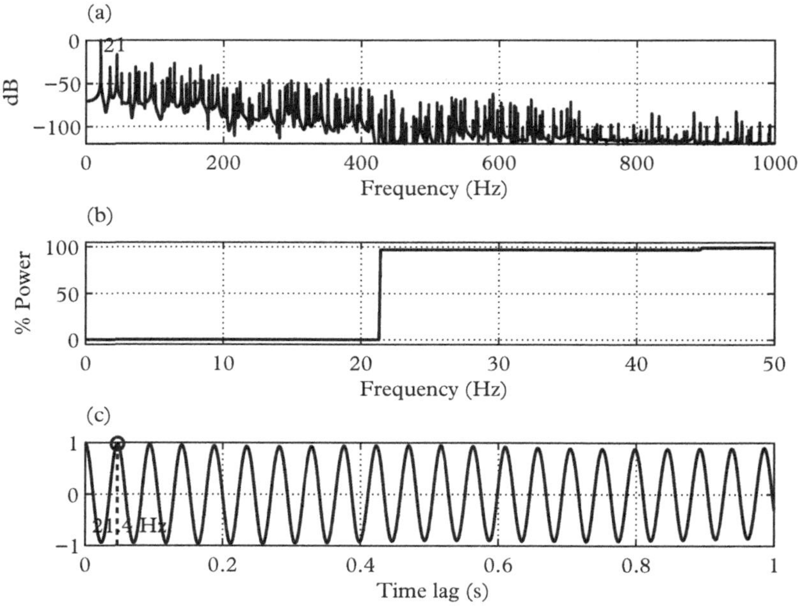

Figure 8.5.2 *Triangular vibrating plate: (a) line spectrum with main peak at 21 Hz, (b) cumulative line spectrum, and (c) autocorrelation*

where the reflections interact. The autocorrelation shows that an estimate of the pitch is 21 Hz. The reader is invited to listen to the rattling sound of the vibrating triangular plate in the file TriangularPlate.mp3 and watch video no. 21 referenced in the book's website.

8.6 Coupling a Vibrating String with a Vibrating Plate

In Section 5.6 an idealization of a soundboard, namely a mass/dashpot/spring system, was coupled with a vibrating string. Here the vibrating string will be coupled with the vibrating plate of Section 8.4, which is a square with a side length of 1.68 m. I picked the string parameters to match those in Giordano and Nakanishi's book (*Computational Physics*, pp. 362–7) where the C4 piano key was struck by a hammer. I also adopted Giordano's numerical method for solving the hammer–string dynamics.

The transverse force of the vibrating string is distributed over the plate or sound board via a smoothed force form shown in Figure 8.6.1, showing how the bridge is placed diagonally across the square board. This smoothing device spreads the applied force from the string and bridge over a larger area than that associated just with the string.

The hammer force is shown in Figure 8.6.2, where it is seen to be similar to that in Giordano and Nakanishi's book (p. 366, Figure 11.6).

Figures 8.6.3 and 8.6.4 show how the soundboard surface evolves in the early going.

The average displacement of the sound board surface, its line spectrum, and cumulative line spectrum are shown in Figures 8.6.5 and 8.6.6.

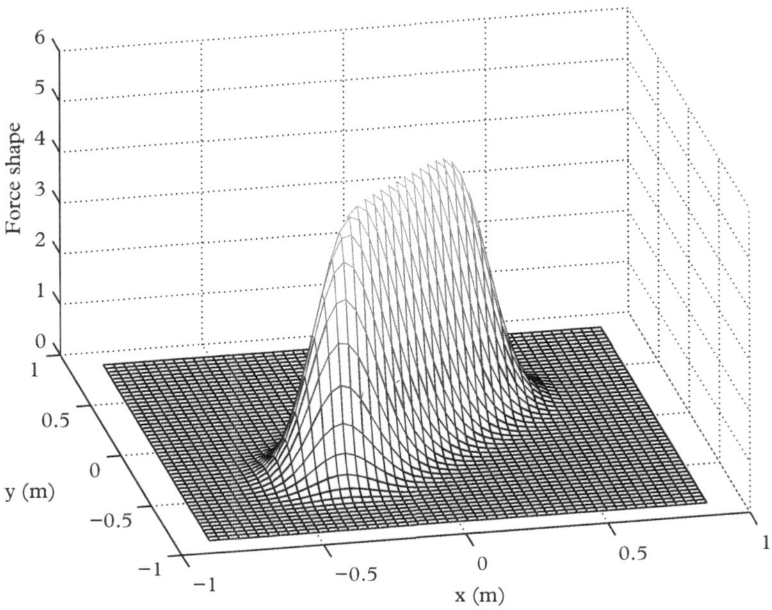

Figure 8.6.1 *Force form for the coupling of the vibrating string to a square sound board*

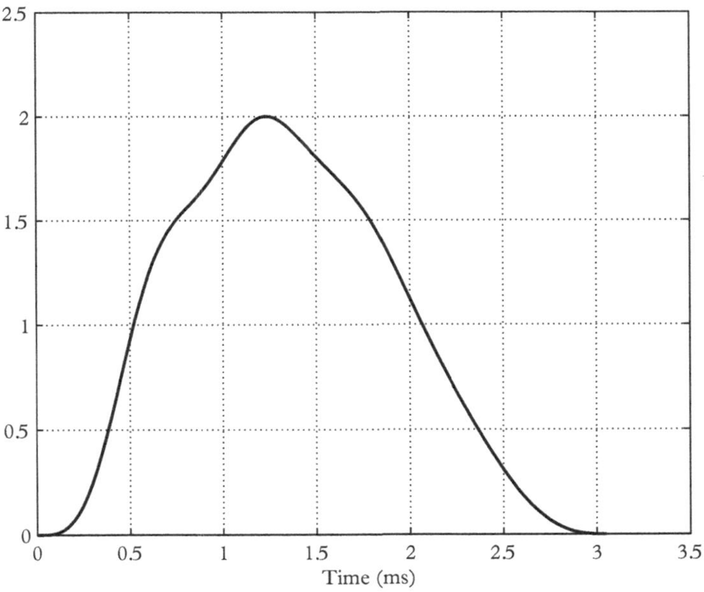

Figure 8.6.2 *Hammer force during the simulation of the coupling of the vibrating string to a square board*

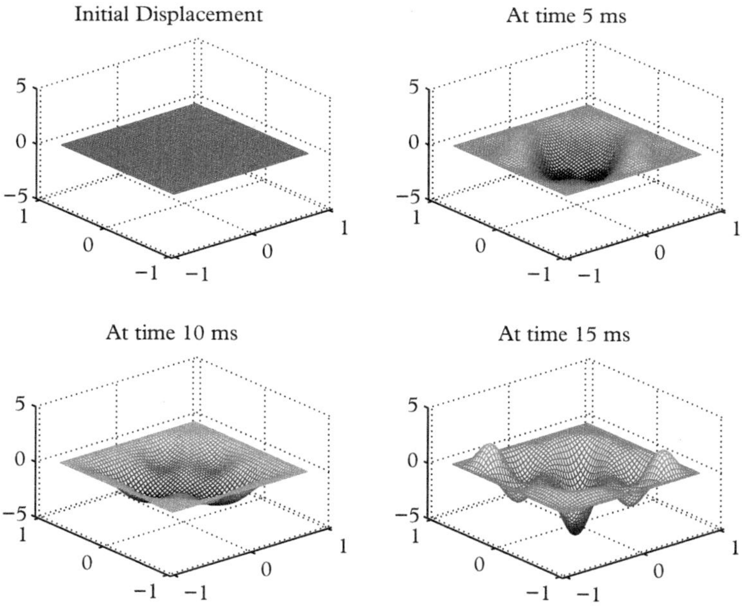

Figure 8.6.3 *Snapshots of simulated soundboard when coupled to a vibrating string at 5 ms, 10 ms, and 15 ms*

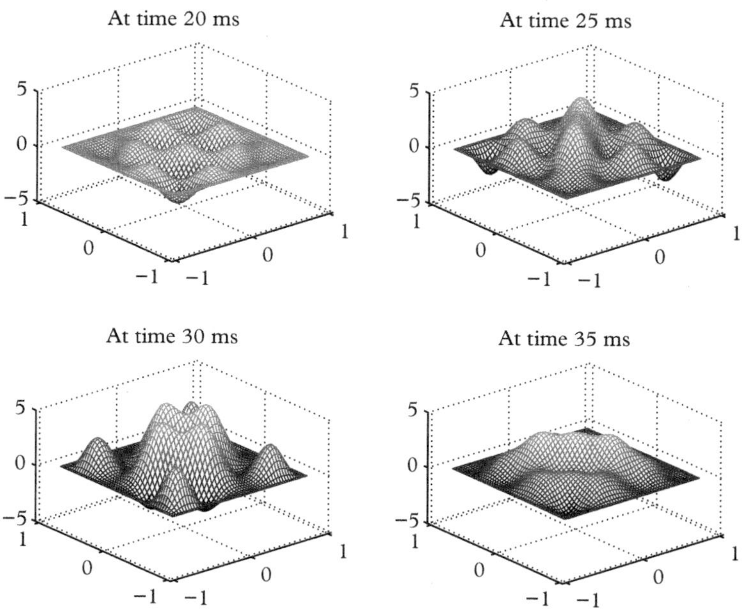

Figure 8.6.4 *Snapshots of simulated soundboard when coupled to a vibrating string at 20 ms, 25 ms, 30 ms, and 35 ms*

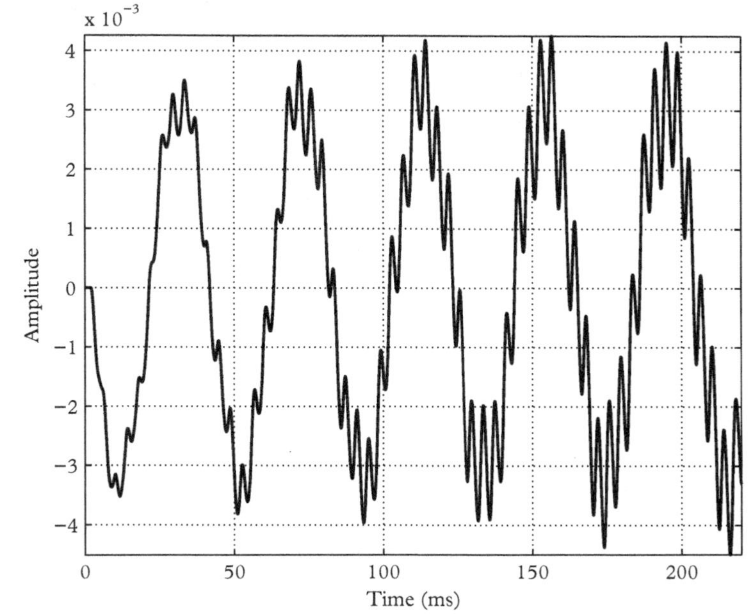

Figure 8.6.5 *Average displacement of plate when coupled to a vibrating string*

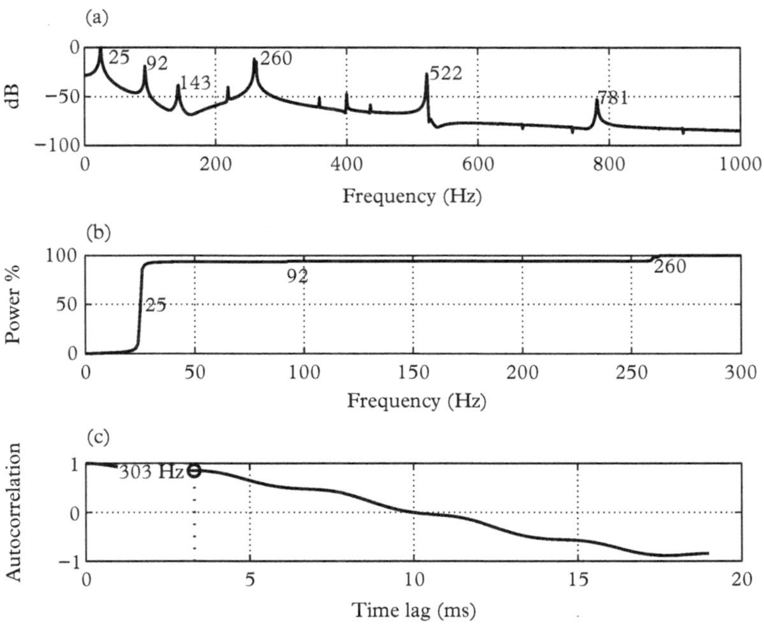

Figure 8.6.6 *Average displacement of plate when coupled to a vibrating string: (a) line spectrum, (b) cumulative line spectrum, and (c) autocorrelation*

The average displacement time domain trace in Figure 8.6.5 shows higher frequency waves riding on an overall wave with a period of approximately 40 ms (a frequency of 25 Hz). The line spectrum in Figure 8.6.6(a) shows that there are spectral peaks at 25 Hz (probably the fundamental frequency of the board), a doublet at 262 Hz (the fundamental frequency of C4), and peaks at 522 Hz (second harmonic of C4) and 781 Hz (third harmonic of C4). The peaks at 92 Hz, 143 Hz, 219 Hz, 358 Hz, 400 Hz, and 436 Hz are probably overtones of the board. The peak at 262 Hz is actually split between 259 and 262 Hz [difficult to see in Figure 8.6.6(a)], suggesting the influence of the board. The autocorrelation in Figure 8.6.6(c) shows a weak peak corresponding to 303 Hz.

The reader with access to Matlab® can run the script in the file VibStrgPlate4.m to investigate rectangular plates and to experiment with changing the parameters.

To develop a reference point, the bridge is tapped once at $t = 0.25$ ms with no influence from the strings. The line spectrum of the plate's average displacement is shown in Figure 8.6.7.

Figure 8.6.7(a) shows that the peaks at 25 Hz, 92 Hz, and 143 Hz show up again, and that there is also one at 259 Hz, which explains the doublet observed in Figure 8.6.6(a).

This numerical experiment was carried out for five sizes of square soundboard. The first two overtones are shown as a function of the length of a plate side in Figure 8.6.8.

These results are consistent with the idea that a bigger soundboard will have a lower fundamental frequency (and lower second partials as well).

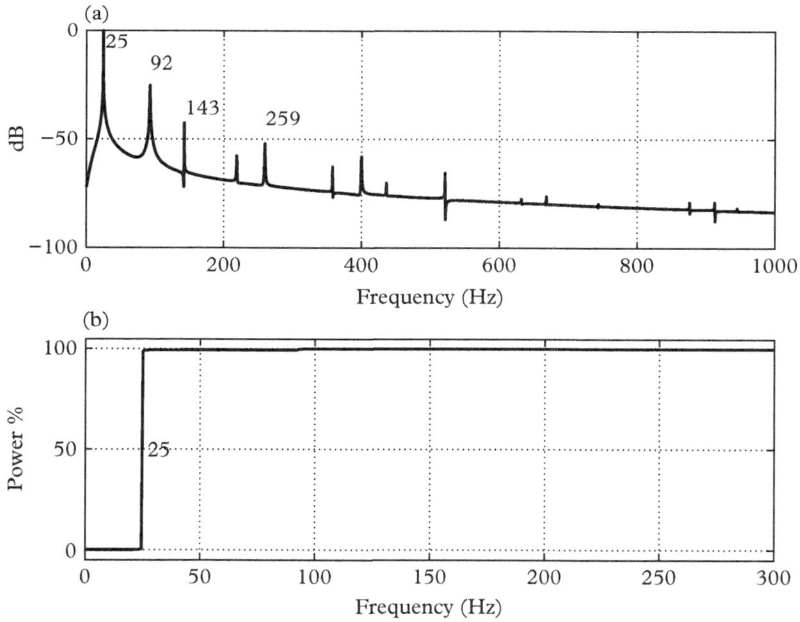

Figure 8.6.7 *Average displacement of tapped soundboard: (a) line spectrum, and (b) cumulative line spectrum*

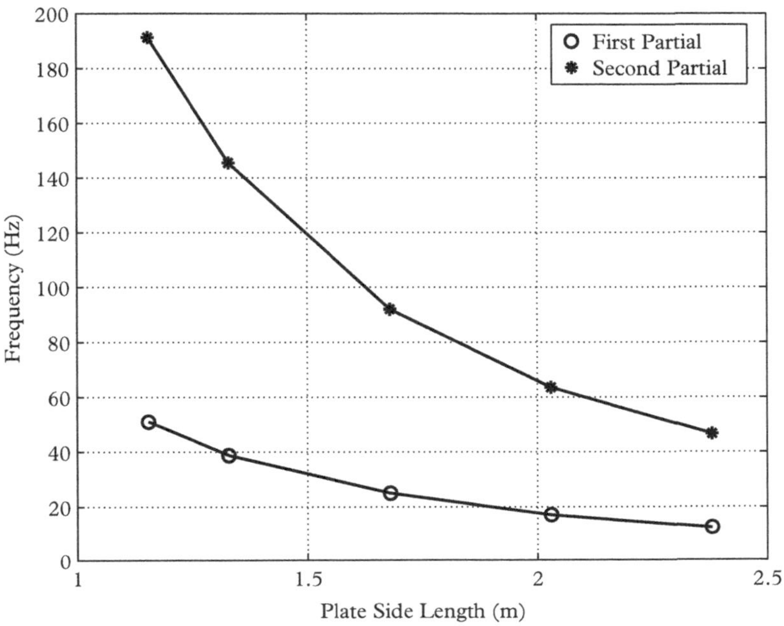

Figure 8.6.8 *Dependence of first and second plate partials as a function of plate side length*

A video showing some of the dynamics of the plate with vibrating string when struck by a hammer is referenced in the book's website (no. 26).

The results of this section can be compared to those of Section 5.6, which came from a much simpler model. In either case one might conclude that the overtones of the board can color the sound coming from the string. Note that many realities of a piano soundboard have not been included in these numerical experiments. For example, a grand piano soundboard is not square and is convex upward. Also, the soundboard's performance depends on the alignment of the grain in the Sitka spruce that makes up the board.

8.7 Knabe Soundboard Experiment

A thump on the treble bridge of my Knabe grand as sensed by a Schatten piezoelectric transducer yielded Figure 8.7.1.

The line spectrum of the real soundboard in Figure 8.7.1(b) shows at least four rather blurred peaks at 73 Hz, 113 Hz, 254 Hz, and 365 Hz. Comparing Figures 8.6.6 and 8.7.1 shows that both the model and the real soundboard have a non-harmonic structure.

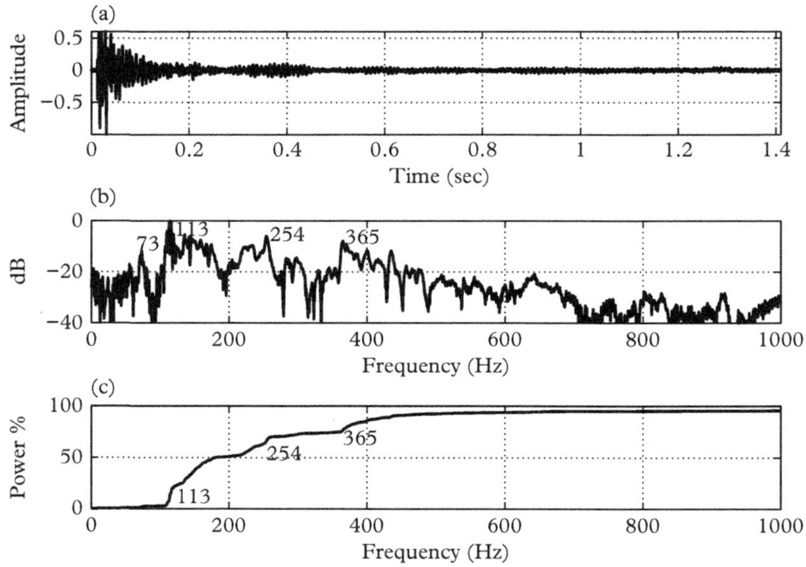

Figure 8.7.1 *Knabe soundboard thump using a Schatten transducer: (a) time domain plot, (b) line spectrum, and (c) cumulative line spectrum*

8.8 Summary of Chapter Eight and Part Two

The flexible vibrating string under tension, fixed at both ends, and the vibrating bar, free at both ends, have been studied because they represent opposite poles in terms of stiffness and flexibility and because they play an important role in generating musical sounds.

To compare the three vibrating objects that have fixed boundaries (the string, the membrane, and the plate), imagine a flexible string, strung across a diameter, embedded in the surface of the plate and the membrane, and watch how it vibrates in response to a Gaussian initial displacement. Figure 8.8.1 shows snapshots of a flexible string, Figure 8.8.1(a), and snapshots of slices made across the diameter of a circular membrane, Figure 8.8.1(b), and plate, Figure 8.8.1(c).

The flexible vibrating string subject only to tension repeats the initial shape. The vibrating membrane, also subject only to tension, does not repeat its initial shape and exhibits a more complex surface. Finally, the vibrating plate, subject to elasticity, develops a host of even more complex movements that also do not repeat.

The flexible string exhibited harmonics defined by integer multiples of the fundamental frequency. On the other hand, the vibrating bar exhibited partials that were non-integer multiples of the fundamental.

In between, the stiff string showed partials that were increasingly greater in frequency than the associated flexible string harmonics in a manner that will lead us to the concept of inharmonicity when we discuss the piano in Part 3. A real string (the C1 string on a piano) exhibited longitudinal vibrations as a consequence of its stiffness, and we were able to quantify them using spectral analysis.

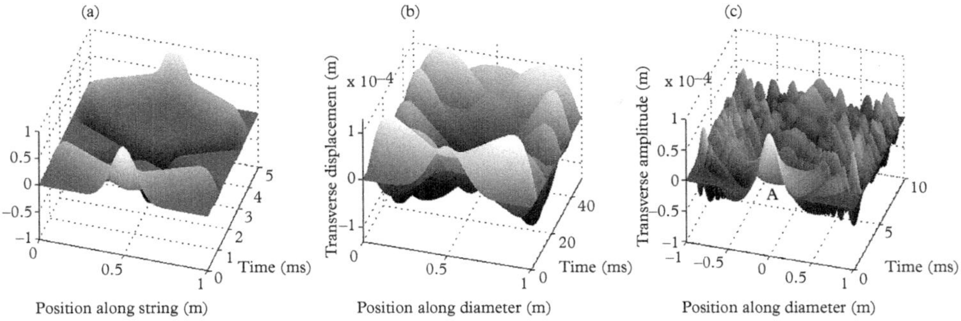

Figure 8.8.1 *Comparison of string, membrane, and plate: (a) snapshots of vibrating string, (b) snapshots of slices over a diameter of a vibrating membrane, and (c) snapshots of slices over a diameter of a circular vibrating plate*

A mass/dashpot/spring system coupled with a vibrating string was used to simulate the behavior of a piano. The impact of different natural frequencies of the soundboard was illustrated.

The flexible circular vibrating membrane under tension and fixed on the circumference was studied because it approximates the behavior of a drum head. Unlike the vibrating flexible string, the flexible membrane does not have an integer-based harmonic structure. Also, the speed of the membrane waves as they move outward from an initial central disturbance is not constant like that of the string.

To address soundboards, whether from a guitar or a piano, we started with a stiff circular board not under tension as a departure from the flexible circular membrane. Next, a square board and a triangular board were studied and compared to the circular one. Finally, the square plate was coupled with a vibrating string that was struck by a hammer. The effect of the string was transmitted through a bridge. For comparison, the numerical experiment was repeated without the string where the soundboard was struck at one instant of time. The mass/dashpot/spring model behaved somewhat like this last model but it did not show any harmonics other than those of the string.

The final part of the book draws on these results and studies the piano.

Part 3

The Piano

9

An Introduction to Pianos

We have studied several real and simulated musical sound sources in previous chapters: horns, pitch pipes, the human voice, a tuning fork, a Didjeribone®, vibrating strings, drums, soundboards, and bars. In Chapter 5 a good bit of time was spent on the vibrating string with special emphasis on how the force of a hammer strike might be transmitted to the string and then to the bridge and then to the soundboard. This chapter and those following will draw on those efforts in an attempt to look at a variety of pianos, both acoustic and digital. The six pianos that will receive the most analysis are introduced. The different stringing arrangements and the locations of the two bridges are discussed. The geometric or exponential relationship between the key numbers at that key's fundamental frequency is presented algebraically and graphically. The nonlinear relationship between the key numbers and the fundamental frequencies motivates the concept of cents, which in effect creates a localized linear relationship for the spacing between semitones.

9.1 The Cast of Pianos

The following pianos appear at various places in the book:

1. Yamaha P120 digital piano
2. Yamaha M1A 43″ console piano
3. 1927 Knabe 7′ 8″ grand piano
4. Steinway Model D from a church in Ithaca
5. Bösendorfer Model 280 from Cunningham Pianos
6. Schimmel 213 from Cunningham Pianos
7. Steinway Model B from a local music school
8. Ithaca College Steinway Model D no. 1
9. Ithaca College Steinway Model D no. 2
10. Ithaca College Steinway Model B (Wapin Conversion)

The first six of these pianos will be studied as a group. In addition, two Model D Steinways in a concert hall were compared, using a couple of the tools to be presented in this part of the book. A Wapin conversion on no. 10 is evaluated in Chapter 17. Thanks to the generosity of Larry Buck

Spectral Analysis of Musical Sounds with Emphasis on the Piano. First Edition.
David M. Koenig, with comments by Delwin D. Fandrich. © David M. Koenig and Delwin D. Fandrich 2015.
Published in 2015 by Oxford University Press.

and Kirk Burgett, I have also sampled Steingräber and Mason Hamlin pianos, but those results will not be presented in this book.

9.2 The Six-Pack

Six of the 88-key pianos to be studied are shown in Figure 9.2.1. This is followed by a brief description of each.

The Knabe Grand Piano

The Knabe semi-concert grand is 7′ 8″ long and appears to have been built in 1927. It was recently rebuilt by Beethoven Pianos in Manhattan. Keys 1 through 8 are single strung (or monochords) and wrapped. Keys 9 through 16 are double strung (bichords) and wrapped. Keys 17 through 24 are wrapped trichords. Keys 25 to 88 are triple strung (trichords) and not wrapped. Keys 1 through 24 make up the bass section (see Figure 9.5.1). This piano has the bass–tenor break between keys 24 and 25. The tenor zone contains keys 25 through 45. The tenor–treble break is between keys 45 and 46. The treble–upper treble break (beyond which where there are no dampers) is between keys 66 and 67.

Yamaha Console

The Yamaha console is 43″ high. Keys 1 through 13 are wrapped monochords. Keys 14 through 32 are wrapped bichords. Keys 33 and higher are unwrapped trichords. Keys 1 through 30 make up the bass section. This piano has the bass–tenor break between keys 30 and 31. The tenor zone contains keys 31 through 56. The tenor–treble break is between keys 56 and 57. There are no dampers on keys 67 and higher.

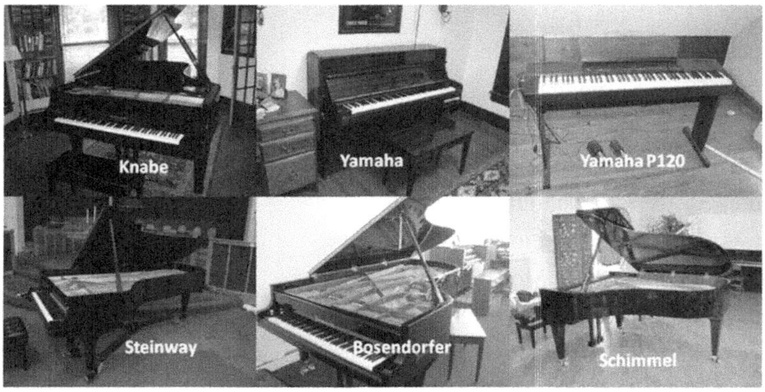

Figure 9.2.1 *Six pianos: 1927 7′ 8″ Knabe grand, 43″ Yamaha console, Yamaha P120 digital piano, 1911 Steinway Model D, Bösendorfer 280, Schimmel 213*

Steinway Model D

Located in a church in Ithaca, NY, the Model D was built in 1911 and rebuilt in the 1990s. The monochord wrapped strings compose A0 to E1 (8 strings). The wrapped bichords are F1 to A1 (5 strings). The remaining keys are trichords.

Yamaha P120

The Yamaha P120 is a digital piano reportedly based on samples of the Yamaha CF IIIS, which is 9′ long. The CF IIIS stringing and break points are similar to the Steinway Model D.

Bösendorfer 280

The Bösendorfer 280 has 7 wrapped monochord keys and 10 wrapped bichord keys up to D2. All the keys beyond this are trichord.

Schimmel 213

The Schimmel 213 has 8 wrapped monochords and 12 wrapped bichords.

9.3 Naming Convention and the Equal Temperament System

The convention for designating the piano keys is rather simple and it appears to be reasonably widely accepted. The lowest-frequency note on the left-hand side of the keyboard, an A, is designated as A0 and it is the first of 88. The next key is A0#. The sequence continues as A0, A0#, B0, C1, C1#, ..., A1, A1#, B1, C2, ..., B2, C3, ..., B3, C4 (middle C), ..., B4, C5, ..., C8 (the highest note on the piano and number 88).

If the notes are numbered 1 to 88 for A0 to C8 then the fundamental frequency of the nth note is given by

$$f(n) = 440r^{n-49} \qquad r = 2^{\frac{1}{12}}. \tag{9.3.1}$$

The 49th note, A4, has a fundamental frequency of 440 Hz. This arrangement of the fundamental frequencies defines the equal temperament scale. Table 9.3.1 gives the note names, key numbers, and fundamental frequencies. Equation 9.3.1 shows that the fundamental frequency depends nonlinearly on the number of the note. This is the basis for our need to occasionally use logarithmic axes in the graphical presentations.

9.4 A Different Measure of Frequency: Cents and Semitones

According to Eq. 9.3.1 the piano's scale is defined in terms of the fundamental frequencies of the 88 notes from A0 (the lowest note on a piano) to C8 (the highest note on a piano). The relationship between the fundamental frequencies and the key numbers is presented using linear axes, Figure 9.4.1(a), and logarithmic axes, Figure 9.4.1(b). The logarithm of the frequencies will be the basis for some of the cumulative line spectrum maps to come later.

Table 9.3.1 *Piano key names and fundamental frequencies*

Key no.	Key name	Fund. freq.
1	A_0	27.5
2	A0#	29.1352
3	B_0	30.8677
4	C_1	32.7032
5	C1#	34.6478
6	D_1	36.7081
7	D1#	38.8909
8	E_1	41.2034
9	F_1	43.6535
10	F1#	46.2493
11	G_1	48.9994
12	G1#	51.9131
13	A_1	55
14	A1#	58.2705
15	B_1	61.7354
16	C_2	65.4064
17	C2#	69.2957
18	D_2	73.4162
19	D2#	77.7817
20	E_2	82.4069
21	F_2	87.3071
22	F2#	92.4986
23	G_2	97.9989
24	G2#	103.8262
25	A_2	110
26	A2#	116.5409
27	B_2	123.4708
28	C_3	130.8128
29	C3#	138.5913
30	D_3	146.8324
31	D3#	155.5635
32	E_3	164.8138
33	F_3	174.6141
34	F3#	184.9972
35	G_3	195.9977
36	G3#	207.6523
37	A_3	220
38	A3#	233.0819
39	B_3	246.9417
40	C_4	261.6256
41	C4#	277.1826
42	D_4	293.6648
43	D4#	311.127
44	E_4	329.6276
45	F_4	349.2282
46	F4#	369.9944
47	G_4	391.9954
48	G4#	415.3047
49	A_4	440
50	A4#	466.1638
51	B_4	493.8833
52	C_5	523.2511
53	C5#	554.3653
54	D_5	587.3295
55	D5#	622.254
56	E_5	659.2551
57	F_5	698.4565
58	F5#	739.9888
59	G_5	783.9909
60	G5#	830.6094
61	A_5	880
62	A5#	932.3275
63	B_5	987.7666
64	C_6	1046.502
65	C6#	1108.731
66	D_6	1174.659
67	D6#	1244.508
68	E_6	1318.51
69	F_6	1396.913
70	F6#	1479.978
71	G_6	1567.982
72	G6#	1661.219
73	A_6	1760
74	A6#	1864.655
75	B_6	1975.533
76	C_7	2093.005
77	C7#	2217.461
78	D_7	2349.318
79	D7#	2489.016
80	E_7	2637.021
81	F_7	2793.826
82	F7#	2959.955
83	G_7	3135.964
84	G7#	3322.438
85	A_7	3520
86	A7#	3729.31
87	B_7	3951.066
88	C_8	4186.009

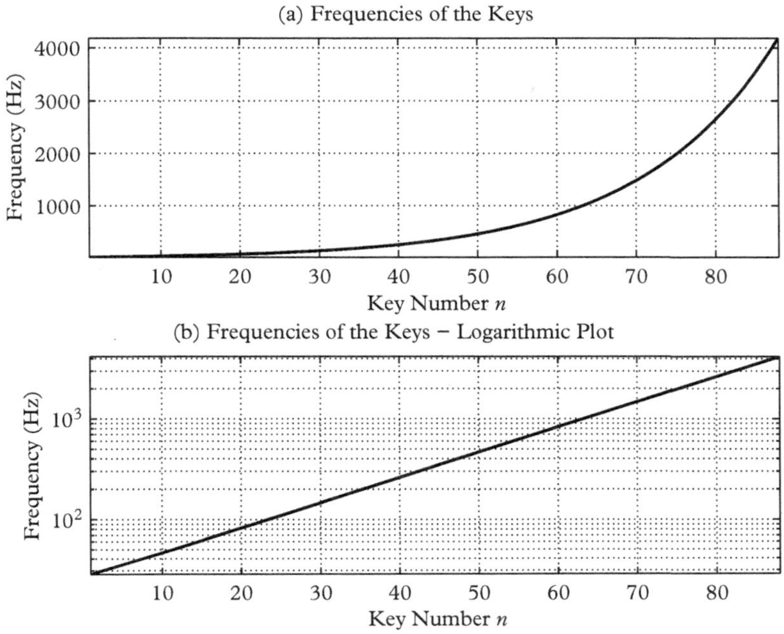

Figure 9.4.1 *Fundamental frequencies of the keys plotted on (a) a linear scale, and (b) a semi-logarithmic scale*

Although it may not be obvious from Equation 9.3.1, C2 (65.4 Hz) is an octave higher than C1 (32.7 Hz) and has a fundamental frequency that is double that of C1. This is the case for all similarly named notes, as in B5 (987.7 Hz) and B6 (1975.5 Hz).

Each of the 88 notes is separated by a so-called semitone. For example, the piano's middle C is C4 and it is separated from B4 by a semitone. It is also separated from C4# by a semitone. By definition there are 100 cents between every two adjacent notes, i.e., each semitone is worth 100 cents. Because the frequency of a piano note increases geometrically with the note number, the spacing in the number of Hz corresponding to 100 cents also varies geometrically. The ratio Hz/cent is given in Figure 9.4.2.

For example, the A0 note frequency is 27.5 Hz and A0# is 29.1352 Hz. Since there are 100 cents between these two notes, the number of Hz per cent will have to be relatively low, namely 0.016352 Hz/cent, as in

$$(29.1352 - 27.5)\ \text{Hz}/100\ \text{cents} = 0.016352\ \text{Hz/cent}.$$

On the other hand, the A7 note is 3520 Hz and A7# is 3729.31 Hz, so 100 cents is spread over approximately 209 Hz and the ratio of Hz to cents is 2.09931 Hz/cent, as in

$$(3729.31 - 3520)\ \text{Hz}/100\ \text{cents} = 2.09931\ \text{Hz/cent}.$$

The unit of cents can sometimes be more useful than Hz. For example, the situation of the C4 key being sharp such that the first partial is at 265 Hz, i.e., over 5 Hz above the fundamental,

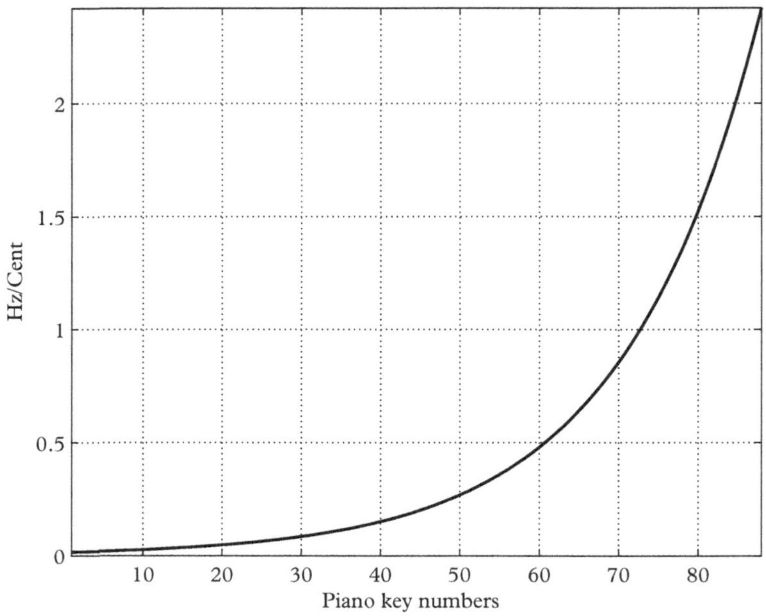

Figure 9.4.2 *Number of Hz in a cent as a function of key number*

is quite different from the C7 key being sharp by 5 Hz at 2098 Hz. In the former case, 5 Hz is approximately 51 cents sharp, while in the latter case 5 Hz is about 4 cents sharp.

9.5 Different Zones in the Stringing Arrangement

Figure 9.5.1 shows the four sections of the Knabe grand piano's strings: the bass, tenor, treble, and upper treble. The bass strings are directed over the tenor strings at an angle. They also are strung over the bass bridge. Within the bass section there are usually three different kinds of strings. The lowest eight (or so) strings are single strings (monochords) and have copper wire wrapped around the center wire. The next eight (or so) strings are also wrapped but there are two strings per key (a two-string unison or a bichord). The remaining keys in the bass section have three wrapped strings (a trichord).

The strings in the tenor, treble, and upper treble sections are all single-strand three-string unisons. All these strings except those in the upper treble have dampers. All of these strings pass over the treble bridge, which is separate from the bass bridge. The strings in the treble and upper treble pass under a V-shaped bar called the capo tasto bar. The strings in the bass and tenor sections pass through guides called agraffes.

Later on we will see that the keys' spectral characteristics differ depending on the number of strings per key, whether they are wrapped or not, and which section they are in. These sections are often defined by the struts or bars that one sees in Figure 9.5.1. These struts, part of the cast-iron plate, are necessary because iron (which behaves admirably under compression) will not tolerate twisting or bending forces.

In the 1890s the Broadwood Piano Company designed and constructed a piano with a steel plate that did not need struts or bars and therefore the different sections were defined only by the

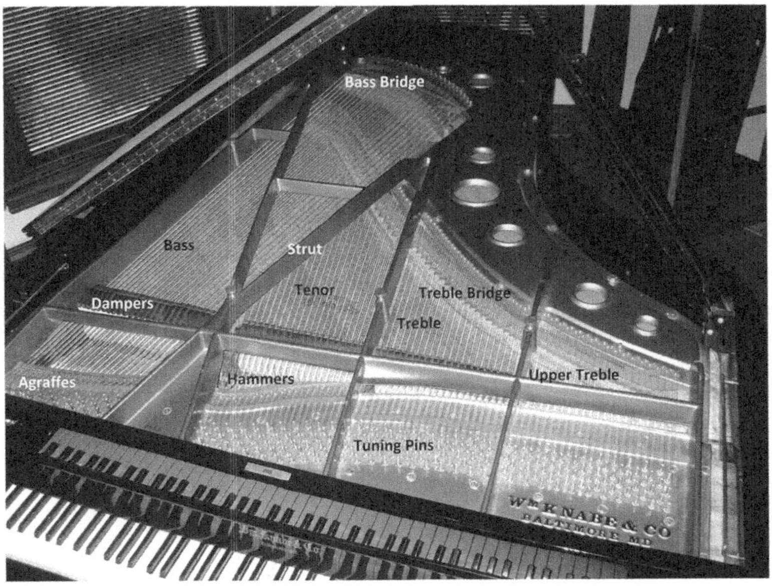

Figure 9.5.1 *Picture of a grand piano showing (1) the bass, tenor, treble, and upper treble zones; (2) the bass and tenor bridges; (3) tuning pins, agraffes, and dampers; and (4) the cast-iron frame including several struts*

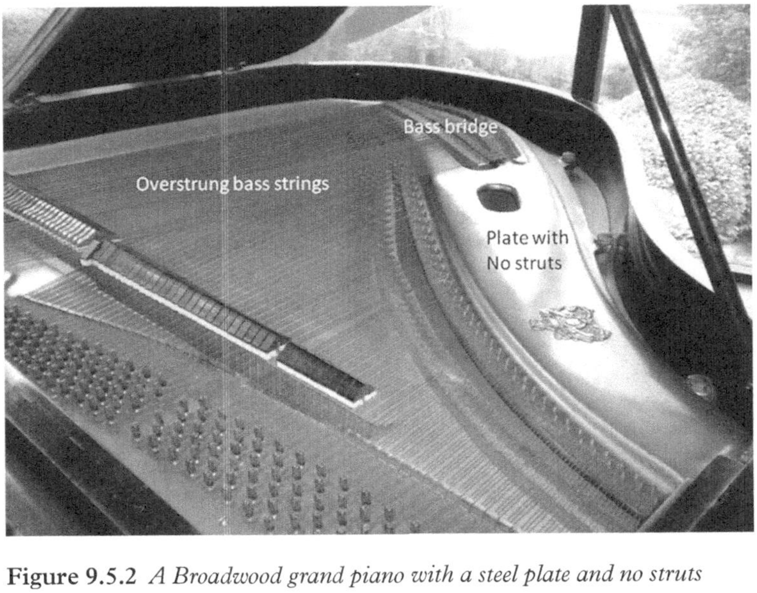

Figure 9.5.2 *A Broadwood grand piano with a steel plate and no struts*

number of strings per key and whether they were wrapped or not. Broadwood discontinued the barless piano in the early 1900s but there was a rebirth around 1990 that was shortlived. Figure 9.5.2 shows a picture of such a piano.

9.6 Summary

This short chapter summarizes some of the physical characteristics of pianos such as the different stringing arrangements and the locations of the two bridges. The geometric or exponential relationship between the key numbers at that key's fundamental frequency is presented algebraically and graphically. The nonlinear relationship between the key numbers and the fundamental frequencies motivates the concept of cents, which in effect creates a localized linear relationship for the spacing between semitones.

10

Some Individual Piano Notes and Their Interactions

The spectral characteristics of several individual piano notes and their interaction will be studied in preparation for the more complicated construction of piano maps in later chapters.

10.1 Aspects of Three Piano Notes: Steinway C3, Knabe C4, Knabe C8

Before discussing the graphs, we should review some terminology. *Harmonic* frequencies are defined as integer multiples of the fundamental frequency. *Partial* frequencies are defined as frequencies near the harmonics that have significant power. We have occasionally used the term *overtones* to describe frequencies higher than the fundamental but not related to its integer multiples. Because of octave stretching (to be discussed in Chapter 12) and tuning preferences, partials are close to the harmonics but not exactly equal to them.

Steinway C3

Let's start with the C3 note on a Model B Steinway (located at a local music school). Figure 10.1.1 shows the time domain trace. The "big picture" is shown in Figure 10.1.1(a), while Figure 10.1.1(b) shows a close-up of the wave where one can see the appearance of the fundamental wave that appears to have a period of 7.5 ms, corresponding approximately to a frequency of 134 Hz. Figure 10.1.2 shows the line spectrum, cumulative line spectrum, and autocorrelation, while Figure 10.1.3 shows a fine grid DFT of the fourth partial. In Figure 10.1.2(a) the first seven harmonics have a fairly strong representation, but starting with the eighth harmonic (1048 Hz) the power drops off. The reader might notice that although the markers on the graphs give the numerical value of the harmonic frequency, the actual location of the peak is at a higher frequency. This deviation grows as the harmonic number increases and is a consequence of inharmonicity where the harmonic frequencies are "stretched." We will discuss this in depth later on, in Chapter 12. The cumulative line spectrum in Figure 10.1.2(b) shows that the first partial is the largest in strength. The autocorrelation in Figure 10.12(c) indicates that the pitch is equal to the fundamental frequency.

A close-up of the neighborhood around the fourth harmonic, using a fine grid DFT centered at 524.5 Hz, is shown in Figure 10.1.3. There is peak splitting, probably due to one of the three

Spectral Analysis of Musical Sounds with Emphasis on the Piano. First Edition.

Published in 2015 by Oxford University Press.

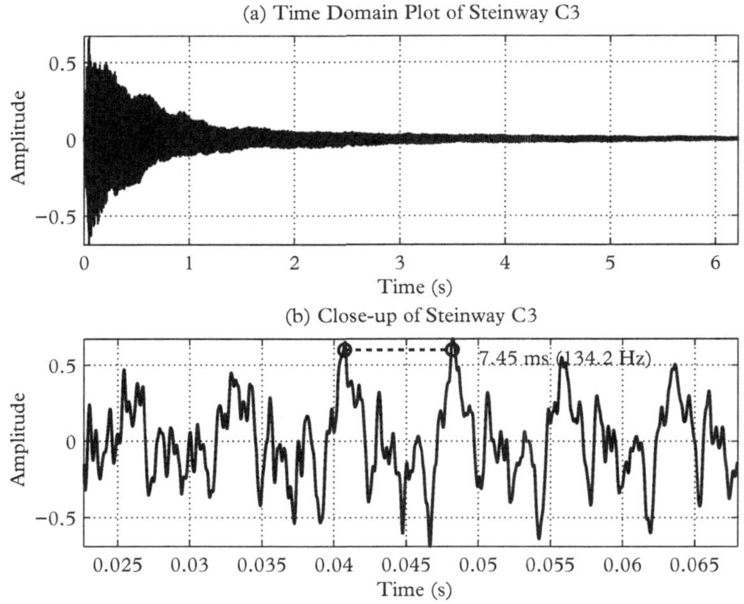

Figure 10.1.1 *Steinway C3 note: (a) entire wave, and (b) a close-up of about 50 ms of the wave*

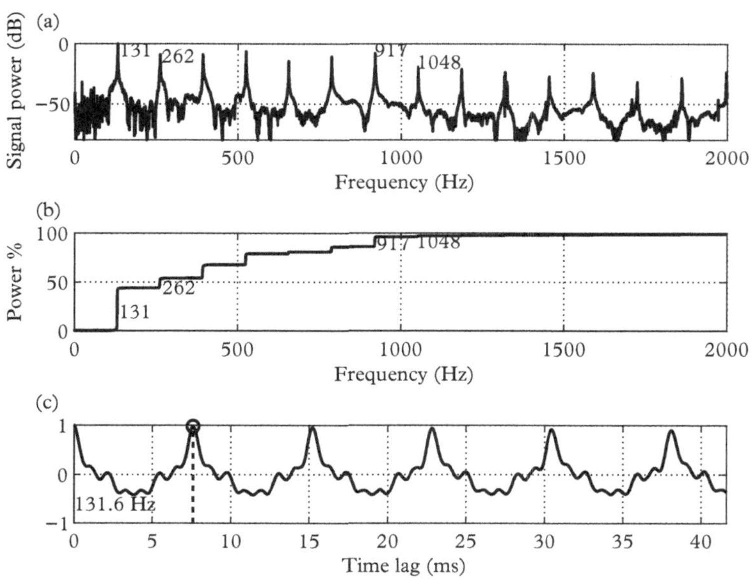

Figure 10.1.2 *Steinway C3 note: (a) line spectrum, (b) cumulative line spectrum, and (c) autocorrelation*

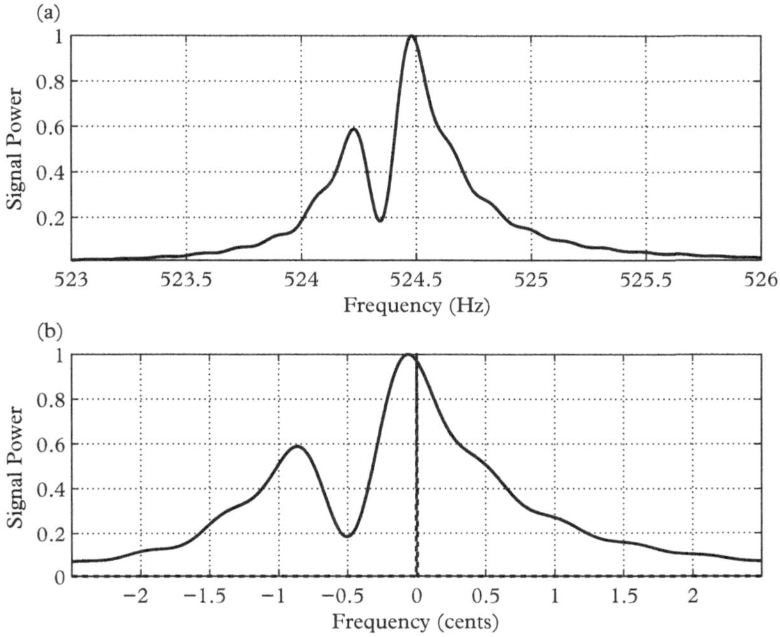

Figure 10.1.3 *Fine grid line spectrum of the fourth partial of the Steinway C3: (a) in Hertz, and (b) in cents*

unison strings for that note not being tuned to match the other two. Note that the frequency axis in Figure 10.1.3(a) is plotted in Hertz while Figure 10.1.3(b) is in cents. The latter graph suggests that the three C3 strings of the Steinway are tuned to within about 1 cent. Some professional piano tuners claim to be able to tune a note to within 0.2 cents, others say it is more like one or two cents. As we progress through this chapter you will see that we tend to side with the latter group.

Knabe C4

Next is the Knabe C4 represented in Figures 10.1.4–10.1.6.

Figures 10.1.4–10.1.6 show that the amplitude attenuates relatively quickly for the first second of time and then fades away rather slowly until the time when the recording stops. There appear to be two modes in the decay, which are elegantly explained in Nicholas Giordano's book *Physics of the Piano* and in Gabriel Weinreich's article "The Coupled Motion of Piano Strings," the latter of which is accessible on the Internet. Essentially, the first mode containing the rapid decay represents the attenuation of the vertical (for a grand piano) transverse vibration mode of the string, which we studied in Chapter 5. The more slowly decaying mode represents the horizontal transverse vibration mode, which is said to be a consequence of the bending vibrations of the soundboard in response to the movement of the string and the bridge. Each partial in the spectrogram in Figure 10.1.6 shows a trough where the transition appears to take place.

The line spectrum in Figure 10.1.5(a) shows strength at the first two harmonics and, consistent with some of the comments in Chapter 5, relatively little at the ninth harmonic (or partial) because

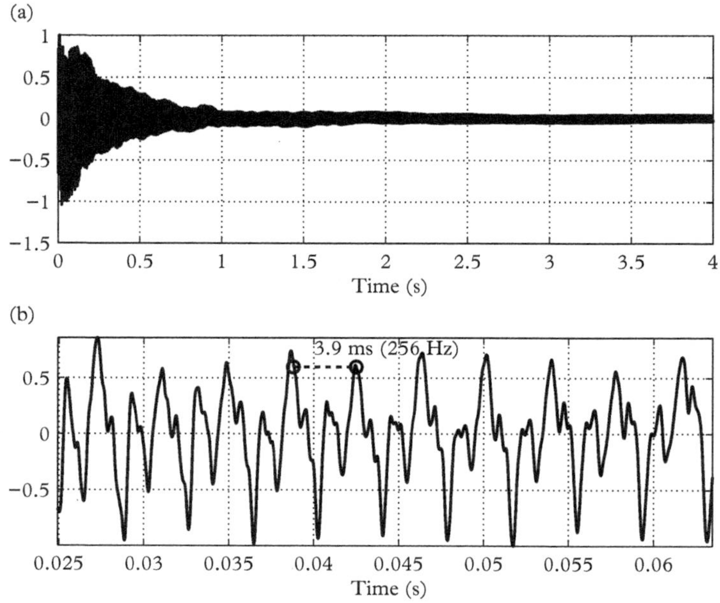

Figure 10.1.4 *Time domain plot of Knabe C4: (a) long view, and (b) close-up view showing a period of 3.9 ms or 256 Hz*

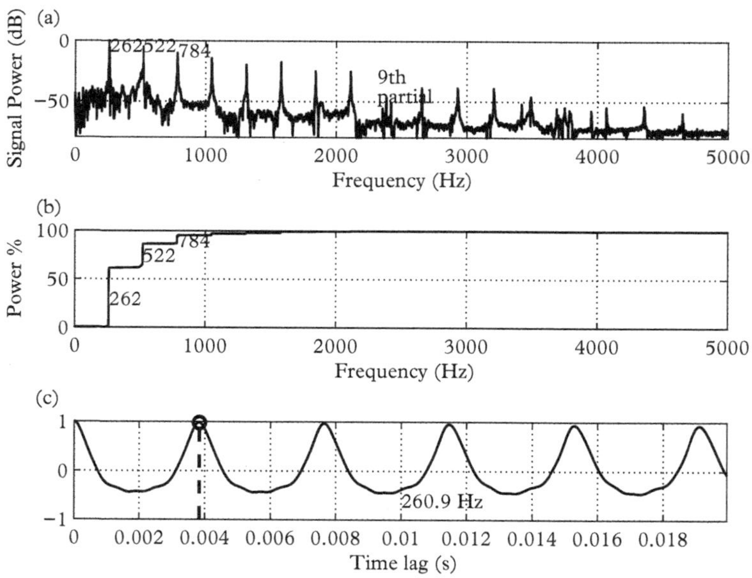

Figure 10.1.5 *Knabe C4: (a) line spectrum, (b) cumulative line spectrum, and (c) autocorrelation*

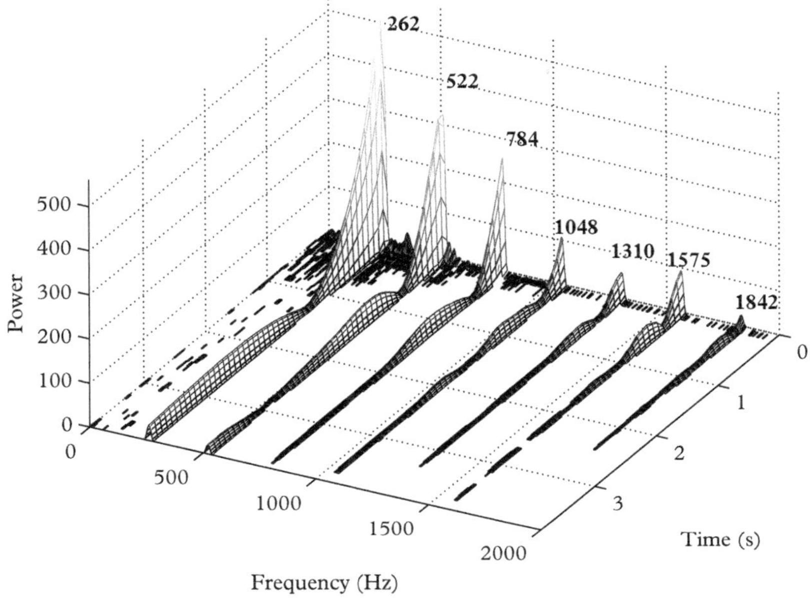

Figure 10.1.6 *Spectrogram of Knabe C4*

of the point where the hammer strikes the string. The autocorrelation in Figure 10.1.5(c) indicates that the pitch is 261 Hz.

A close-up of the of the neighborhood of the second partial centered at 523 Hz, using a fine grid line spectrum, is shown in Figure 10.1.7, where one sees two peaks separated by about 2 cents. This is not surprising because the C4 note consists of three unison strings and it is generally difficult to get them to match each other exactly.

Knabe C8

For the sake of comparison, Figure 10.1.8 shows two views of the time domain plot of C8 (4186 Hz fundamental frequency) where the note has lost all of its intensity after 0.5 s and has an apparent half-time decay of about 0.05 s. Figure 10.1.9 shows the line spectrum, cumulative line spectrum, and autocorrelation. We have commented on C8 before but, at the risk of overdoing it, notice the sound power below the fundamental in the line spectrum, Figure 10.1.9(a), and the cumulative line spectrum, Figure 10.1.9(b). We will talk more about this effect (hammer knock) in Chapter 16. The autocorrelation in Figure 10.1.9(c) suggests that the pitch is 4410 Hz; however, as pointed out in Chapter 2 and Appendix 2, the conventional autocorrelation calculation is sensitive to the time grid for high frequency signals. A fine grid autocorrelation (not shown here) produces a peak at 4225 Hz which is about 16 cents sharp of the fundamental at 4186 Hz and more in line with the line spectrum peak in Figure 10.1.9(a).

The three notes analyzed in this section have touched on several characteristics of piano notes. First, the decay of the amplitude in the time domain sometimes shows the transfer of power between the vertical and horizontal transverse modes. Second, the line spectrum often shows split

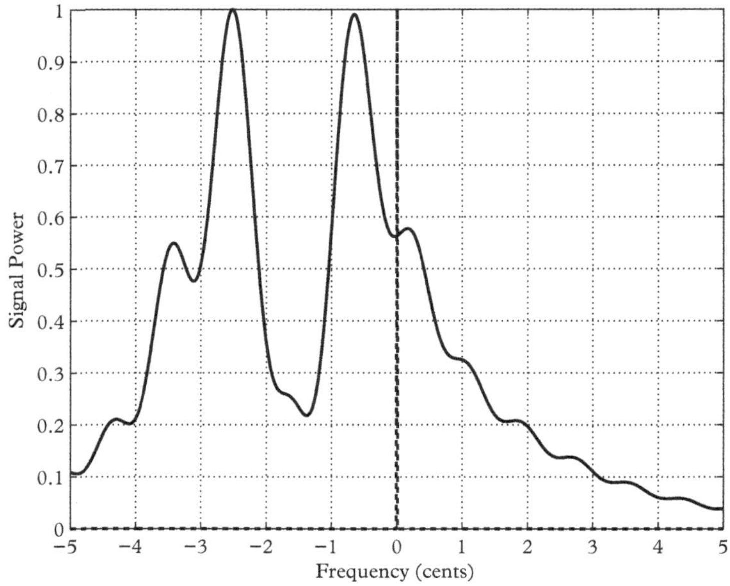

Figure 10.1.7 *Fine grid line spectrum of the second partial of the Knabe C4*

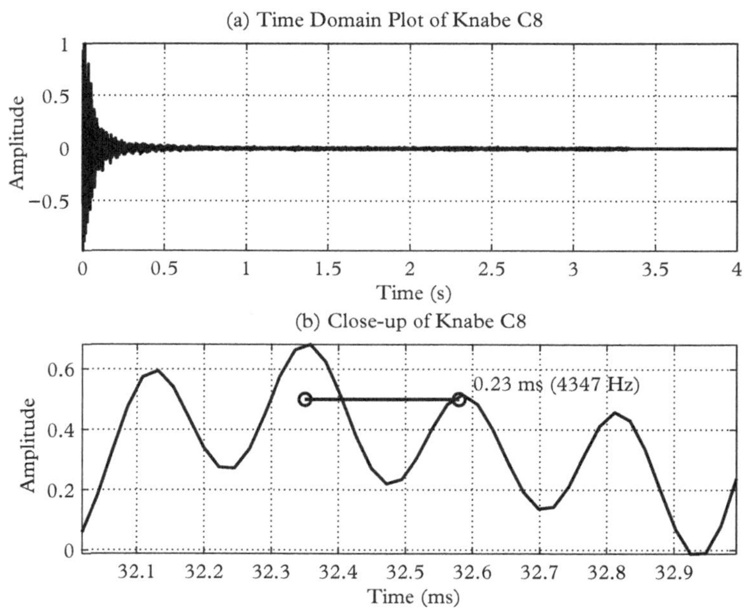

Figure 10.1.8 *Knabe C8: (a) entire wave, and (b) a close-up showing a period of approximately 0.23 ms which relates to 4347 Hz*

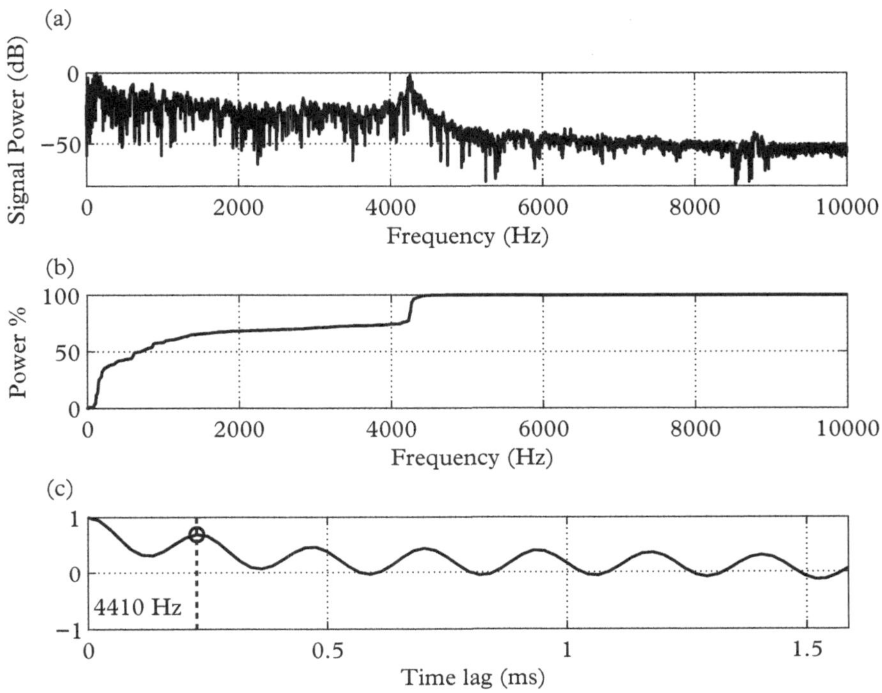

Figure 10.1.9 *Knabe C8: (a) line spectrum, (b) cumulative line spectrum, and (c) autocorrelation*

peaks due to slight deviations in the tuning of three string unisons. Third, high-numbered notes exhibit power below the fundamental called hammer knock. Fourth, the location of the hammer strike point, usually at ⅟₉ th of the length of the string, can cause the line spectrum peaks associated with the eighth and ninth harmonics to be attenuated. Fifth, high-numbered notes decay more rapidly than low-numbered ones. Sixth, the conventional autocorrelation calculation appears to be relevant for the low-numbered notes but not for the high-numbered ones where one must resort to a fine grid approach.

10.2 A Chromatic Progression

Figure 10.2.1 shows the spectrogram of a 13-note chromatic progression from C4 to C5.

On the back wall of Figure 10.2.1, where time is zero, the various partials have been denoted as (1), (2), . . ., (5). The first note of the sequence is struck at time zero and the strength of each harmonic dies away within about one second, at which time the second note in the sequence is struck and the harmonics for that note show their strengths. With each succeeding note the frequencies of the harmonics increase. Note how each note has a slightly different distribution of signal power between the fundamental and the harmonics. We will see this again when piano maps are discussed in Chapter 15.

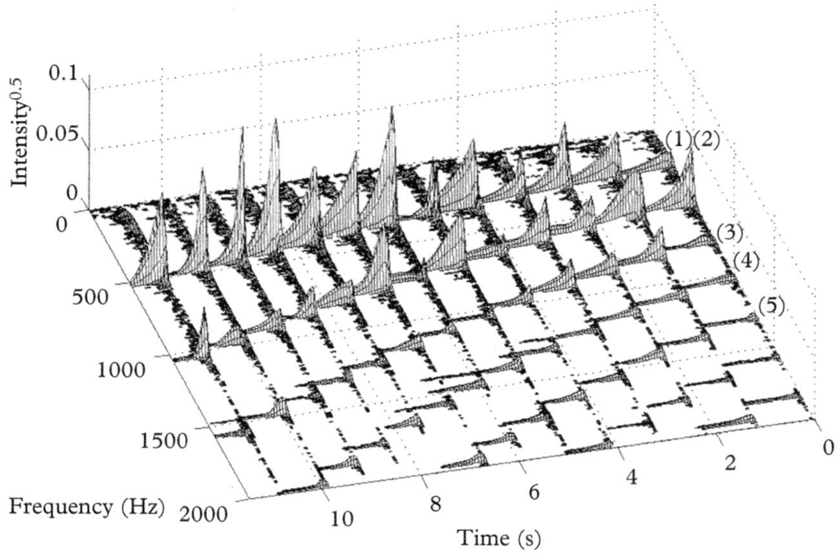

Figure 10.2.1 *Spectrogram of a chromatic run*

10.3 Interaction between Piano Notes

In Section 4.1 we showed how the harmonics of various piano notes line up. In this section we will show how striking a key can activate certain harmonics or partials of another key even if it has not been struck. Consider two notes: F2 and C3. Table 10.3.1 shows that C4, C5, G5, and C6 are harmonics for both F2 and C3. These notes are in bold print. For example, C4 has a fundamental frequency of 261.6 Hz and is the second harmonic of C3. On the other hand, C4 is also the third harmonic of F2. Likewise, C5 is the fourth harmonic of C3 and the sixth harmonic of F2.

Now, consider the following experiment. Slowly depress F2 such that no sound results and then hold the key down so that the F2 damper is off the strings. Strike C3 firmly but briefly. Figure 10.3.1 shows the time trace of the resulting sound wave. At a time of about 0.9 s the C3 key is released so that the damper is lowered on the C3 strings while the F2 key is still depressed and its damper is still off the F2 strings. At a time of 3.0 s a second set of data is collected. By this time it is assumed that the C3 string has been completely damped.

Figure 10.3.2 shows snapshots of the line spectrum during the course of the whole experiment. In the period from 0.0 s to about 1.0 s, C3 and its harmonics dominate the spectrum. Note that most of the peaks along the back wall of the graph correspond to entries in column 2 of Table 10.3.1.

After 1.0 s, there are five ridges in Figure 10.3.2 due to activity in the F2 string. Two of the ridges denoted by the point A appear to have small power at 88 Hz, which is F2, and at 120 Hz, which does not correspond closely with another key—at best, it is between A2# and B2—and it may be related to background noise, noticeable because the overall audio level from the key is so low due to C3's damper having been lowered. Point B denotes a ridge that corresponds to C4, which is the third harmonic of F2. There is a fourth weak ridge stemming from C5, the sixth harmonic of F2, denoted by C. The ninth harmonic of F2, namely G5, has a ridge denoted by D.

Table 10.3.1 *Harmonics of C3 and F2 (frequencies in bold type denote matches between C3 and F2)*

Note Name	Equal Tempered Fundamental Frequency (Hz)	C3 Harmonic Number	Note Name	Equal Tempered Fundamental Frequency (Hz)	F2 Harmonic Number
C3	130.8128	1	F2	87.30705786	1
C4	**261.6256**	2	F3	174.6141157	2
G4	391.9954	3	C4	**261.9211736**	3
C5	**523.2512**	4	F4	349.2282314	4
E5	659.2551	5	~A4	436.5352893	5
G5	**783.9909**	6	C5	**523.8423471**	6
A5#	932.3275	7		611.149405	7
C6	**1046.5022**	8	F5	698.4564629	8
D6	1174.6591	9	G5	**785.7635207**	9
E6	1318.5102	10	~A5	873.0705786	10
				960.3776364	11
			C6	**1047.684694**	12
				1135	13
				1222.3	14
			~E6	1309.6	15
				1397	16

There is also a ridge at about 1317 Hz, denoted by E, which corresponds approximately to E6, the 15th harmonic of F2.

These results show that the notes can talk to each other via the bridge and soundboard even when they are not struck.

10.4 Effect of the Damper

The interaction between the two notes discussed in the previous section begs the question about the impact of lifting the dampers on all the notes. Figure 10.4.1 shows that effect on C2, C3, and C4 via the cumulative line spectrum and the spectral centroid.

For these three notes lifting the damper does not change the location of the strengths on the frequency axis but it does allow power from the lower harmonics to color the tone as evidenced by both the cumulative line spectra and by the values of the spectral centroids. That is, the spectral centroid for the C2 note when the dampers are down on all the other notes is 330.9 Hz (remember that the C2 fundamental frequency is 65.4 Hz). When the C2 note is struck with the dampers up

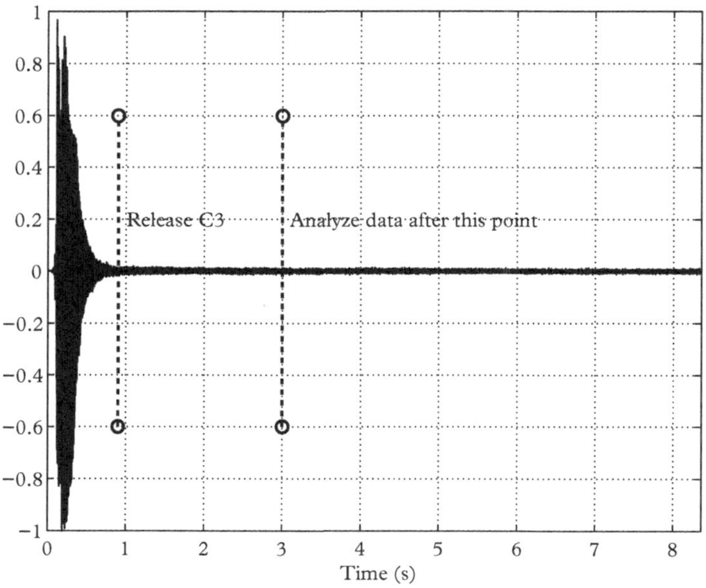

Figure 10.3.1 *Time domain plot showing results of striking C3 having depressed F2 earlier*

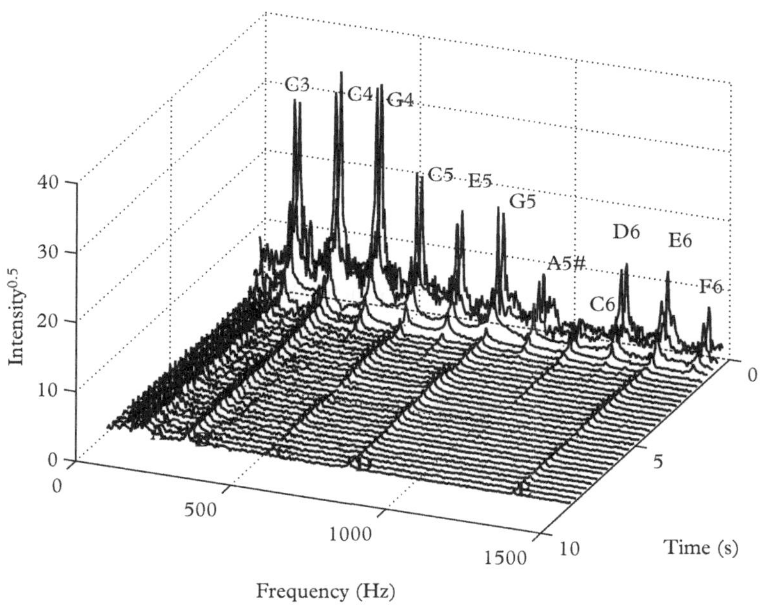

Figure 10.3.2 *Snapshots of the line spectrum for striking C3 having depressed F2 earlier*

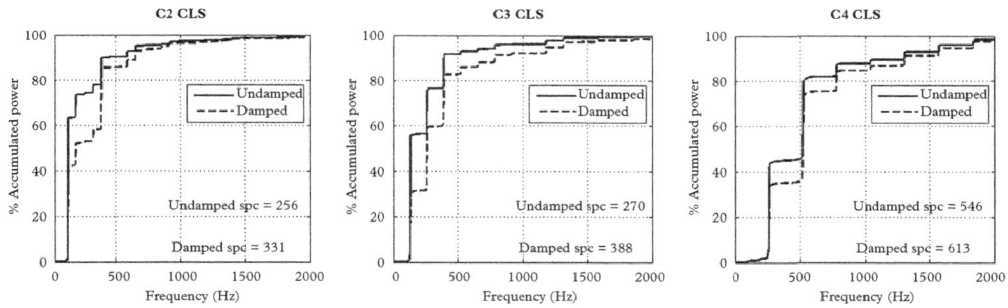

Figure 10.4.1 *The effect of using the damper on C2, C3, and C4*

on all the other notes, the spectral centroid is 255.7 Hz, suggesting that sympathetic vibrations from lower notes have brought the average weighted frequency down. The same thing happens for the C3 and C4 notes, namely, the average weighted frequency is decreased when the dampers are off the strings.

10.5 Summary

In this section we looked at several single notes. First came the C3 note from a Steinway B located at a nearby music school, followed by the C4 and C8 notes from the Knabe piano. A fine grid line spectrum was used to closely look at some of the peaks, where we saw some splitting as a consequence of not all the multiple strings of the note of interest being exactly in tune. The spectra of several of the notes showed weakness at the eighth and ninth harmonics corresponding to where the hammer strikes the strings. All the notes, to one degree or another, showed the amplitude envelope decreasing quickly at first and then more slowly. This phenomenon is purported to be the result of energy being transferred from the vertical plane of vibration to the horizontal. A chromatic sequence of notes next got our attention where we used the spectrogram to show how the notes and their partials exhibited different fade-away times. An experiment demonstrated how energy from a strike of one key can stimulate certain harmonics of another key whose damper has been raised. Finally, the effect of raising all the dampers was shown to emphasize the lower harmonics of a single struck note.

11 The Missing Fundamental

In previous chapters we have touched on the strength (or lack of it) of the fundamental tones for those waves that have harmonic structure. In this chapter we study the variation of the fundamental's strength as the key number increases. This is followed by a simulation where an analysis is made of when various harmonics are removed and/or added. After all the dust has settled, it appears that a low note can be sounded effectively by combining its higher harmonics without using its fundamental frequency.

The concept of removing sinusoidal components is extended to removing portions of the frequency spectrum and studying the resulting sound of an actual musical instrument.

11.1 Empirical Evidence—The Low Notes

Consider the two spectra and the autocorrelation of the C1 note in the Knabe grand in Figure 11.1.1.

The line spectrum in Figure 11.1.1(a) has markers for the 1st, 2nd, 8th, and 16th partials or harmonics. The cumulative line spectrum in Figure 11.1.1(b) has markers for the 3rd, 8th, and 16 partials. There is a periodicity in the shape of the harmonics. Every eighth spike is attenuated, probably due to the location of the hammer strike. There is a slender spike at 32 Hz in Figure 11.1.1(a) that is almost 40 dB down. This spike probably represents the fundamental of 32.7 Hz. If one looks closely at the line spectrum one can see spikes at 64 Hz, 96 Hz, 129 Hz, 162 Hz, 194 Hz, and so on. The *differences* between the frequencies of these spikes are 35 Hz, 32 Hz, 43 Hz, 33 Hz and 32 Hz. Therefore, although the fundamental itself is weakly represented (more about this later), the partials are spaced as though the fundamental existed at 32.7 Hz.

The cumulative line spectrum of C1, Figure 11.1.1(b), shows that the signal intensity is spread over the harmonics with 25% of the power supplied at the fourth harmonic (129 Hz) and another 17% at the 13th harmonic (426 Hz). There is no noticeable contribution from the fundamental. One must remember that the power percentage is plotted on a linear scale that does not emphasize small levels.

The autocorrelation in Figure 11.1.1(c) shows that, although the two spectra did not show any significant strength at the fundamental, there is a contribution to the pitch estimation at 32.4 Hz (the first isolated but not strongest peak). This suggests that the impact of the structure of the harmonics, even though they contain little strength below the third harmonic, combine to generate a pitch at the fundamental frequency.

The C1 note on a Steinway B grand piano (at a local music school) has a similar pattern, and the cumulative line spectra of the C1 notes for it and the Knabe are compared in Figure 11.1.2.

Spectral Analysis of Musical Sounds with Emphasis on the Piano. First Edition.
David M. Koenig, with comments by Delwin D. Fandrich.
Published in 2015 by Oxford University Press.

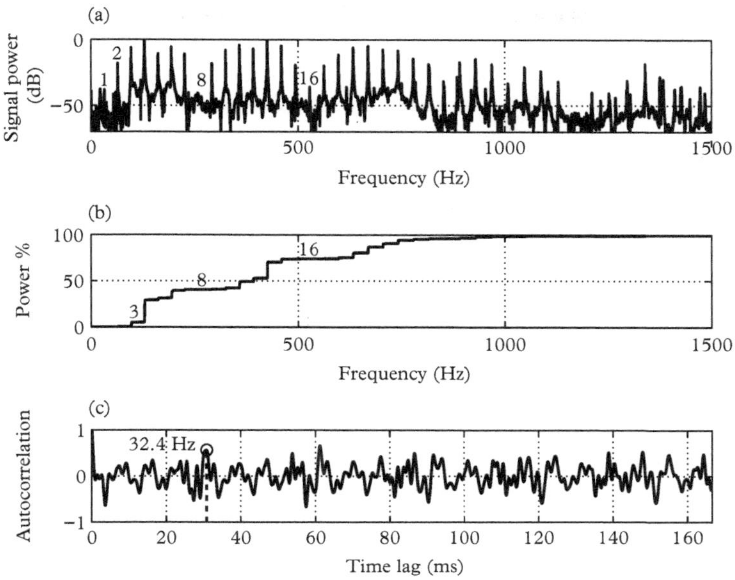

Figure 11.1.1 *Knabe C1: (a) line spectrum, (b) cumulative line spectrum, and (c) autocorrelation*

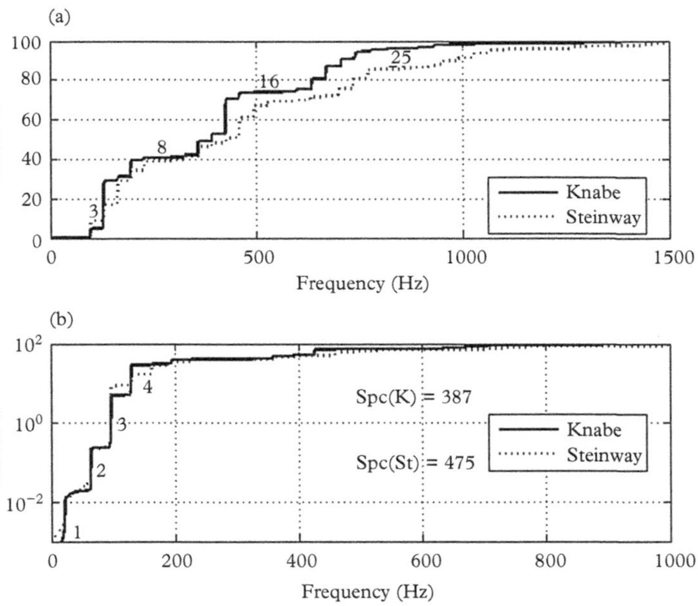

Figure 11.1.2 *Cumulative line spectra of Knabe and Steinway C1s with percent power plotted (a) linearly, and (b) logarithmically (showing the spectral centroids)*

The distributions of the signal power in Figure 11.1.2(a) for the two pianos look similar but one could argue that there is more power in the upper harmonics on the Steinway because by the 25th harmonic the Steinway has still only produced about 85% of its total signal power while the Knabe has already produced almost 96%. The spectral centroids support this contention, the spectral centroid (or average weighted frequency) of the Steinway being significantly higher than that of the Knabe.

In Figure 11.1.2(b) the power is plotted on a *logarithmic scale* where the low levels of the first two partials do in fact appear (note the different span of the frequency axis). Although these strengths at the fundamental and first harmonic are now discernable, it is the logarithmic scale that makes them so. On the other hand our ears are sensitive to sound on a logarithmic scale so perhaps we can indeed hear these weak components.

These two versions of the cumulative line spectrum suggest that the sound of the C1 note does not come from a strong contribution at the fundamental frequency.

To support our contention that the fundamental component is not particularly evident, consider Figure 11.1.3, which shows the Knabe C1 wave in two visualizations. The "big picture" graph in Figure 11.1.3(a) shows a conventional waveform with a sustain time of several seconds. The close-up graph in Figure 11.1.3(b) shows an attempt to locate the fundamental which would have a period of 1000/(32.7 Hz) = 30.6 ms. The graphs show that it is not easy to find peaks with the requisite spacing and that there are a lot of higher-frequency components that dominate the wave. In fact, the two peaks that I found were 25 ms apart, corresponding to a frequency of 40 Hz. Compare this to Figure 10.1.1 in the previous chapter where the same analysis was performed on the C3 note.

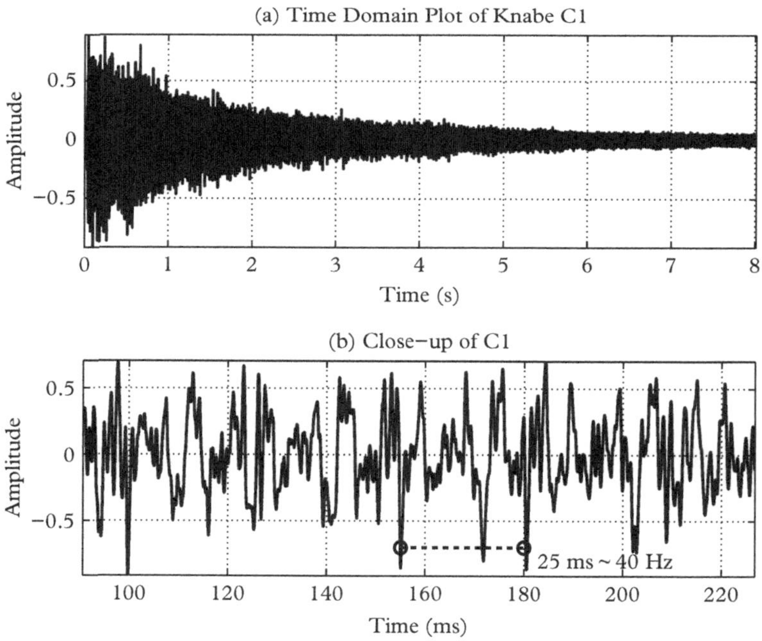

Figure 11.1.3 *Knabe C1: (a) entire wave, and (b) a close-up showing a period of ~25 ms, which relates to 40 Hz*

At the end of this chapter there are a couple of exercises. Use one of them to compare the sounds of the Steinway C1 and C2 with the Knabe C1 and C2. I think you will agree that they are significantly different and that there is a slight high-frequency component to the Steinway notes that is not present in the Knabe notes.

11.2 The Appearance of the Fundamental in the Piano's Higher Octaves

In this section we will go up an octave at a time and watch how the presence of the fundamental changes. Figure 11.2.1 shows the C2 note on the Knabe.

The fundamental at 65 Hz is still over 20 dB down but, compared to the C1 note, it has more of a presence and it is supported by the autocorrelation where the first isolated peak is also the strongest. Once again, the 8th and 16th partials are diminished compared to the others.

Figure 11.2.2 shows C3, where the fundamental at 130.8 Hz is now the third strongest harmonic.

Figure 11.2.3 shows the Knabe C4. The fundamental at 262 Hz is the largest spike. Also, notice how the eighth harmonic is stronger than the ninth.

The Knabe C5 note is shown in Figure 11.2.4. The fundamental at 523 Hz is the strongest harmonic and there is negligible power beyond the third harmonic, at least as indicated by the cumulative line spectrum.

The C6, C7, and C8 notes are shown in Figures 11.2.5–11.2.7.

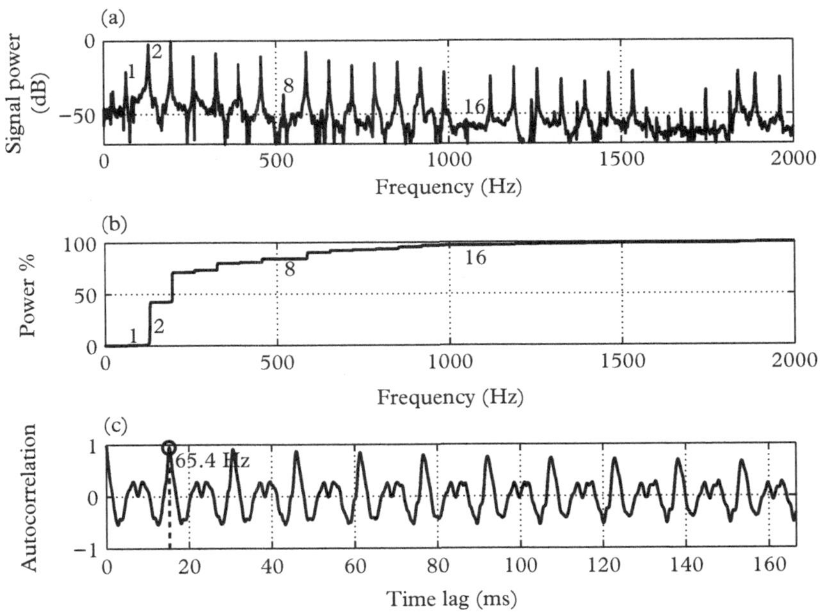

Figure 11.2.1 *Knabe C2: (a) line spectrum, (b) cumulative line spectrum, and (c) autocorrelation*

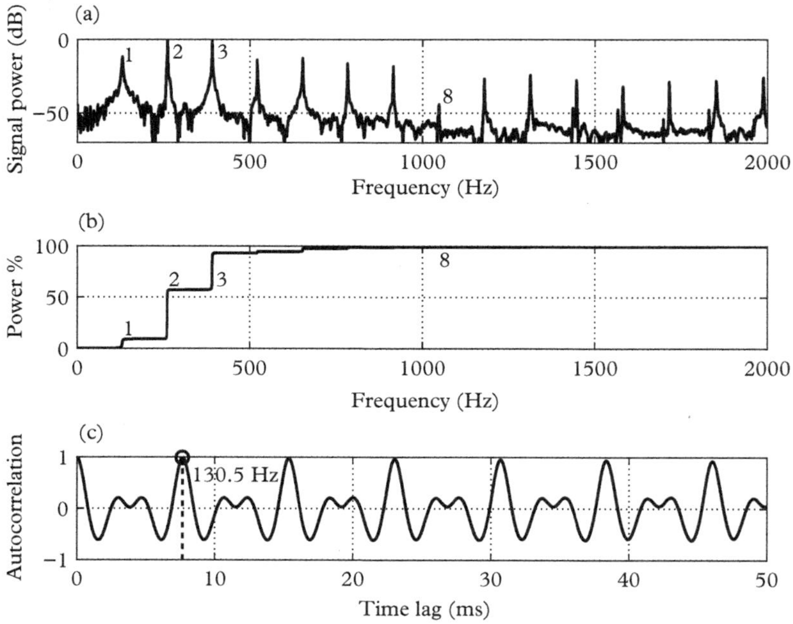

Figure 11.2.2 *Knabe C3: (a) line spectrum, (b) cumulative line spectrum, and (c) autocorrelation*

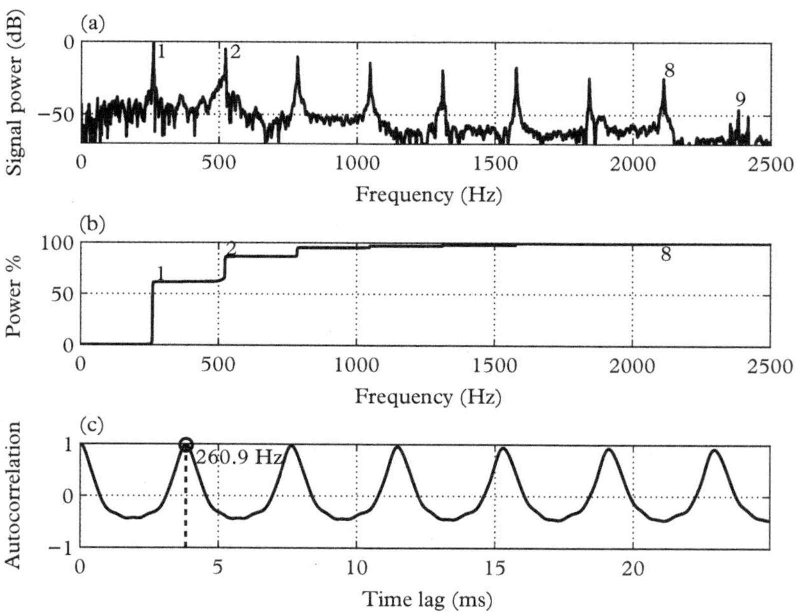

Figure 11.2.3 *Knabe C4: (a) line spectrum, (b) cumulative line spectrum, and (c) autocorrelation*

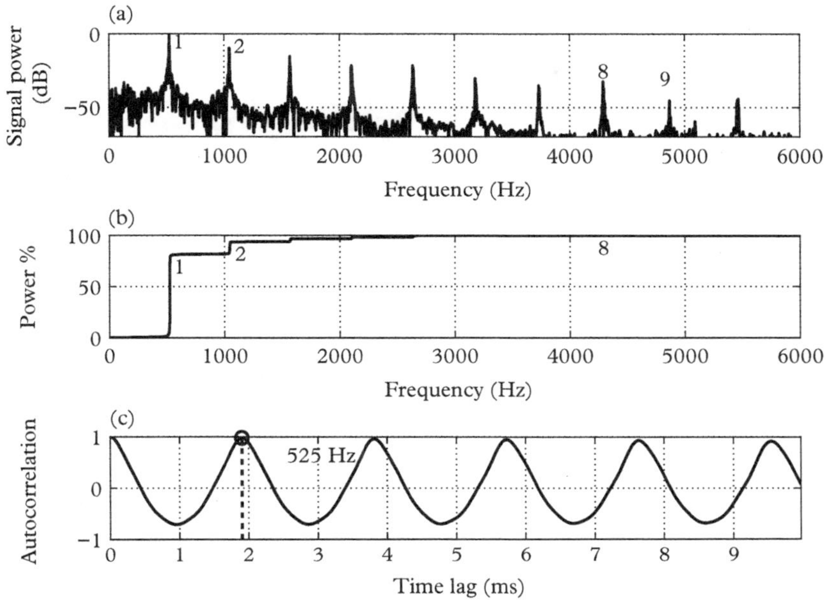

Figure 11.2.4 *Knabe C5: (a) line spectrum, (b) cumulative line spectrum, and (c) autocorrelation*

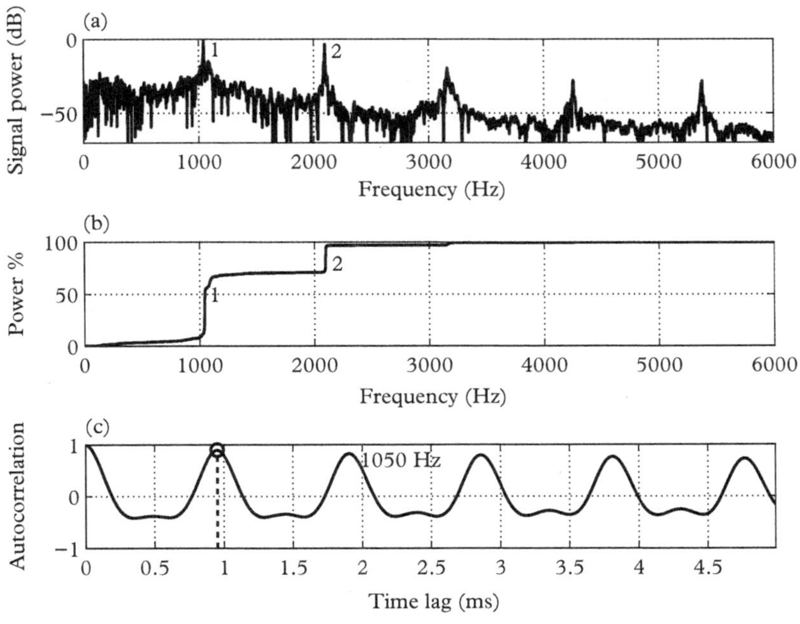

Figure 11.2.5 *Knabe C6: (a) line spectrum, (b) cumulative line spectrum, and (c) autocorrelation*

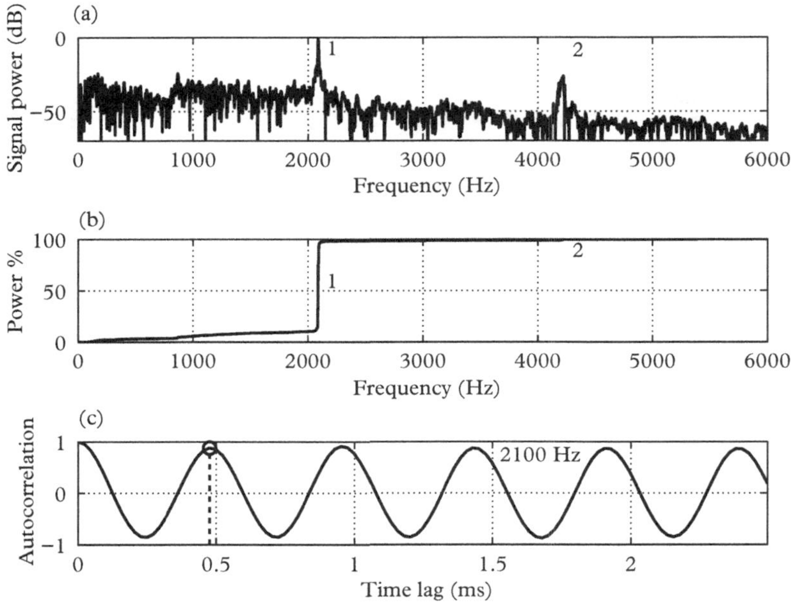

Figure 11.2.6 *Knabe C7: (a) line spectrum, (b) cumulative line spectrum, and (c) autocorrelation*

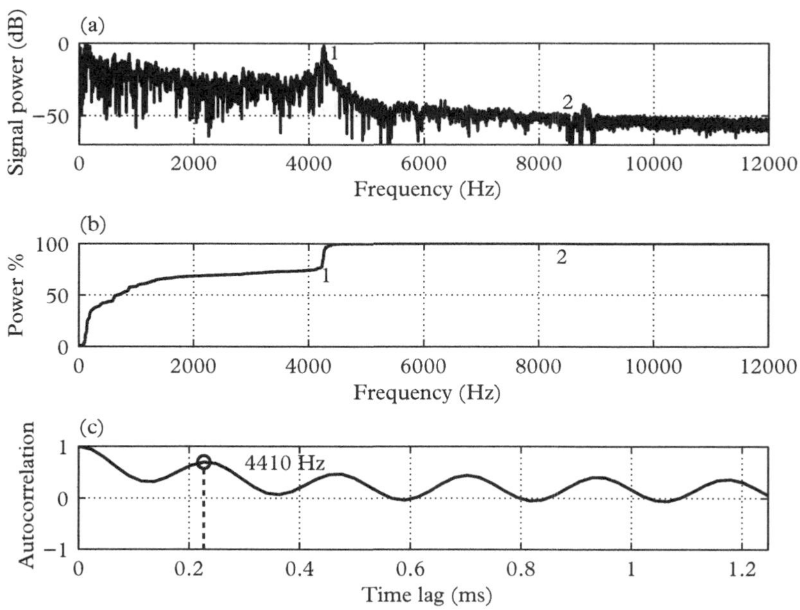

Figure 11.2.7 *Knabe C8: (a) line spectrum, (b) cumulative line spectrum, and (c) autocorrelation*

In these last three notes there is significant signal power below the fundamental frequency, especially for C8. This is evidence of the infamous "hammer knock" that we will come across many times in subsequent chapters in Part 3. Also, the C8 shows a first partial at 4250 Hz (theoretical fundamental at 4186 Hz) and a weak partial at 8780 Hz, which is 40 dB down. The autocorrelation estimate for C8 is not particularly close to the fundamental, probably because, as mentioned in Chapter 2, the error in the estimate depends on the number of samples per cycle, which for the high frequencies is marginally low. An alternative calculation of the autocorrelation using the Wiener–Khinchin theorem fine grid approach (discussed in Chapter 2 and in Appendix 2) yields a pitch estimate of 4226 Hz, which is closer to the maximum in the line spectrum than the 4410 Hz estimate given in Figure 11.2.7(c).

11.3 Some Support Based on Videos and Calculations

Videos

The reader is invited to watch the YouTube video "TheMissingFundamental" and "The Missing Fundamental, Second Attempt", both of which are referenced in the book's website (no. 10), which will provide audio and visual support for the material in this section. In those videos we construct a sine wave having a frequency of 70 Hz and send it to the Matlab sound-playing function *soundsc* where it is barely audible. Links to these videos, and others, are on the book's website.

Removing Harmonics, One by One

To address the problem graphically, a composite signal is composed of a sum of sine waves having frequencies of 70 Hz, 140 Hz, 210 Hz, 280 Hz, 350 Hz, 420 Hz, and 490 Hz, i.e., the fundamental and its harmonics up to the seventh. Each harmonic's amplitude is divided by the harmonic number, e.g., the third harmonic's amplitude is divided by three. In mathematical terms the signal is

$$Y(t) = \sin(2\pi 70t) + \frac{1}{2}\sin(2\pi 140t) + \frac{1}{3}\sin(2\pi 210t) + \cdots + \frac{1}{7}\sin(2\pi 490t)\,. \qquad (11.3.1)$$

In other words, the fundamental is the strongest component and the harmonics are attenuated by their harmonic number. The resulting Matlab *soundsc* response is a note sounding like 70 Hz but with considerable strength.

Figure 11.3.1 shows the time trace, line spectrum, and autocorrelation of the sum of seven harmonics. The time trace in Figure 11.3.1(a) shows a period of 14.3 ms, which corresponds to 70 Hz. The line spectrum in Figure 11.3.1(b) shows seven spikes at the seven harmonics with each spike decreasing in height with increasing frequency. The autocorrelation in Figure 11.3.1(c) shows that the estimated pitch is 70 Hz. The sound can be heard in the file Harms1_7.mp3.

Next, the fundamental is removed, leaving the second through seventh harmonics:

$$Y(t) = \frac{1}{2}\sin(2\pi 140t) + \frac{1}{3}\sin(2\pi 210t) + \cdots + \frac{1}{7}\sin(2\pi 490t)\,. \qquad (11.3.2)$$

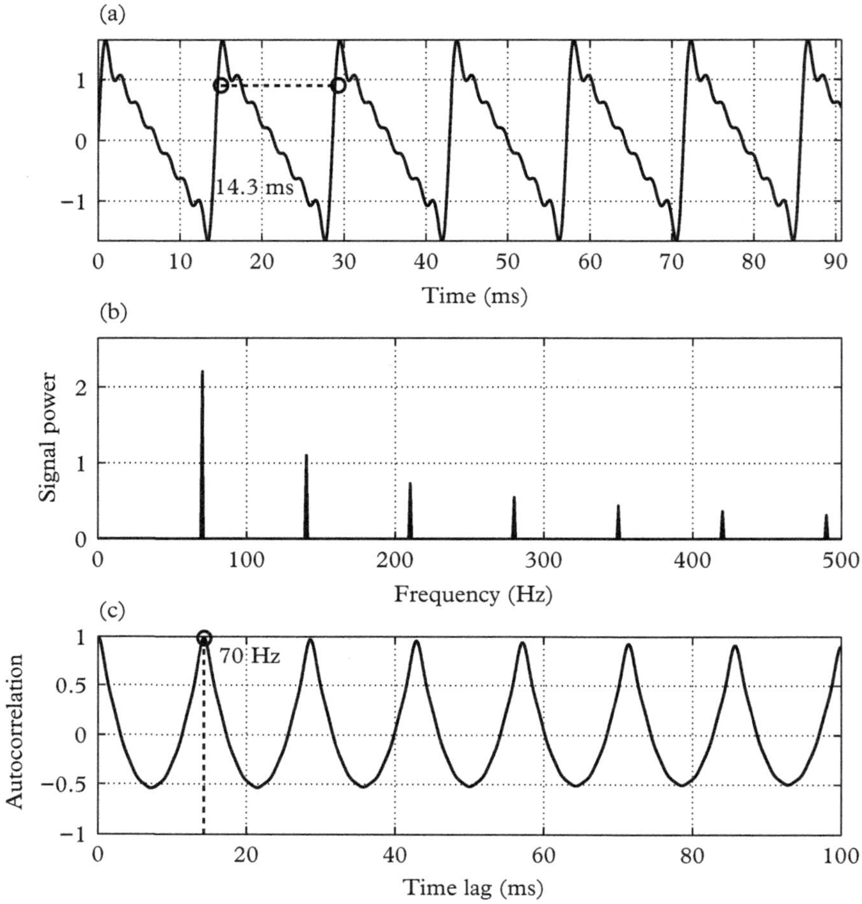

Figure 11.3.1 *A sum of seven harmonics that produces a 70 Hz tone: (a) time domain plot, (b) line spectrum, and (c) autocorrelation*

Remarkably, the sound generated from Eq. 11.3.2 is aurally indistinguishable from that generated from Eq. 11.3.1. Likewise, the autocorrelation in Figure 11.3.2(c) shows that the pitch is still 70 Hz, although there are secondary peaks pointing to higher harmonics. It appears that it is the harmonic structure that the ear perceives rather than the individual components. The sound can be heard in the file Harms2_7.mp3.

Let's repeat this exercise and take away the second harmonic as well, keeping the third through the seventh harmonics. Figure 11.3.3(c) shows that the pitch is still 70 Hz. The sound can be heard in the file Harms3_7.mp3.

Figure 11.3.4 shows the result of keeping only three harmonics, the fifth, sixth, and seventh. The autocorrelation shows more competing peaks but the first isolated, strongest peak still points to 70 Hz as the pitch. The wave still has a sound like 70 Hz.

We can continue this experiment, but we will leave it to a reader with access to Matlab and skip to the case of keeping just the top three harmonics, the results of which are presented in

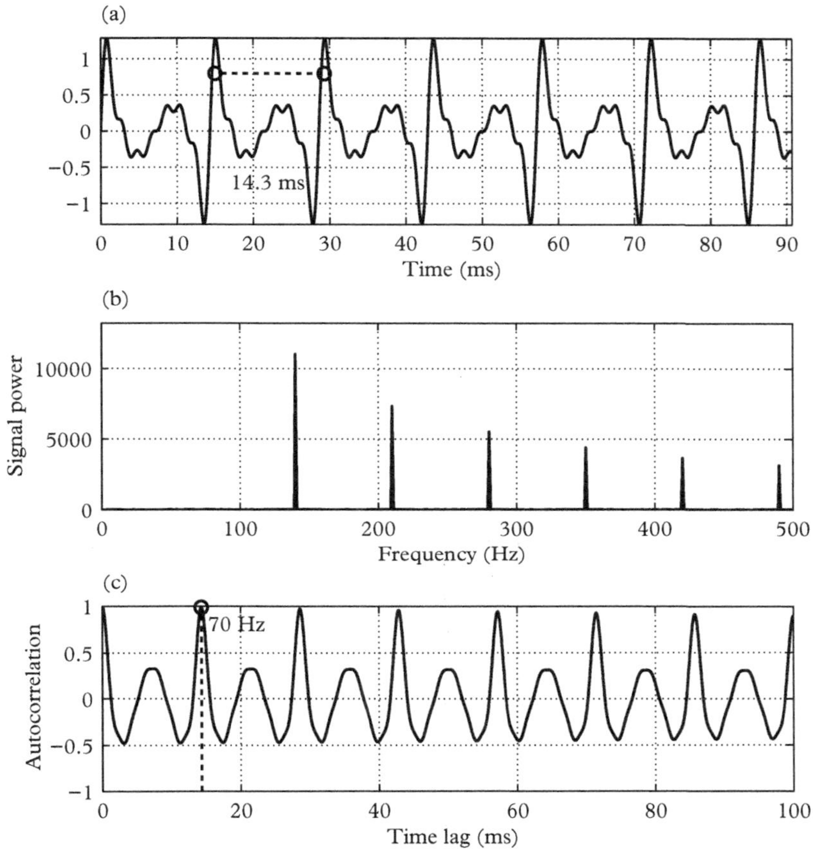

Figure 11.3.2 *A sum of six harmonics that produces a 70 Hz tone: (a) time domain plot, (b) line spectrum, and (c) autocorrelation*

Figure 11.3.4. The wave now only has a hint of 70 Hz in its sound even though the autocorrelation shows 70 Hz as the first and highest peak. The sound can be heard in the file Harms5_7.mp3.

As shown in Figure 11.3.5, when we get down to the last two harmonics the pitch estimate starts to be questionable, although the autocorrelation peak in Figure 11.3.5(c) is still at 70 Hz. Also, the sound from this two-term sum is not close to 70 Hz. Note how the time domain trace and the autocorrelation trace change as we remove the harmonics. The sound can be heard in the file Harms6_7.mp3.

Putting the Harmonics Back, One by One

In the following discussion, the various components will be added graphically in an attempt to show that the sum is periodic with a frequency equal to the fundamental even though the fundamental is not present. First, Figure 11.3.6 shows the second harmonic all by itself.

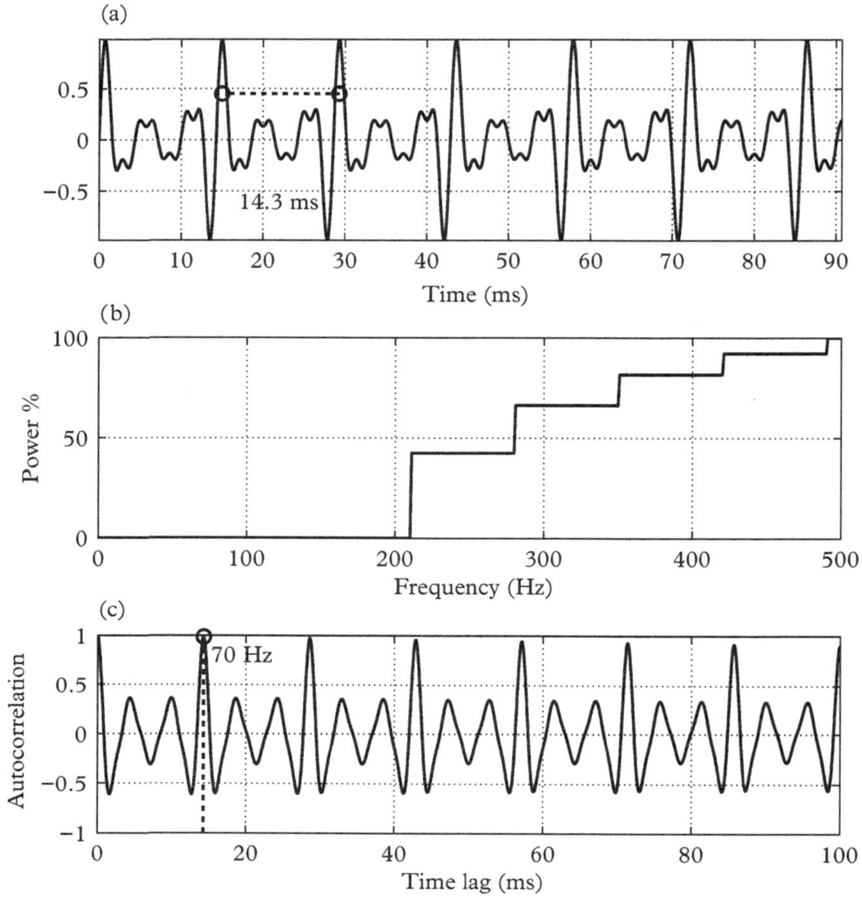

Figure 11.3.3 *A sum of five harmonics that produces a 70 Hz tone: (a) time domain plot, (b) line spectrum, and (c) autocorrelation*

Figure 11.3.6 shows that the second harmonic has a period of about 7.1 ms, half that of the fundamental, whose period is 14.2 ms. Therefore, the second harmonic by itself can not produce a 70 Hz sound. The sound of the second harmonic can be heard in the file AddHarms2.mp3.

Figure 11.3.7 shows the individual second and third harmonics and their sum weighted by the harmonic number. The sum is denoted by the thickest trace, the second harmonic by the solid line, and the third harmonic by the dotted line.

The sum of these two harmonics by definition contains two components with periods too short to produce a 70 Hz tone, but the overall sum has a period of 14.2 ms (the period of the missing fundamental). The autocorrelation in Figure 11.3.7(b) supports this concept with a peak at a time lag corresponding to 70 Hz. The sound can be heard in the file AddHarms2_3.mp3.

If this summing is continued to the point where the second through the seventh harmonics are added together, one gets Figure 11.3.8 where the sum now shows a strong period of

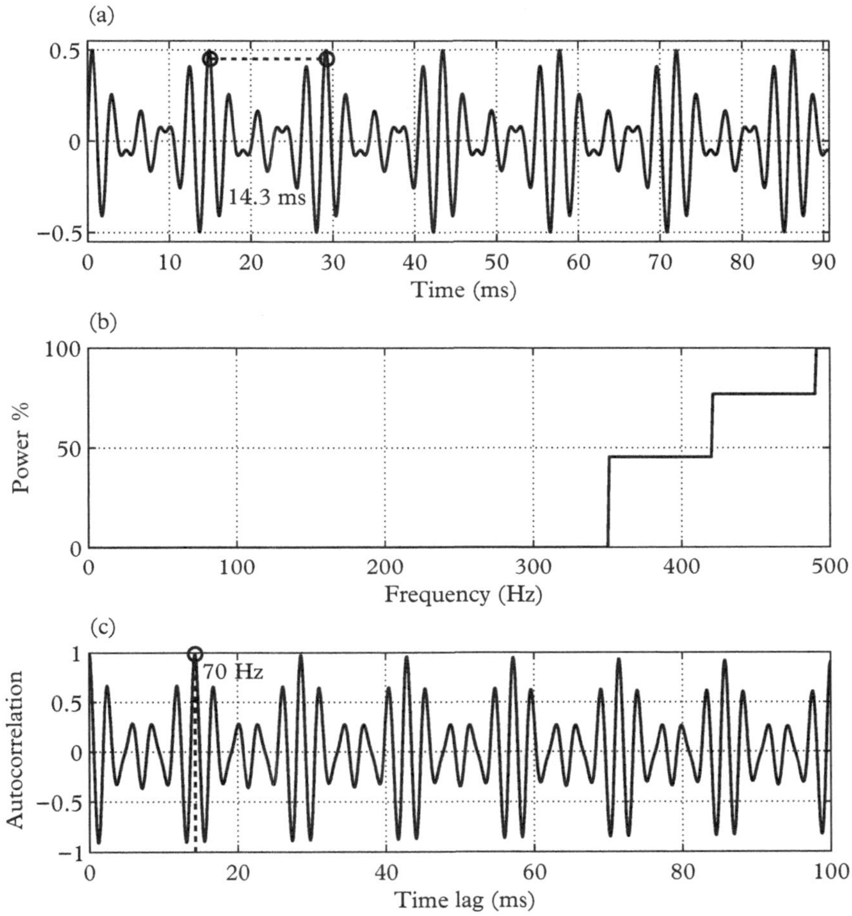

Figure 11.3.4 *A sum of three harmonics that produces a 70 Hz tone: (a) time domain plot, (b) line spectrum, and (c) autocorrelation*

14.2 ms, which is supported once again by the autocorrelation. The sound can be heard in the file AddHarms2_7.mp3.

These graphs suggest that the sum of the second through the seventh harmonics produces a wave that has a repeating structure with a period equal to that of the missing fundamental. This result is somewhat reminiscent of the approximation of the square wave by a sum of sine waves in Chapter 2. In effect, the higher harmonics are combining to provide an approximation to the fundamental.

As an example of the utility of this idea, consider an inexpensive speaker system (such as the one in my laptop) that can not produce a low frequency sound. If, instead of a single wave having the low frequency, one sends a properly configured sum of harmonics to the speaker, a strong low frequency sound will result.

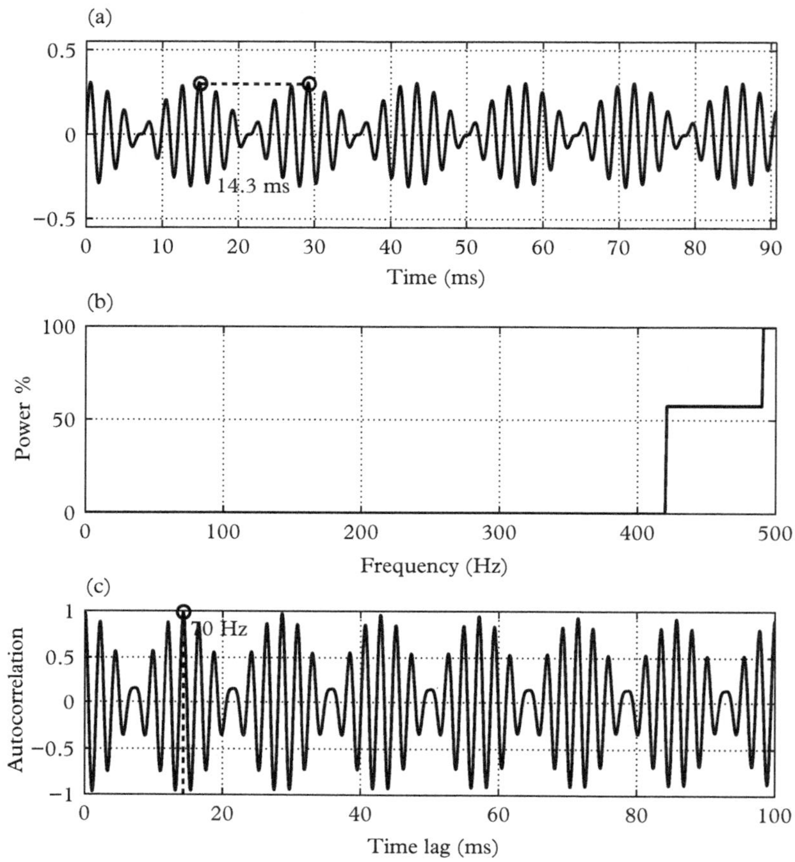

Figure 11.3.5 *A sum of two harmonics that produces a 70 Hz tone: (a) time domain plot, (b) cumulative line spectrum, and (c) autocorrelation*

11.4 Changing the Sound of a Note by Frequency Domain Filtering

This section requires that you go to the YouTube videos associated with this book. Since they are available on the Internet I will not bore you with pictures but instead I will just give an introduction. The first video looks at the effect of removing parts of the spectrum for the C4 note on my Knabe piano. Using frequency domain filtering, sections of the spectrum are removed one at a time. For example, all of the power just below the fundamental is removed and the sound is played. Next, all the power just above the fundamental and below is removed. This continues with more sections around higher harmonics being removed.

Next, we start at the high frequency end of the spectrum and remove all the power above a certain harmonic. The removal point is decreased one harmonic at a time until we are left with just the fundamental. These steps are demonstrated on the YouTube video available at

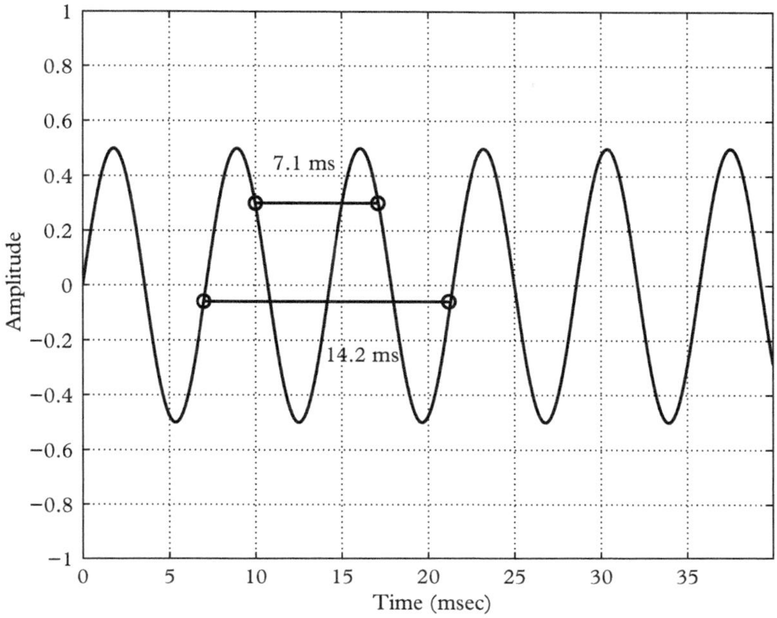

Figure 11.3.6 *The second harmonic all by itself, exhibiting a period of 7.1 ms*

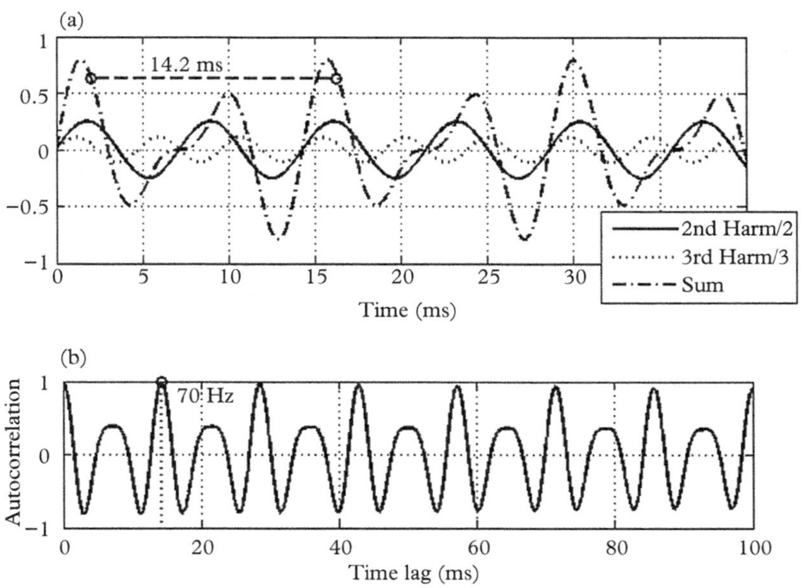

Figure 11.3.7 *Sum of (2nd harmonic)/2 and (3rd harmonic)/3: (a) time domain plot, and (b) autocorrelation*

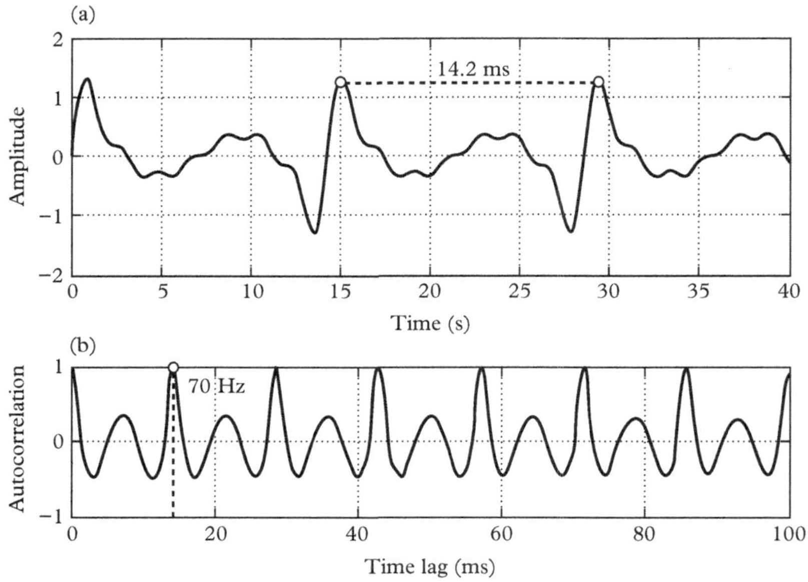

Figure 11.3.8 *Sum of second through seventh harmonics (all divided by their harmonic number): (a) time domain plot, and (b) autocorrelation*

no. 10 on the book's website. The reader with access to Matlab can see and listen to the above actions by running the scripts in the file C4AnalysisQ.m.

The same procedure is carried out on the G4 note of my cornet (played by me, unfortunately). The YouTube video is available at no. 9 on the book's website, and the sounds and pictures of the spectra can be seen if you run the scripts in the file Cornet.m.

Both of these videos were made without the aid of CamStudio and are rather amateurish in appearance, but hopefully the reader will still be able to see what I am getting at. I plan to redo the videos using CamStudio.

As mentioned above, there is more frequency domain filtering in the exercises at the end of the chapter.

11.5 Summary

This chapter has dealt with adding and removing periodic components from a signal and studying the resulting sound. The low frequency notes on the piano and on several other instruments that we looked at in this book have a harmonic structure wherein the separation of the harmonics (or partials) is approximately equal to the fundamental frequency. However, there is virtually no strength at the fundamental frequency. We showed that the sound of the fundamental frequency could be created by a sum of the higher harmonics even though there was no strength at that fundamental frequency.

Using frequency domain filtering to remove portions of the line spectrum (and therefore periodic components) we were able to manipulate the sound of the piano's C4 note and the cornet's G4 note.

11.6 Exercises

Download the Matlab script file GenKeyAnalysis.m. Use it to look at the effect of filtering out parts of the spectrum. The filtered sound is followed one second later by the unfiltered sound and there is an overlap of several seconds. Try to see what part of the spectrum is important to the sound and what part is not. For example, when does the "buzz" go away when more and more high partials are removed? When the lowest harmonics are removed does the sound noticeably change? Start with the Knabe and Steinway C1s and compare the sounds.

12

Octave Stretching, Inharmonicity, and the Railsback Curve

The equal temperament system defines the fundamental frequencies of the 88 notes on a piano. An actual piano exhibits deviations from these theoretical frequencies because of inharmonicity and because of planned tuning profiles. The Railsback curves present these deviations graphically. Other measures of inharmonicity are also illustrated in this chapter.

12.1 Inharmonicity or Octave Stretching

In Chapter 5, Section 5.10, we looked at inharmonicity and string stiffness via simulation. It was suggested there that the elastic stiff string reacts with the fixed connections at the termination points. The string exerts a force on the connections as a consequence of the reflection and inversion of the traveling wave. There is an equal and opposing force on the string that, because of the elasticity, causes the wave to change shape and to accelerate. Using snapshots of two strings, one flexible and one with stiffness, we showed that the stiff string increases its lead over the flexible string with each reflection and inversion. The abruptness of the reflection and inversion at the connection points generates higher harmonics that also feel the force and experience the acceleration. This feature of pianos is perhaps best illustrated by looking again at the first 36 partials and harmonics in the line spectrum of a Steinway model B C1 note in Figure 12.1.1. Figure 12.1.1(a) shows the first 12 harmonics represented by the numbered dot-dashed vertical lines. All 12 harmonics line up closely with the partials. In Figure 12.1.1(b), showing the next 12 harmonics, we start to see the partials move to the right of the harmonic's dot-dashed lines. In Figure 12.1.1(c), showing harmonics 25 through 36, it is difficult to make an association between the harmonics and the partials because of this increased movement. This is the phenomenon of inharmonicity or octave stretching due to the stiffness of the strings. One would expect this for the C1 note because it is usually a single wrapped string exhibiting much more stiffness than the single-strand strings in the treble section of the piano.

Figure 12.1.2 shows the relationship between the theoretical harmonics and the observed partials for the C1 note on the Steinway model B. The shape of the curve in Figure 12.1.2 is supported by H. Fletcher in "Normal Vibration Frequencies of a Stiff Piano String" (*Journal of the Acoustical Society of America*, vol. 36, no. 1) who showed that

$$f_n = nf_1\sqrt{(1 + bn^2)}, \tag{12.1.1}$$

Spectral Analysis of Musical Sounds with Emphasis on the Piano. First Edition.
David M. Koenig, with comments by Delwin D. Fandrich.
Published in 2015 by Oxford University Press.

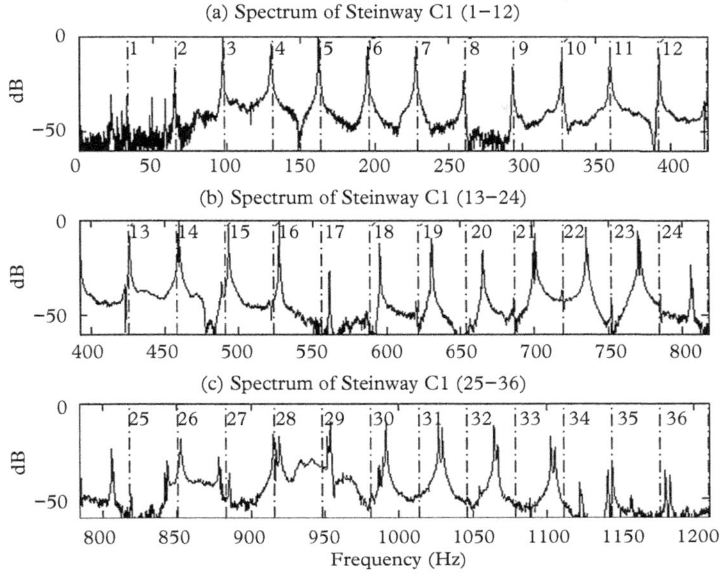

Figure 12.1.1 *Line spectrum of Steinway C1 for three ranges: (a) 0 to 400 Hz, (b) 400 to 850 Hz, and (c) 800 to 1200 Hz, showing the locations of the partials and of the harmonics, which are numbered*

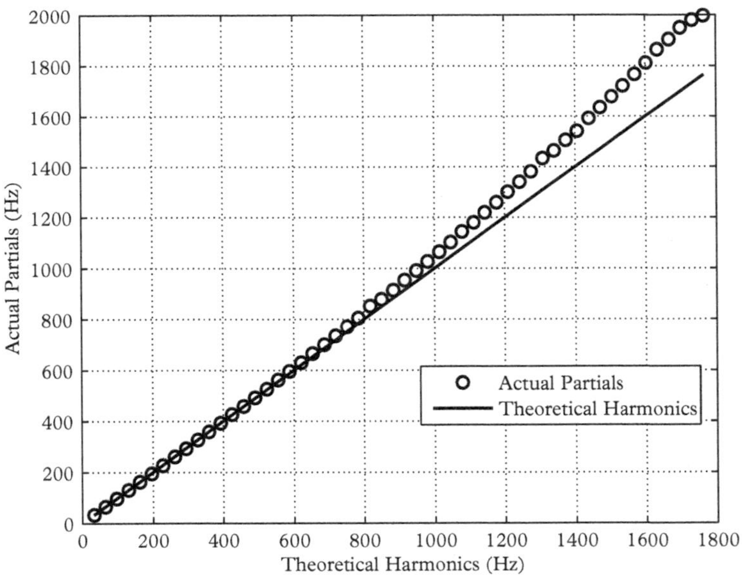

Figure 12.1.2 *Partial frequencies versus harmonic frequencies for the Steinway C1 note*

where f_n is the nth partial frequency, which is shown to be a function of the harmonic number n, the fundamental frequency f_1, and a coefficient b. Fletcher showed that the coefficient b is proportional to the Young's modulus which in turn is a measure of elasticity and therefore stiffness. It is inversely proportional to the fourth power of the string length.

If the coefficient b is zero then Eq. 12.1.1 becomes

$$f_n = nf_1,$$

which is the definition of the harmonic frequencies; that is to say, the n th harmonic frequency is n times the fundamental frequency.

Figure 12.1.3 shows the quantity (which appears in Professor Nicholas Giordano's book *Physics of the Piano*)

$$f_n - nf_1$$

plotted against f_n. If there were no inharmonicity, each partial, f_n, would equal nf_1 and the plot would have zero slope.

As mentioned above, this inharmonicity or frequency stretching is usually attributed to the stiffness of the piano strings. This is consistent with the findings of Section 5.10 where we looked at the force on a fixed connection of a vibrating string. In addition, Section 6.3 (Figure 6.3.3) showed that the partials of a vibrating bar (which is stiff to the extreme) are significantly greater than the harmonics expected for a flexible string.

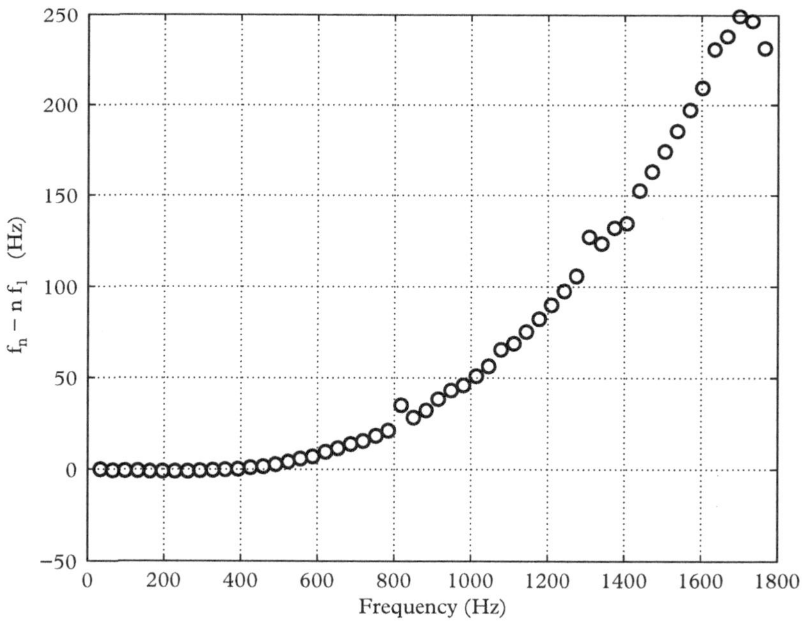

Figure 12.1.3 *The Giordano measure for the Steinway C1 note*

12.2 Inharmonicity of the C1 Note for Several Pianos

In Section 12.1 inharmonicity was introduced using the Steinway C1 note. The stretching (or Giordano) measure defined by

$$f_n - nf_1$$

was introduced and plotted for the C1 note in Figure 12.1.3. In this section, Figure 12.2.1 shows the same metric plotted for all six of the pianos mentioned at the beginning of Part 3.

The 43″ Yamaha console, symbolized in Figure 12.2.1 by *YCon*, has the greatest degree of inharmonicity, followed by the Knabe. The Bösendorfer 280, symbolized by *Bos*, shows the least, followed by the Yamaha P120, symbolized by *YP120* (which is supposedly based on a Yamaha concert grand). The Schimmel, symbolized by *Sch*, is in the center of the pack. Later on we will refer to this graph when we attempt to differentiate between European and American pianos.

The Bösendorfer has the longest strings, therefore the thickness of the wrapping need not be as great as, say, for the 43″ Yamaha console which has the shortest strings. The thicker the string, the more stiff it is and therefore it will exhibit more inharmonicity. Consequently, it is not surprising that the Bösendorfer has the least inharmonicity and the Yamaha console the most.

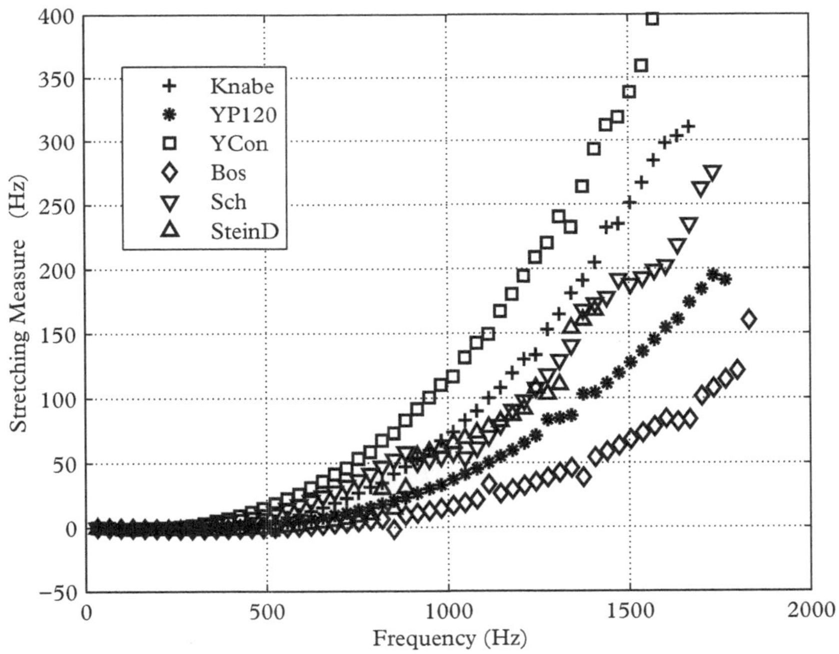

Figure 12.2.1 *The Giordano measure for C1 of six pianos*

12.3 Inharmonicity of the Strings from Chapter Five

In Section 5.2 of Chapter 5, the results of striking and plucking a single-strand piano wire were presented. Also, the results of striking and plucking a wrapped bass string were presented. These results will be reprised now that we have introduced the stretching, or Giordano, measure. Figure 12.3.1 shows the stretching measure for all four wires.

The legend on the graph is a bit cryptic and bears some explanation. "SS P @ $\frac{1}{2}$" means a single-strand wire plucked at $\frac{1}{2}$ of its length. "SS S @ $\frac{1}{2}$" means the same thing except the wire was struck. "W P @1/9" means a wrapped wire was plucked at $\frac{1}{9}$ th of its length. Finally, "W S @1/9" means a wrapped wire was struck at $\frac{1}{9}$ th of its length. The single-strand wire struck at $\frac{1}{2}$ th of its length appears to have the least inharmonicity while the wrapped wire plucked at $\frac{1}{9}$ th of its length appears to have the most inharmonicity. Also, it appears that the wrapped strings have more inharmonicity than the single-strand strings.

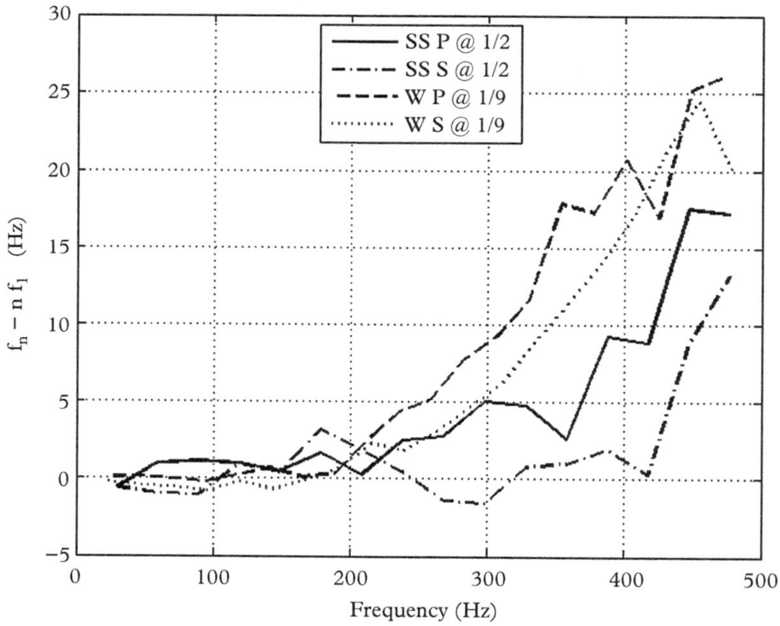

Figure 12.3.1 *The Giordano measure for the four wires studied in Section 5.2 of Chapter 5*

12.4 The Inharmonicity Coefficient for the Knabe Piano

The value of the coefficient b in Eq. 12.1.1 (repeated here as Eq. 12.4.1) can be determined by least squares for each of the 88 keys of the Knabe piano. Therefore, instead of constructing a graph such as that in Figure 12.2.1, one could use the quantity b as a more compact representation. The equation theoretically relating the frequency of the nth partial has already been given as

$$f_n = nf_1\sqrt{(1 + bn^2)}, \tag{12.4.1}$$

where the inharmonicity coefficient b depends on the Young's modulus. As was pointed out in Section 12.1, if b is zero then the partial frequencies become harmonics with frequencies that are integer multiples of the fundamental frequency f_1.

The following paragraph goes through some algebra to develop an equation that can be easily used to provide a numerical estimate of the inharmonicity coefficient. The reader can skip to Figure 12.4.1 and the associated discussion if so desired. However, the following manipulations simply rearrange Eq. 12.4.1 into a formula that has the form of a straight line, with the slope of the line being the coefficient b. The coefficient can then be determined by finding the slope of the straight line either graphically or by least squares. I chose the latter approach.

A simple algebraic manipulation of Eq. 12.4.1 yields, for the nth partial,

$$\left[\frac{\left(\frac{f_n}{n}\right)^2 - f_1^2}{f_1^2 n^2}\right] = b\ \left[n^2\right], \tag{12.4.2}$$

which looks a bit complicated until one realizes that the elements in the square brackets on the left-hand side, which we will denote as Y_n, namely

$$\left[\frac{\left(\frac{f_n}{n}\right)^2 - f_1^2}{f_1^2 n^2}\right] = Y_n,$$

are all measurable for the n th partial. Likewise, the elements in the bracketed factor on the right-hand side, which we will denote as X_n, namely

$$\left[n^2\right] = X_n,$$

are also measurable. The only unknown is the coefficient b.

Since Eq. 12.4.2 has the form of a straight line, as in

$$Y_n = bX_n \quad n = 1, 2, ..., \tag{12.4.3}$$

one can apply least squares and the coefficient b can be found by solving one equation. The result of the calculation for each of the first 39 Knabe grand keys is shown in Figure 12.4.1.

The inharmonicity coefficient is estimated for only the first 39 keys because for the higher-numbered keys there are fewer detectable partials and the calculation becomes erratic (one might argue that Figure 12.4.1 exhibits quite a bit of variation, too).

Figure 12.4.2 shows how well Eq. 12.4.1 fits the data for the Knabe grand C1 key.

In a 2008 University of Edinburgh master's thesis by Simon Hendry, available at <http://www.simonhendry.co.uk/wp/wp-content/uploads/2012/08/inharmonicity.pdf>, there is a development of estimates of the inharmonicity coefficient using graphical means for the A1, A2, A3, A4, A5, and A6 keys from a "well-tuned" Steinway Model D. The values in Figure 12.4.1 are roughly similar to those found in the thesis.

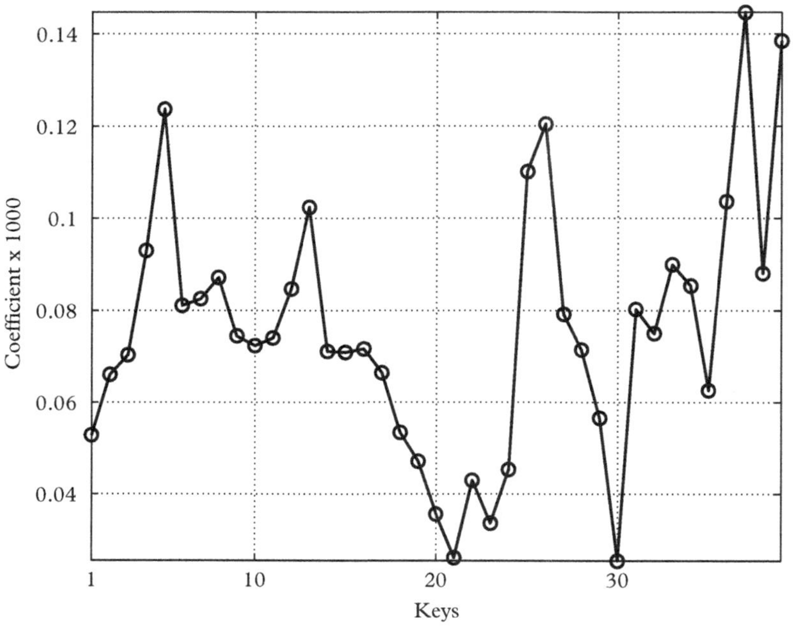

Figure 12.4.1 *Inharmonicity coefficient for the first 39 keys on the Knabe grand piano*

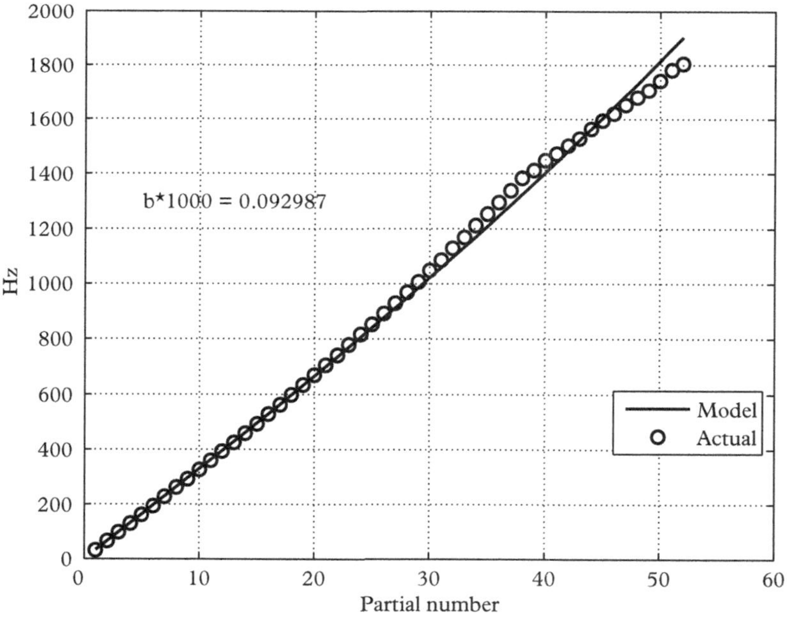

Figure 12.4.2 *Equation 12.4.1 versus actual data for Knabe C1*

12.5 The Inharmonicity Coefficients for the Six Pianos

Figure 12.5.1 shows estimates of the inharmonicity coefficient for each of the first 39 keys for the six pianos that we are studying in detail. All six show basically the same shape except for the Yamaha console. Remember that Figure 12.2.1 referred to C1 or the fourth key. Figure 12.5.1 shows that for this key the Bösendorfer indeed has the lowest value, but for, say, keys 20 to 30, the Steinway Model D has the lowest values.

An overall metric for the degree of inharmonicity could be the average value of the inharmonicity coefficient. Figure 12.5.1 shows that the average inharmonicity coefficient (multiplied by 1000 and denoted by <*b*>) is the least for the Steinway Model D and the greatest for the Yamaha console.

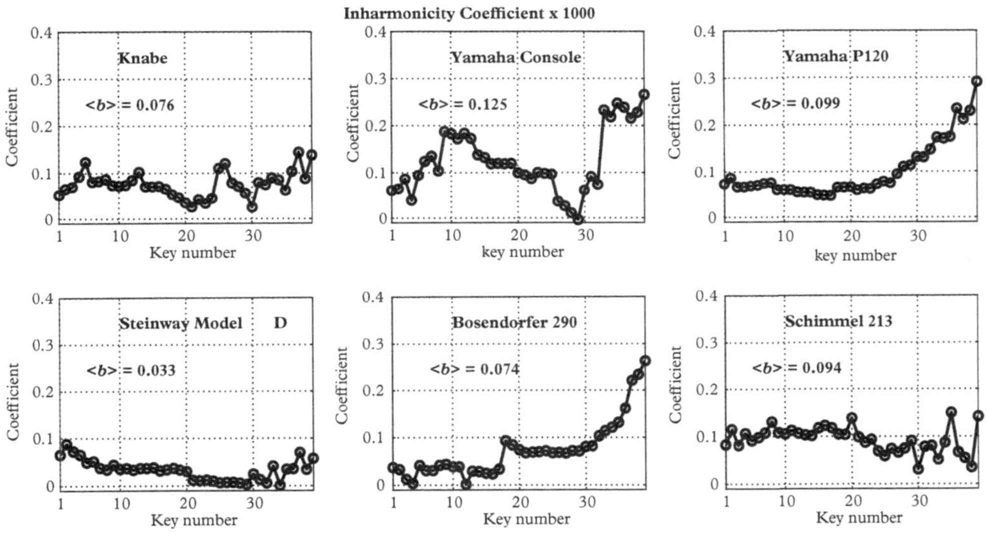

Figure 12.5.1 *Inharmonicity coefficients for the six pianos*

12.6 The Difference between the Fundamental and First Partial Frequencies

The Knabe Grand Piano

A measure of how a piano is tuned relative to the equal temperament scale can be estimated by the difference between the theoretical fundamental frequency and the observed first partial, as is shown in Figure 12.6.1. This difference can be determined by searching the line spectra of each key for the appearance of the peak associated with the first partial. Keep the shape of this graph in mind because we will be discussing it later when the Railsback curve is presented.

Note that the lower-numbered keys in Figure 12.6.1 appear to be flat relative to equal temperament while the opposite is true for the higher-numbered keys. For example, the first key,

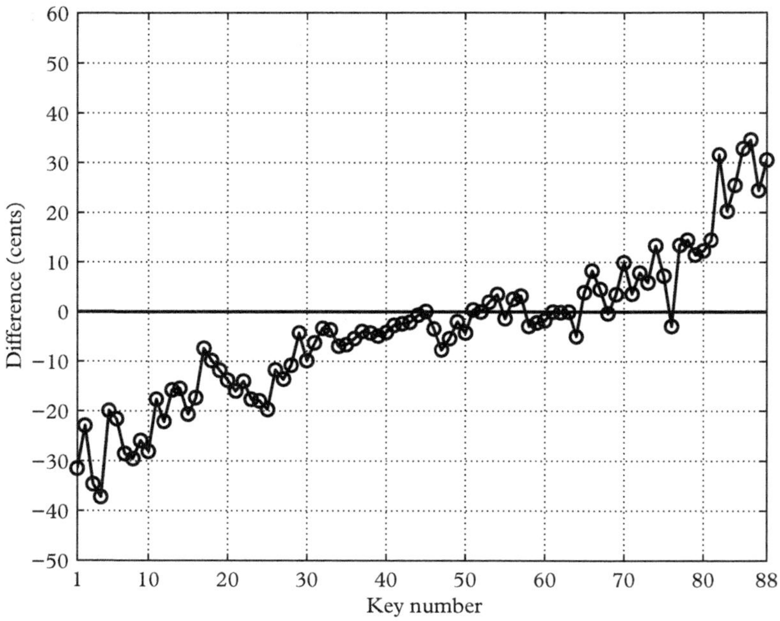

Figure 12.6.1 *Difference (in cents) between the fundamental frequency and the first partial for the Knabe grand piano*

A0, appears to be flat relative to equal temperament by approximately 30 cents. In the middle of the piano's compass (keys 35 to 70), the difference is effectively zero, indicating that those notes are tuned to equal temperament. The higher-numbered notes from 70 on up are sharp relative to equal temperament, with C8 being sharp by approximately 30 cents. Thus, if the piano were tuned such that the frequency of each key's first partial equaled the equal temperament fundamental frequency, then the curve in Figure 12.6.1 would be flat, that is, the same as the solid horizontal line in the graph.

Piano tuners apparently choose to tune their pianos in this manner to compensate for the inharmonicity of the partials. This makes sense because, due to the harmonic stretching exhibited by the low-numbered keys (and illustrated for C1 in Figure 12.1.1), the higher harmonics of these keys would be sharp relative to the equal temperament fundamental frequencies of the high-numbered notes. Therefore, if the low-numbered notes are tuned slightly flat of equal temperament and the high numbered notes are tuned slightly sharp of equal temperament, there will be less clashing between the higher partials of the low-numbered notes and the first partials of the higher-numbered notes.

Difference of Frequencies for All Six Pianos

Figure 12.6.2 shows the difference between the fundamental and first partial frequency for the six pianos.

The composite graph showing an overlay of all six pianos will be presented next in conjunction with the discussion of the Railsback curve in the next section.

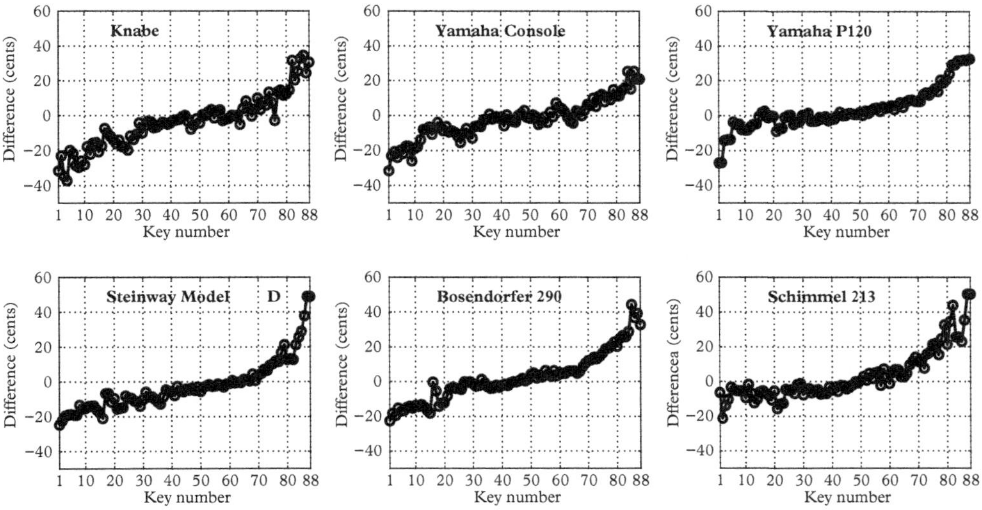

Figure 12.6.2 *Difference (in cents) between the fundamental frequency and the first partial for the six pianos*

12.7 The Railsback Curve

Figure 12.7.1 shows the Railsback curve, information about which is widely available on the Internet (for example, <http://en.wikipedia.org/wiki/Piano_acoustics>).

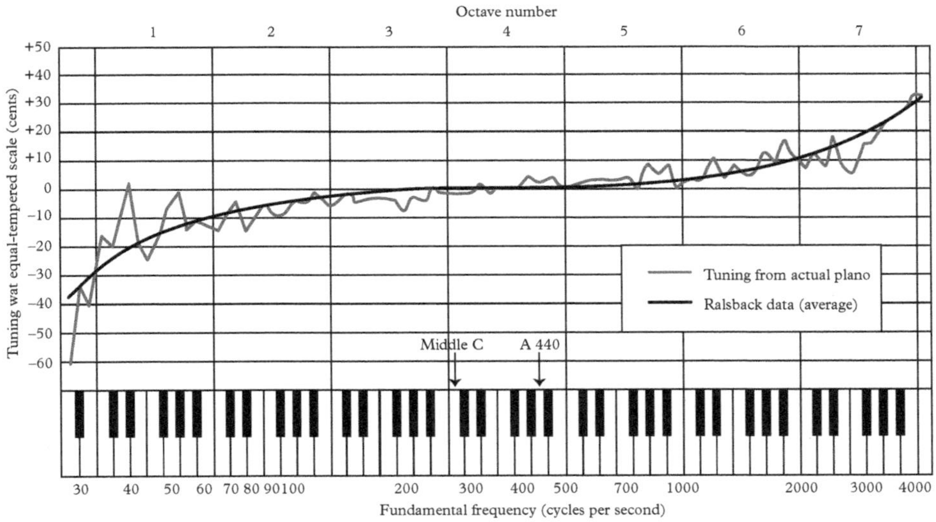

Figure 12.7.1 *The Railsback curve*

This curve is purportedly based on data from the work of many piano tuners collected by O. L. Railsback and first referenced in 1943 in an article by O. H. Schuck and R. W. Young (J. Acoust. Soc. Am. 15, 1–11, 1943). The shape of the plot in Figure 12.7.1 appears to have had its debut in 1961 in an article by D. W. Martin and W. D. Ward (*J. Acoust. Soc. Am.* Vol. 33, 5, 582–5, 1961). The curve suggests that the low notes are purposely tuned flat and the high notes sharp.

The Railsback curve can be compared to the curves for each of the six acoustic pianos shown in Figure 12.7.2. It appears that each of the six pianos has been tuned a little differently, but they all seem to have some flatness in the extreme bass and some sharpness in the extreme treble.

As mentioned in the previous section, the shape of the curves in Figures 12.7.1 and 12.7.2 is related to inharmonicity. For example, the partials of C1 in Figure 12.1.1 become increasingly sharp relative to its theoretical harmonics. These partials will clash with the partials of the higher frequency notes such as C5. To address this one might purposely tune the low notes a few cents flat and the high notes a few cents sharp—hence the shapes of the two curves.

The reader is invited to read the article by Schuck and Young to get an idea of the apparatus required in 1943 to collect, analyze, and present the data for the Railsback curve. In doing so, the reader will probably appreciate the luxury of having the Zoom H2 and the Matlab system at hand (as do I).

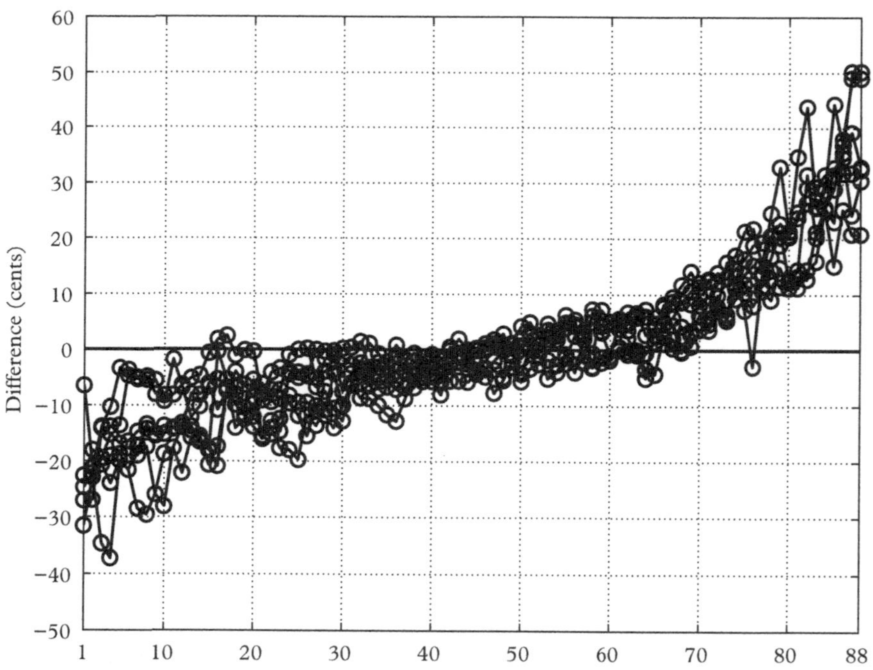

Figure 12.7.2 *Difference (in cents) between the fundamental frequency and the first partial for the six pianos in the Railsback format*

12.8 Summary

There have been isolated references to inharmonicity and octave stretching in previous chapters. This chapter concentrated on six fairly diverse pianos and computed and compared the various metrics for inharmonicity. We started by simply plotting the partials versus the harmonics. Professor Giordano's measure was presented. Then we spent some time estimating the so-called inharmonicity coefficient for all 88 keys for the six pianos. Another measure is the difference between the observed first partial and the fundamental frequency as specified by equal temperament. This was presented as a complement to the Railsback curve.

13

Beating, Unisons, and Tuning

The line spectrum can be used to monitor and even assist in tuning a piano. In fact, for those like me, who do not have the experience or expertise to tune by ear, the line spectrum approach is a handy aid. We begin with spending a few moments on beating, since aural tuning is based on being able to analyze beating sounds resulting from striking two notes simultaneously. Next, we display some time domain and line spectrum plots associated with an amateurish attempt to tune a three-string unison. Because we are trying to discern differences in frequencies, a short discussion on the limits of detectability is presented next. A spectral analysis of the results of an intervallic tuning completes the chapter.

13.1 Beating

Consider the sum of two sine waves having frequencies of 440 Hz and 441 Hz. The difference in the two waves' frequencies is 1.0 Hz. These two sine waves when added together form periods of addition and cancellation and generate a beating wave having a period of 1.0 s. This is evident in Figure 13.1.1(a). The fine grid DFT line spectrum in Figure 13.1.1(b) is able to visually separate the two signals. The lobes in the fine grid line spectrum are a consequence of the boxcar window and are explained in Appendix 1.

13.2 Using Spectral Analysis to Follow a Tuning Exercise

With the exception of a handful of the lowest notes on the piano, most of the keys on a piano use two or three strings. The A4 note at 440 Hz has three strings. These three strings form a "unison" and they should be tuned to have the same frequency or an audible beating will result.

Figure 13.2.1 shows the time domain plot and a fine grid DFT of the A4 note on my Yamaha 43″ console.

There are three peaks in Figure 13.2.1(b), apparently one for each string. Of the three strings, Figure 13.2.1 shows that only one is satisfactorily close to the target frequency of 440 Hz—the worst one is about 3 Hz off. The time domain plot in Figure 13.2.1(a) shows the beating resulting from the presence of one or more frequencies. The sound of this note can be heard in the file A4YamCon0.mp3. Note that what follows is not an accepted method of tuning and no mutes were used.

Spectral Analysis of Musical Sounds with Emphasis on the Piano. First Edition.
David M. Koenig, with comments by Delwin D. Fandrich.
Published in 2015 by Oxford University Press.

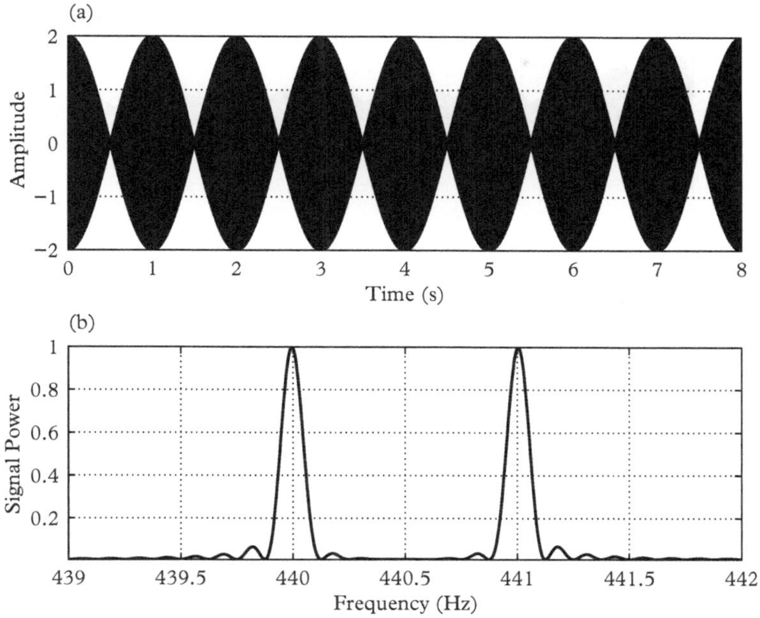

Figure 13.1.1 *A signal composed of two sine waves with frequencies of 440 Hz and 441 Hz: (a) time domain plot, and (b) fine grid line spectrum*

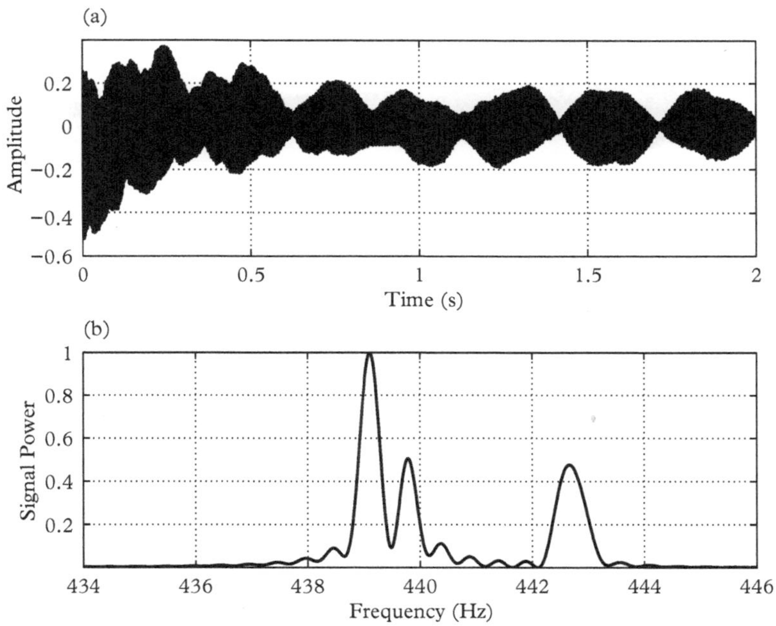

Figure 13.2.1 *First attempt at tuning the Yamaha console A4: (a) time domain plot, and (b) fine grid line spectrum*

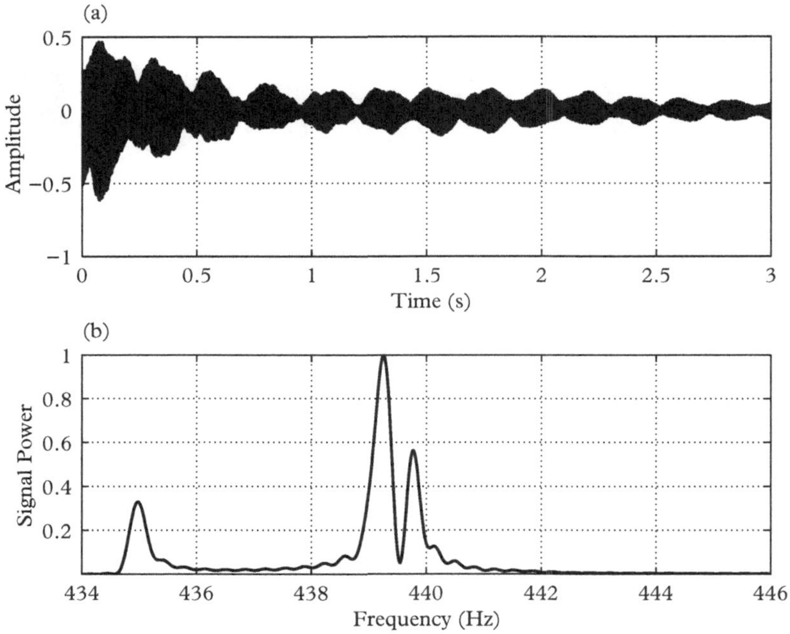

Figure 13.2.2 *Second attempt at tuning the Yamaha console A4: (a) time domain plot, and (b) fine grid line spectrum*

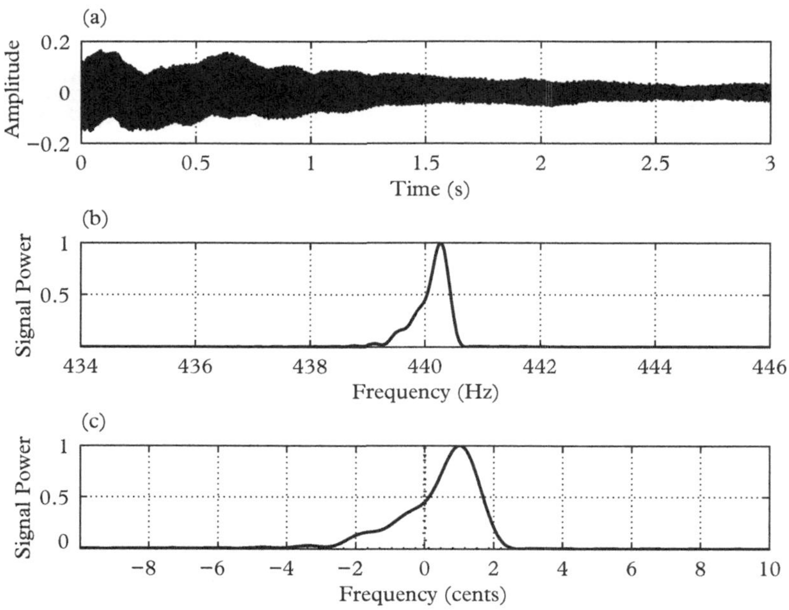

Figure 13.2.3 *Third attempt at tuning the Yamaha console A4: (a) time domain plot, (b) fine grid line spectrum in Hertz, and (c) fine grid line spectrum in cents*

First Adjustment

After some adjustments of the tuning pin for one of the three strings with a tuning hammer, I obtained Figure 13.2.2. Since I adjusted only one tuning pin, two of the peaks in Figure 13.2.2(b) remain unaffected while the one at 443 Hz moves over to 435 Hz.

The sound of the note can be heard in the file A4YamCon5.mp3. Clearly, I made the situation worse with my adjustments. The time domain plot is most telling. The difference in frequencies has generated a low-frequency beat that is audible on the recording.

After Many Adjustments

Finally, after several more adjustments I obtained the results shown in Figure 13.2.3, which indicate that all three strings are within a 5-cent window that is near the target frequency of 440 Hz. The sound of this note can be heard in the file A4YamConFinal.mp3. This exercise confirmed my respect for professional piano tuners.

13.3 Limits of Frequency Detectability

At this point it might be relevant to point out the inherent limits in the detectability of two waves of similar frequency by means of spectral analysis. In the following experiment we carry out a spectral

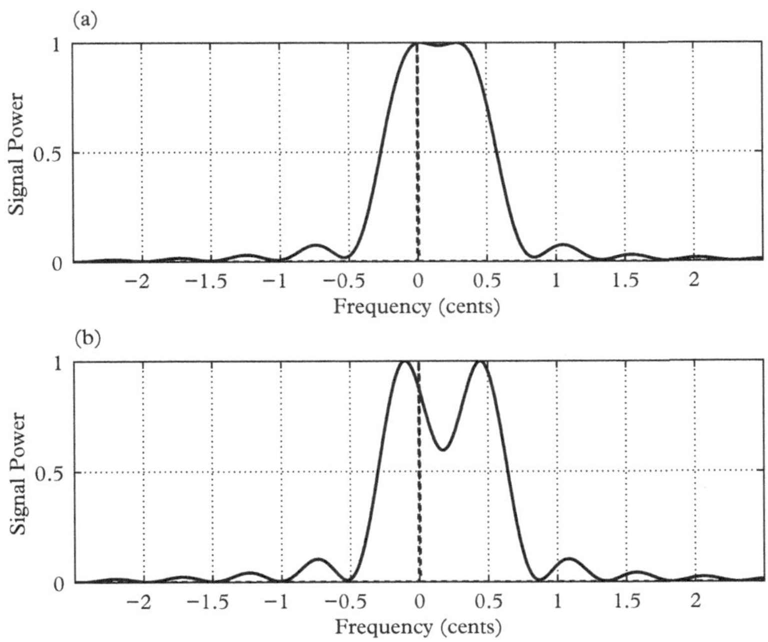

Figure 13.3.1 *Fine grid line spectra of waves containing two sines at (a) 440 Hz and 440.08 Hz, and (b) 440 Hz and 440.09 Hz, both with a duration of 8 s*

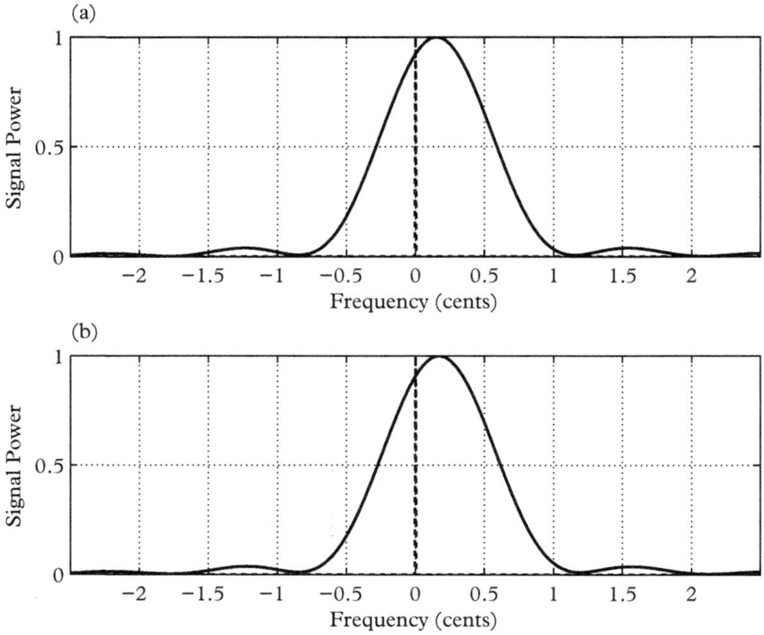

Figure 13.3.2 *Fine grid line spectra of waves containing two sines at (a) 440 Hz and 440.08 Hz, and (b) 440 Hz and 440.09 Hz, both with a duration of 4 s*

analysis of the sum of two waves having frequencies of 440.00 Hz and 440.08 Hz and a duration of 8 s. The difference in frequency is 0.305 cents (not Hz). We then repeat the experiment with a difference of 0.09 Hz, which is 0.343 cents. Figure 13.3.1 shows the results of the experiment.

The two components of this signal that are 0.305 cents apart cannot be identified—Figure 13.3.1(a). However, if the separation of the frequencies in the two components is 0.09 Hz (0.343 cents), the sum can be barely detected, as shown in Figure 13.3.1(b).

Note that this detection exercise was carried out under ideal conditions. The sample duration of 8.0 s with no decay is quite long, especially when a note like C8, shown in Figures 10.1.8 and 10.1.9, is considered.

When the same experiment is carried out but with a duration of 4.0 s and no decay, the results are different, as shown in Figure 13.3.2 where the fine grid DFT can only produce a single peak that is approximately 0.2 cents displaced from 440 Hz. The reader can find more on this subject in Appendix 1.

13.4 A Spectral Analysis of Intervallic Tuning

Based on the analyses presented earlier in the book, the reader should not be surprised that for any two notes on the piano there are likely to be several near coincidences between the partials of those notes.

This section will use spectral methods to study intervallic tuning of the E4 note using the E3 note, where E3 has already been tuned by other means. E3 is an example of a *bass note* that is an octave below the *treble note*, which in this case is E4. The tuning approach is based on the beating that results when the bass note and a third note, called the *test* note, are struck simultaneously as compared to the beating that results when the treble note and the same test note are struck simultaneously.

The tuning algorithm is therefore:

```
Bass note + test note → beating frequency #1
Treble note + test note → beating frequency #2
Tuning = OK if beating frequency #1 "equals" beating frequency #2
```

The beating will manifest itself as a vibrato about a frequency near a coincident partial; that is, at a frequency where a harmonic of the test note and one of either the bass note or the treble note are close. Because the treble note is a harmonic of the bass note, the beating rate should be the same in both cases if the treble note is tuned properly. In practice the treble note may be a bit sharp relative to the exact harmonic of the bass note because of inharmonicity.

Using the First Coincident Partial

The *test* note is C2 (65.4 Hz fundamental frequency). Let E3 (164.89 Hz fundamental frequency) be the *bass* note, which is specified to be already tuned. The *treble* note is E4 and its tuning is to be checked. The first coincidence of partials for C2 and E3 occurs near the fifth harmonic of C2 (327.0 Hz) and the second harmonic of E3 (329.6 Hz).

Table 13.4.1 shows the harmonics for C2, E3, and E4. Note the coincidences at the fifth harmonic of C2 and the second harmonic of E3. Also, note the coincidence at the first harmonic of E4, the second harmonic of E3, and the fifth harmonic of C2.

Table 13.4.1

Test Harmonic Number C2	Note Frequency for C2	Bass Harmonic Number E3	Note Frequency for E3	Treble Harmonic Number E4	Note Frequency for E4
1	65.4064				
2	130.8128				
		1	164.8128		
3	196.2192				
4	261.6256				
5	327.032	2	329.6256	1	329.6256
6	392.4384				
7	457.8448				
		3	494.4384		
				2	659.2551

It turns out that the treble note is also the first coincident partial for C2 and E3. The coincidence of the partials will not be perfect due to octave stretching or inharmonicity and because of the equal temperament scale system. This slight difference will result in a beating signal at or near E4 when C2 and E3 are struck simultaneously. The beating tone will sound like a vibrato about E4.

As mentioned above, to use this information for tuning by intervals one assumes that E3, the bass note, is already properly tuned by some other means. We will attempt to tune E4 by comparing the simultaneous striking of C2 and E4 with that of C2 and E3. We would strive to make the beating nearly the same in both cases. That is, when E4 is in tune, the beat frequency from striking C2 and E3 will be essentially the same (but not slower) as that resulting from striking C2 and E4. This method does not require that C2 (the test note) be properly tuned.

To illustrate this approach I used felt and rubber mutes to allow only one of the two C2 strings to vibrate and only one each of the three E3 and E4 strings to vibrate on my Yamaha 43″ console piano.

A broad look at the result of striking C2 and E3 simultaneously is shown in Figure 13.4.1.

For the C2 key, the first partial is labeled as "C2," the second partial as "C3," the third partial as "G3," the fourth partial as "C4," the fifth partial as "E4," and the sixth partial as "G4." The seventh partial of C2 does not line up with any of the key note frequencies. The eighth and ninth partials of C2 line up with C5 and D5, respectively.

There are two groupings in Figure 13.4.1. The notes E3, E4, and B4 are related as harmonics of E3. The notes C2, C3, G3, C4, E4, G4, and C5 are related as harmonics of C2. The coincidence occurs at E4.

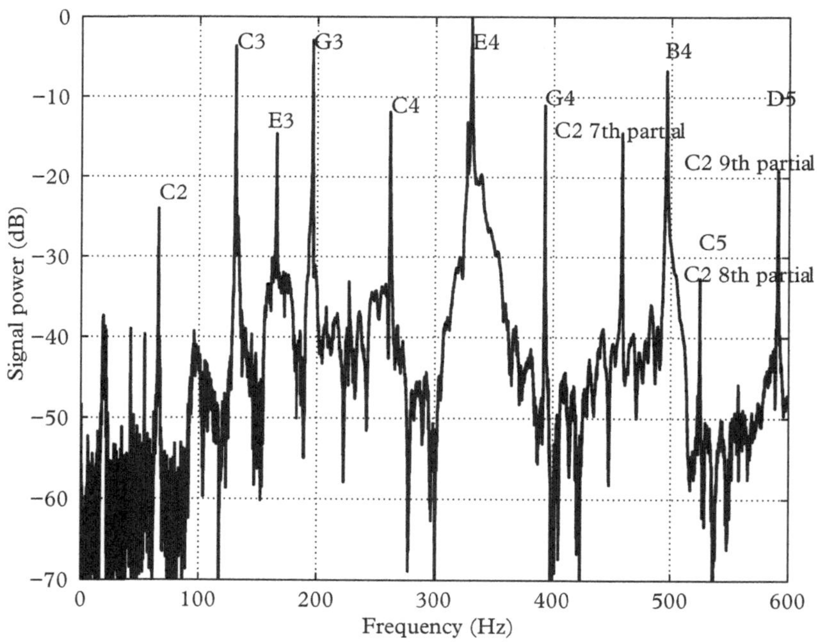

Figure 13.4.1 *Line spectrum of C2 and E3 when struck simultaneously*

Therefore, the first partial coincidence occurs near the peak labeled "E4" at 329 Hz. Because of the plotting range of frequencies shown in Figure 13.4.1 one cannot see the splitting of the peaks. This will be remedied below where we will take close-up looks at C2 by itself, E3 by itself, and C2 and E3 together. The fine grid DFT line spectrum will be centered about 329 Hz, which is near the first coincident partial of C2 and E3.

Figure 13.4.2 shows fine grid spectra for C2 by itself at its fundamental frequency and E3 by itself at its fundamental frequency. The vertical dotted lines represent the fundamental frequency location.

Figure 13.4.3(a) shows the spectrum of C2 at its fifth harmonic, which will be the coincidence location. The vertical dotted line denotes the location of this harmonic, namely 329.6 Hz. This spectrum is for the case when C2 is struck alone.

Figure 13.4.3(b) shows the spectrum of E3 at its second harmonic located near 329.6 Hz, which is denoted by the dotted vertical line. The fact that the peak is located at a frequency greater than that of the second harmonic is a consequence of inharmonicity. This spectrum is for the case when E3 is struck alone.

Figure 13.4.3(c) shows the spectrum for the simultaneous striking of C2 and E3 in the neighborhood of the coincidence location of 329.6 Hz, which again is denoted by the vertical dotted line. Both the fifth harmonic of C2 and the second harmonic of E3 are present, with the difference of 2.7 Hz denoted by the horizontal dotted line. The strength of the fifth harmonic of C2 is significantly less than that of the second harmonic of E3 but that difference will generate a weak beating tone or vibrato-like signal near 329 Hz that varies approximately at three beats per second. To hear this beating signal takes some practice (but do try this at home).

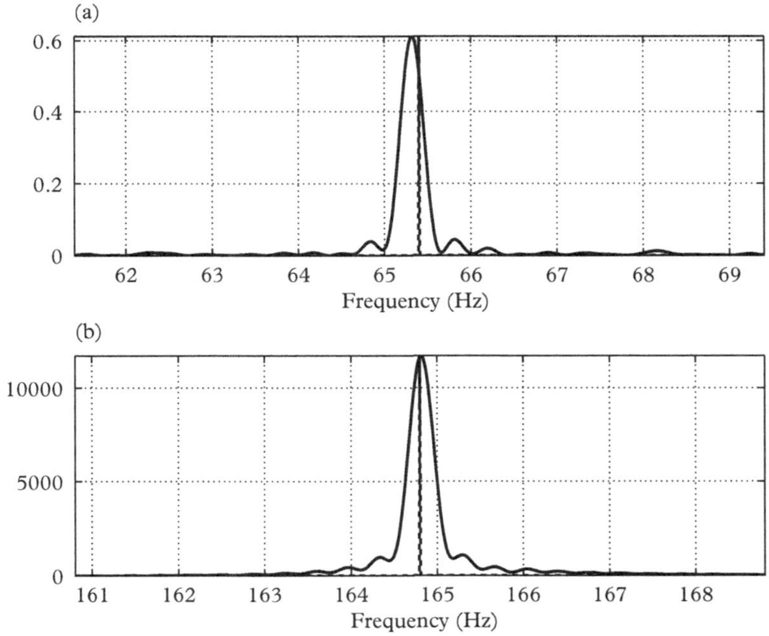

Figure 13.4.2 *Fine grid line spectra of (a) C2 by itself at 65.4 Hz, and (b) E3 by itself at 164.8 Hz*

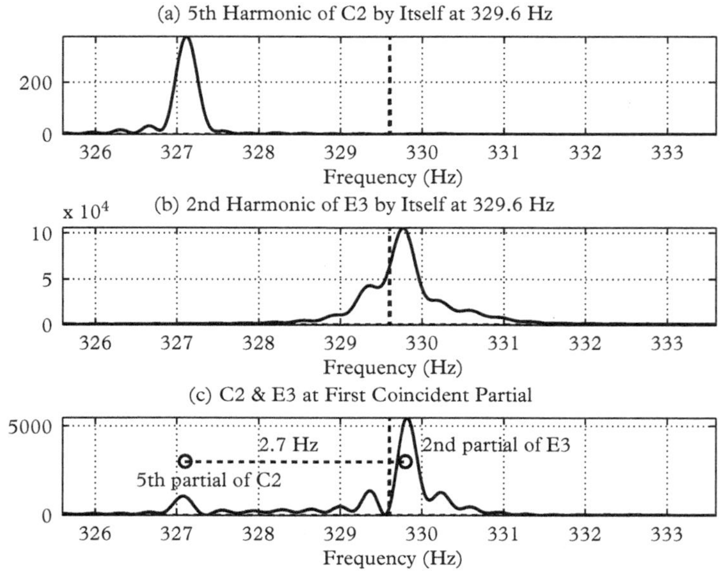

Figure 13.4.3 *C2 and E3: (a) fifth harmonic of C2 by itself, (b) second harmonic of E3 by itself, and (c) C2 and E3 together at the first coincident partial*

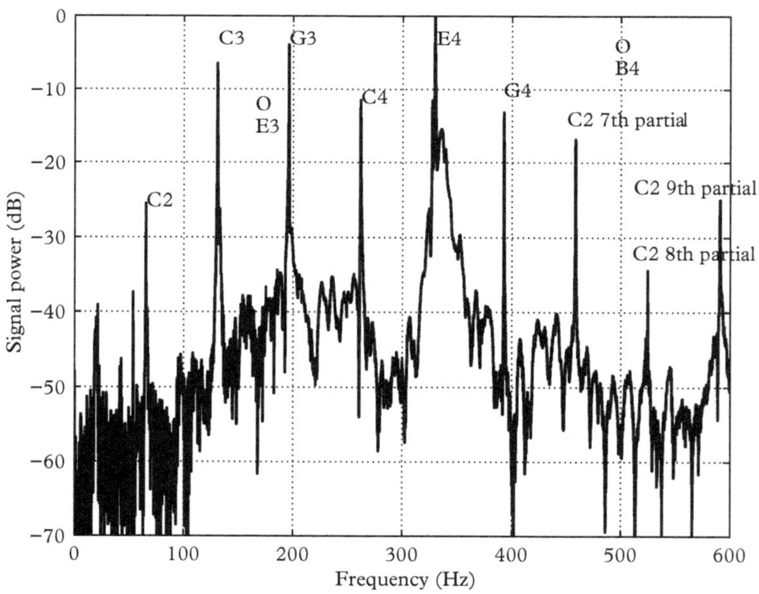

Figure 13.4.4 *Line spectrum of C2 and E4 when struck simultaneously*

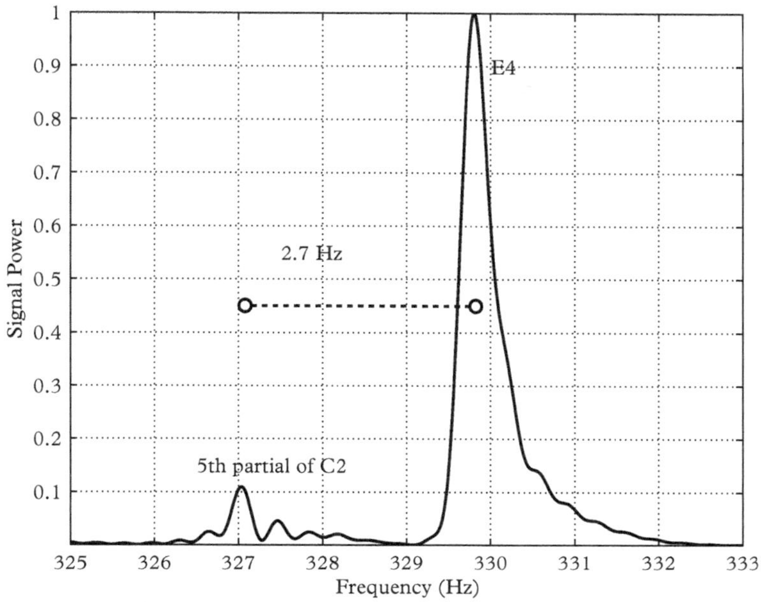

Figure 13.4.5 *Fine grid line spectrum of C2 and E4 at first coincident partial*

Now, move on to the analysis of C2 and E4. A broad view of the spectrum of C2 and E4 when struck simultaneously is shown in Figure 13.4.4.

Figure 13.4.4 is similar to Figure 13.4.1 except that peaks near E3 (165 Hz) and B4 (494 Hz) are not present because they are partials of E3, which has not been struck. These two missing peaks are denoted by small circles.

The fine grid DFT spectrum for C2 and E4 (when struck simultaneously) centered around 329 Hz is shown in Figure 13.4.5. This graph is larger so as to give a more detailed picture of the separation.

The fifth partial of C2 is weaker than the first partial of E4 by approximately 21 dB. The difference in frequencies is approximately 2.7 Hz. Therefore, striking C2 and E4 simultaneously will generate a weak vibrato-like signal with a base near 329 Hz that has approximately three beats per second. Since the beat rates are effectively the same, this suggests that E4 is properly tuned (if one assumes that E3 is properly tuned).

13.5 Summary

When two sound waves of slightly different frequencies are activated simultaneously, a beating signal having a frequency equal to the difference is generated. This phenomenon was illustrated for a 440 Hz and a 441 Hz signal. This idea of beating was then applied to tuning the A4 note on a Yamaha console. Spectral analysis and time domain plots showed three events during the tuning process. The impact of sample length on the ability to detect two signals of slightly different frequency was demonstrated by using the line spectrum. Finally, spectral analysis was used to illustrate the features of intervallic tuning.

13.6 Exercises

1. Download the Matlab script file Chorus.m from the book's website. One user should change the three frequencies and see if another user can determine how much one of the frequencies has to be off from 440 Hz to make the tone sound flat or sharp and see if you can tell whether it is flat or not. For example, one might choose 440, 440, and 440.5, and see if the other player can tell that one of the unisons is sharp.
2. Modify the script in the file Chorus.m by commenting some lines and uncommenting others so that there are 20 sources (as in a chorus) with tones each randomly distributed about 440 Hz. How does the distribution affect the sound, the autocorrelation, and the spectrum?

14

Two-Dimensional Graphical Metrics

Spectral analysis provides a wealth of information that can be manipulated into a variety of forms for gaining insight. The emphasis in this book has been on graphical means and will continue to be so in this chapter. However fancy the graphical methods are, there are some basic questions that underpin our efforts, namely, how is the sound energy distributed over the frequency domain differently for different instruments? What are the differences between instruments? What is the difference between a Knabe and a Bösendorfer piano? Are European pianos, as a class, different from American pianos?

For example, in Larry Fine's *The Piano Book*, fourth edition, p. 43, there is the following comment (emphasis added):

> Some European pianos, for instance, *emphasize the fundamental tone* rather than the harmonics, and probably because of their style of construction, have a tone that sustains at a lower volume. This results in a much less powerful sound that Steinway enthusiasts would likely find dull, but one which has been traditional in Europe since Steinway was founded. Many Asian pianos have a bright, brittle tone, also short on sustain, which is often preferred by jazz pianists for its crispness.

This begs the question: is there a metric that can be used to address this contention? As we go through the many metrics in this chapter keep this question in mind.

We have already compared the six pianos with regard to the inharmonicity coefficient in Chapter 12. The metrics in this section will be applied to those six pianos.

The metric that is the primary subject of this chapter, the spectral centroid, was introduced in Chapter 2 and is two dimensional in that it will be plotted against the piano key number. Here it will be used to compare pianos, especially the American versus the European ones. It will also be used to evaluate the effect of key striking force.

14.1 A Variety of Spectral Centroids

The spectral centroid was defined in Chapter 2 as a weighted average of the frequencies in a spectrum of a wave. There are a variety of ways to define that average.

The simplest is to carry out the average over *all* of the frequencies from zero to the Nyquist frequency. We will call this average the "unrestricted spectral centroid," or SpC_u. This range

Spectral Analysis of Musical Sounds with Emphasis on the Piano. First Edition.
David M. Koenig, with comments by Delwin D. Fandrich.
Published in 2015 by Oxford University Press.

of frequencies is extreme because the lowest note on the piano has a fundamental frequency of 27.5 Hz and, since the sampling frequency is 44,100 Hz, the highest frequency will be the Nyquist frequency at 22,050 Hz. However, there is not much sound power above 10,000 Hz for even the highest-frequency keys.

A second spectral centroid would be a combination of the SpC_u and the fundamental frequency, and it will be defined as the "spectral centroid minus the fundamental," or SpC_{-f}. It will be useful in measuring the degree to which the spectral centroid drifts above and below the fundamental frequency as the key number varies. We will use this in our comparison of pianos.

A third spectral centroid would be the unrestricted spectral centroid, which is denoted as SpC_N, and is SpC_u divided or normalized by the fundamental frequency (the subscript N denotes the spectral centroid being normalized):

$$SpC_N = \frac{SpC_u}{f_{fund}}.$$

It is useful to plot the logarithm of this ratio versus key or note number. This metric will complement the cumulative line spectrum in Chapter 19.

So, in summary, here are the three spectral centroids:

1. The unrestricted spectral centroid (SpC_u). The average is taken over the entire frequency range from 0.0 Hz to the Nyquist frequency. It will include the hammer knock power present for piano keys in the upper two octaves. It will therefore include any low note power below the fundamental.
2. The spectral centroid minus the fundamental (SpC_{-f}). This is the same as the SpC_u except that there is an offset due to the subtraction of the fundamental frequency.
3. A logarithmic ratio of the unrestricted spectral centroid to the fundamental frequency (SpC_N).

European pianos are said by Larry Fine to emphasize the fundamental. If that is the case then a plot of the spectral centroid SpC_{-f} versus key number for such a piano would show a curve closer to zero than a similar plot for American pianos.

Two plots of the spectral centroid SpC_u for the six pianos are given in Figure 14.1.1. The fundamental frequency (the circles) is included as a reference point in Figure 14.1.1(a). There is no attempt to separate the values for each piano. In Figure 14.1.1(a) the unrestricted spectral centroid is plotted against key numbers. In general, at the lower-numbered keys the spectral centroid is above the fundamental for all six pianos. In the middle region, say from key 45 to 70, there is a great deal of variability and it is difficult to discern significant differences, but all six pianos have a spectral centroid above the fundamental. At about key 70 the spectral centroid becomes more erratic and goes below the fundamental frequency due to the hammer knock. In other words, the average weighted frequency of the top 18 keys is less than the fundamental frequency.

In Figure 14.1.1(b) the ratio of the unrestricted spectral centroid to the fundamental frequency (SpC_N) is plotted on logarithmic scale against the key number. On average, there appears to be a linear decrease in the logarithm of the ratio until about key number 50 at which point the curve is essentially constant at unity. At key 75 the logarithm of the ratio decreases and becomes erratic due to hammer knock. Therefore, from this perspective something dramatic happens for all of the pianos at about key number 55 and again at about key number 75.

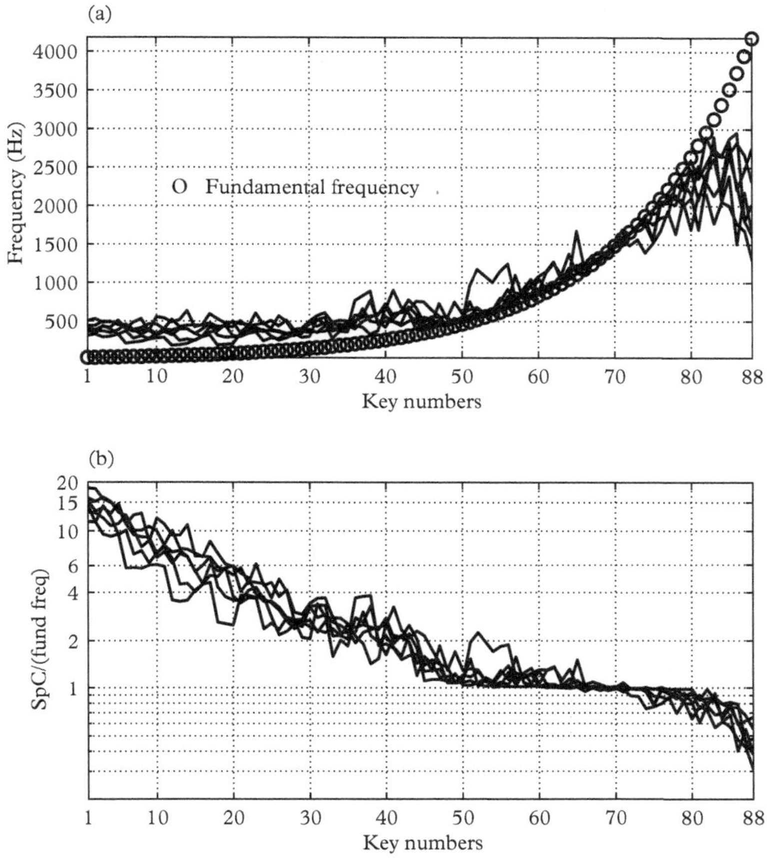

Figure 14.1.1 *Two ways to plot the spectral centroid: (a) unrestricted (along with the fundamental frequency), and (b) the ratio of the spectral centroid to the fundamental frequency*

14.2 Bimodal Comparisons of American and European Pianos

Knabe (American) versus Bösendorfer (European)

To address these differences we will compare two pianos at a time. For the two pianos, in this case the Knabe and the Bösendorfer, the two spectral centroids SpC_{-f} (with the fundamental frequency subtracted) are plotted in Figure 14.2.1(a). Note that the spectral centroids are plotted in cents, and consequently the impact of hammer knock at the high-numbered keys is not as discernible. This is because at the high-numbered notes there are many Hertz between semitones compared to the low-numbered notes. Also, note that the spectral centroid trace for the Bösendorfer (the small circles) lies above that for the Knabe.

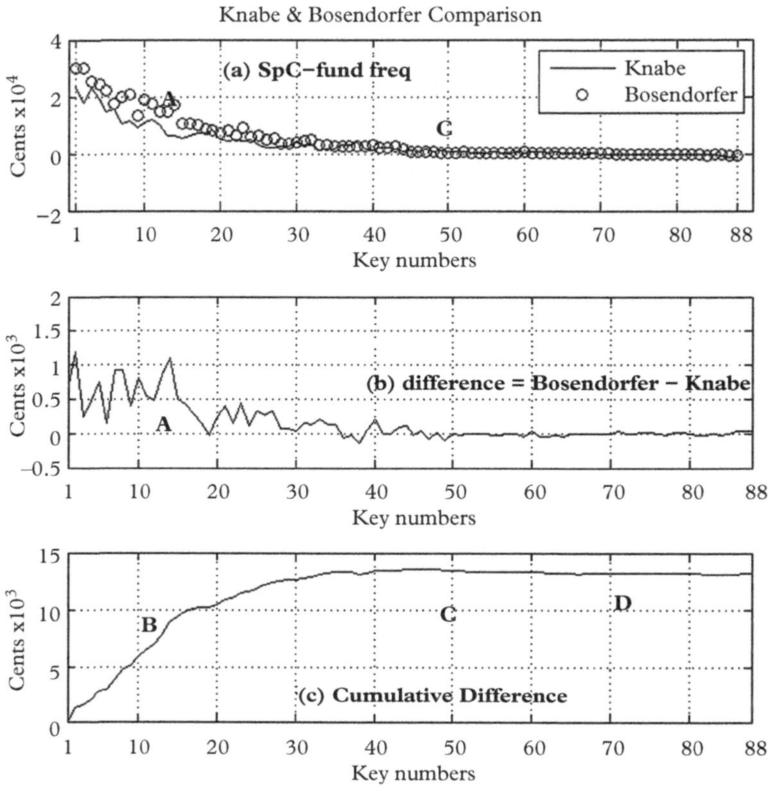

Figure 14.2.1 *Knabe and Bösendorfer comparison: (a) spectral centroid minus fundamental frequency for each, (b) difference of the spectral centroids, and (c) cumulative difference of the spectral centroids*

The difference (in cents) between the two spectral centroids is plotted against the key number in Figure 14.2.1(b). An alternative but complementary visualization is given by the cumulative sum of the difference between the spectral centroids plotted against the key number in Figure 14.2.1(c).

In Figure 14.2.1(a) it appears that for the lower-numbered keys, the Knabe curve lies closer to the fundamental frequency and below the spectral centroid of the Bösendorfer. Because the Bösendorfer spectral centroid is greater than that of the Knabe for most of the keys, the difference in Figure 14.2.1(b) is positive for most of the keys.

Figure 14.2.1(c) shows the cumulative difference. As an aside, the cumulative difference is an accumulation of the differences of two centroids just as the cumulative line spectrum is an accumulation of the line spectrum. The cumulative line spectrum tended to smooth local variations in the line spectrum and the same thing occurs with the cumulative difference. Appendix 3 deals with the cumulative difference in more detail. The cumulative difference curve rises for the lower keys, meaning that the difference is, *on average*, positive and that the Knabe emphasizes the higher harmonics more. By the 30th key the cumulative difference has leveled out, indicating that the difference of the spectral centroids is, on average, zero. For the higher notes the cumulative difference is flat, again indicating that the two pianos exhibit no difference.

Hence one might conclude that for lower-numbered keys, the Knabe emphasizes the fundamental more strongly than the Bösendorfer. On the other hand, for the keys in the middle and upper part of the pianos, the Knabe and Bösendorfer are effectively the same with regard to emphasis on fundamental and harmonic frequencies. Remember that this is a sample of one pairing of two quite dissimilar pianos so any attempts to draw general conclusions about American versus European pianos from this analysis is fraught with danger.

Knabe (American) versus Steinway (American)

The two American pianos, the Knabe (vintage 1927) and New York Steinway (1991) are closer with regard to the spectral centroid, as evidenced in Figure 14.2.2—note especially Figure 14.2.2(c). The Knabe versus Steinway cumulative difference is never above 3.0, whereas it is on the order of 13.0 for the Knabe versus Bösendorfer.

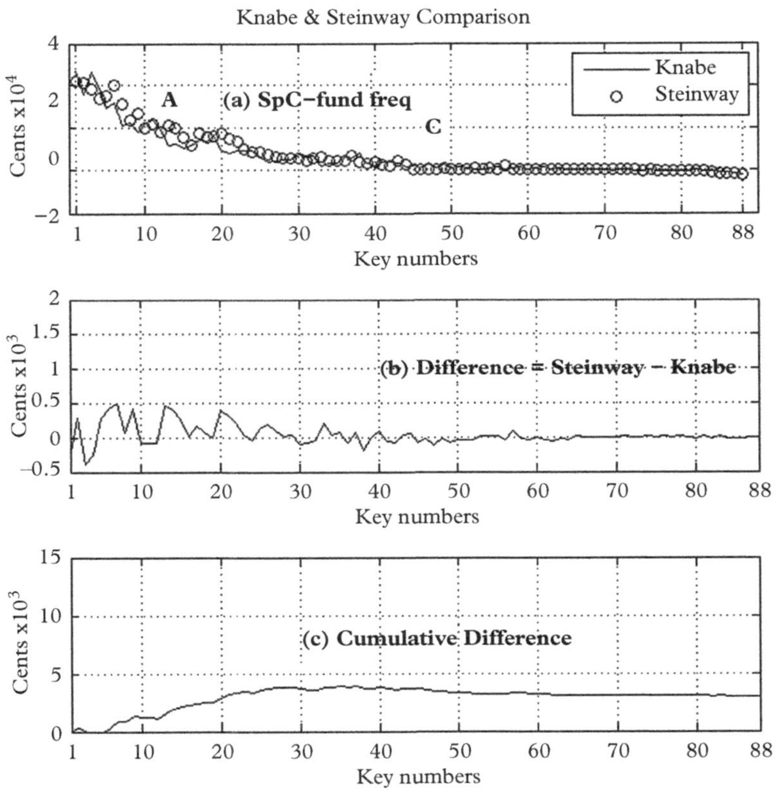

Figure 14.2.2 *Knabe and Steinway comparison: (a) spectral centroid minus fundamental frequency for each, (b) difference of the spectral centroids, and (c) cumulative difference of the spectral centroids*

Although we could look at 13 other comparisons of pairs from the six pianos we will leave it to the above two pairs and suggest that there *may* in fact be something to Larry Fine's assertion.

Are these comparisons conclusive? Not at all. The data come from single samples of one piano from different eras. Perhaps if I had access to ten pianos of each kind—for example, ten Steinway Model D's of the same vintage, ten Knabe 7′ 8″ grands from 1927, and ten Bösendorder 280's from 2010—and carried out the same tests, then I might be able to firmly stand behind any conclusions.

Having access to ten of any model of the same vintage would be an experimenter's dream (at least mine). For example, how would the ten Steinway Model D's compare? Would there be more variation *amongst* the Steinways than, say, between an average Steinway and an average Bösendorfer? How would the dealer preparation affect the distribution of the spectral centroid among the keys? One could go on . . .

14.3 The Effect of Strike Force

The foregoing analyses require striking each of the 88 keys on each piano by some repeatable means. One should question the impact of varying strike force. To see whether the inevitable variation in the strike force is significant, I manually struck the C2 and A4 keys with what I considered a range of forces from weak to strong. Figure 14.3.1 shows the time domain plots of the strikes.

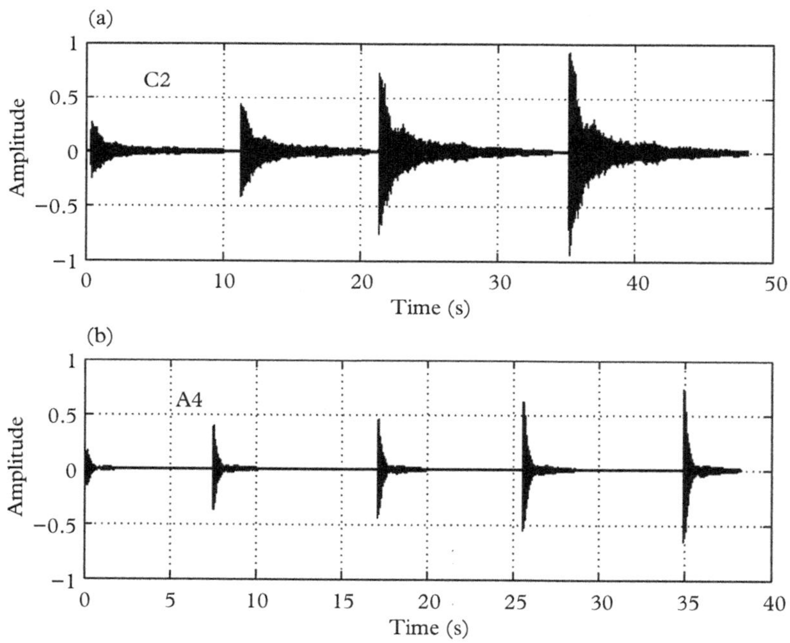

Figure 14.3.1 *Amplitudes resulting from strokes of different strengths: (a) C2, and (b) A4*

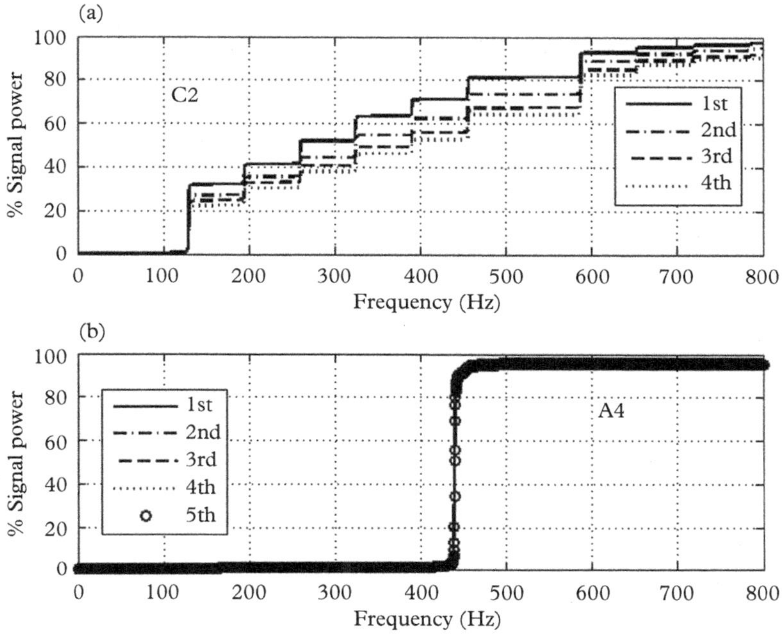

Figure 14.3.2 *Cumulative line spectra from strokes of different strengths: (a) C2, and (b) A4*

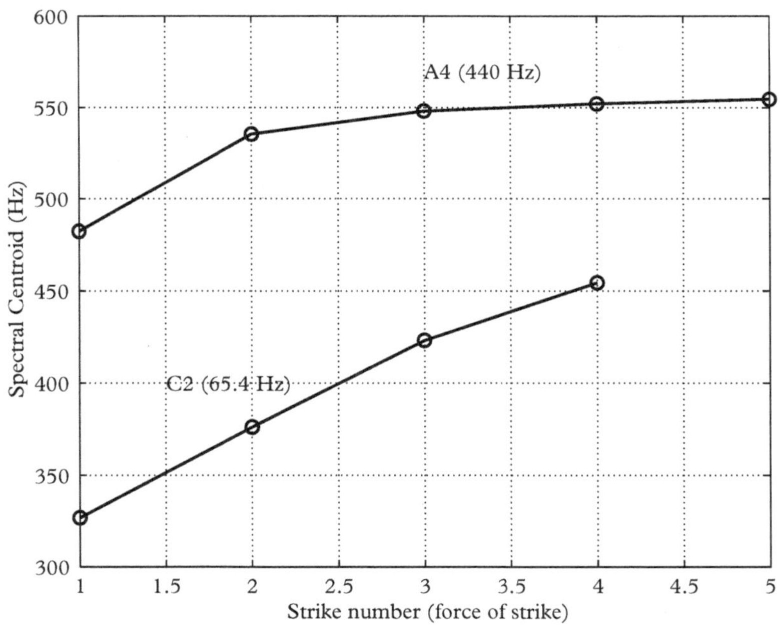

Figure 14.3.3 *Spectral centroids for strokes of different strengths for C2 and A4*

The cumulative line spectra of the strokes of the two notes is shown in Figure 14.3.2. In Figure 14.3.2(b) there appears to be but one significant partial and all of the strikes on A4 appear to have the same effect. In Figure 14.3.2(a), at the partial near C2's second harmonic (130.8 Hz), which is the first to have any power, the weakest strike (strike number 1) appears to have the most power, nearly 10% more than the strongest strike (strike number 4). The strongest strike has the most power near the ninth harmonic (587 Hz).

Figure 14.3.3 shows how increasing the strike force pushes the spectral centroid (the average weighted frequency) to higher frequencies. For the C2 key even the lightest strike still causes the spectral centroid to be significantly above the fundamental frequency because there is so much power in the higher harmonics. For the A4 key the spectral centroid is closer to the fundamental for all the different strikes. This means that although care was taken to maintain a consistent strike force there will be variability in the results. There will be more comments on the repeatability of these kinds of measurement in Chapter 18.

14.4 Summary

Three versions of the spectral centroid were presented. Two pairings of pianos were compared, using the spectral centroid. The Knabe and Bösendorfer appear to be significantly different, while the Knabe and Steinway are similar. The spectral centroid and the cumulative line spectrum were used as a metric for evaluating key strike force.

It appears that the spectral centroid can be a compact and useful metric for comparing pianos with regard to the distribution of the spectral power among the partial frequencies. It also appears that it can be useful in comparing the effect of striking force.

14.5 Exercise

Read Appendix 4 and check out the book's website to figure out how to use the scripts in the two files CollectAndSaveSyncTalk.m and AnalyzeSyncdDataOffline2.m (or CollectPianoNotes.m and AnalyzePianoNotes.m if using Octave). Use them to collect data for a piano of your choice. Use CalcSpecCenOct.m to construct a spectral centroid map. Refer to the book's website for further help.

15

Three-Dimensional Graphical Metrics

The cumulative line spectrum map is three-dimensional because the cumulative line spectrum (first dimension) is plotted against frequency (second dimension) for each of the piano's 88 keys (third dimension). There are other three-dimensional metrics. For example, the line spectrum and its dependence on frequency can be plotted against key number. Likewise, to get a feel for the variation in the sustain times, the envelopes of the time domain traces and their dependence on time can be plotted against key number.

15.1 The Time–Keys–Amplitude Envelope Map

A plot of the wave versus time versus key number would be an indecipherable blur because the wave consists of, at minimum, approximately 30 oscillations per second. Therefore, an envelope of the positive part of the Knabe A0 wave is constructed as shown in Figure 15.1.1.

If this is done for all 88 keys on the Yamaha console and the Knabe grand, one gets Figure 15.1.2.

The reader can see the key number versus the time axes in the plane of the page. At an angle out of the page one can see the axis for the amplitude envelope. Therefore, this is a plot of the amplitude envelope versus time for all 88 keys. Key number 1 is A0 while key number 88 is C8. To aid in the visual presentation, the order of the keys is reversed from that on a piano.

Figures 15.1.2(a) and (b) also show that there is unevenness in the response among the keys—one key's envelope might fade quickly while its adjacent neighbor might have a significantly longer time constant. There are several envelopes that do not have the exponential decay but a "squiggly" shape perhaps indicative of energy moving back and forth between the two transverse dimensions or the result of poor tuning of the unisons.

Finally, Figures 15.1.2(a) and (b) show that the notes in the upper treble region die out quickly, well within one second, and the amplitudes are significantly smaller. A closer look at the higher-numbered keys for the Knabe grand is given in Figure 15.1.3.

This data could perhaps be used as a diagnostic tool or a quality control tool. For an ideal piano, would you want the envelope curves for each key to be more consistent?

Cut-Off Times

The cut-off time is the time required for the envelope of the wave to decrease below 10% of its maximum value. Beyond this time the wave starts to consist more and more of background noise. Therefore, a spectrum should probably be based on that portion of the wave before the cut-off

Spectral Analysis of Musical Sounds with Emphasis on the Piano. First Edition.
David M. Koenig, with comments by Delwin D. Fandrich. © David M. Koenig and Delwin D. Fandrich 2015.
Published in 2015 by Oxford University Press.

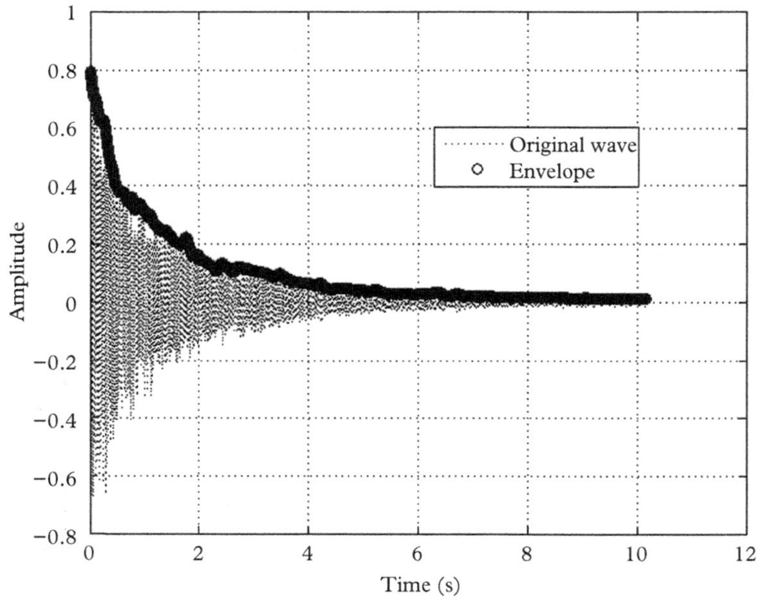

Figure 15.1.1 *A0 wave and its envelope for the Knabe grand piano*

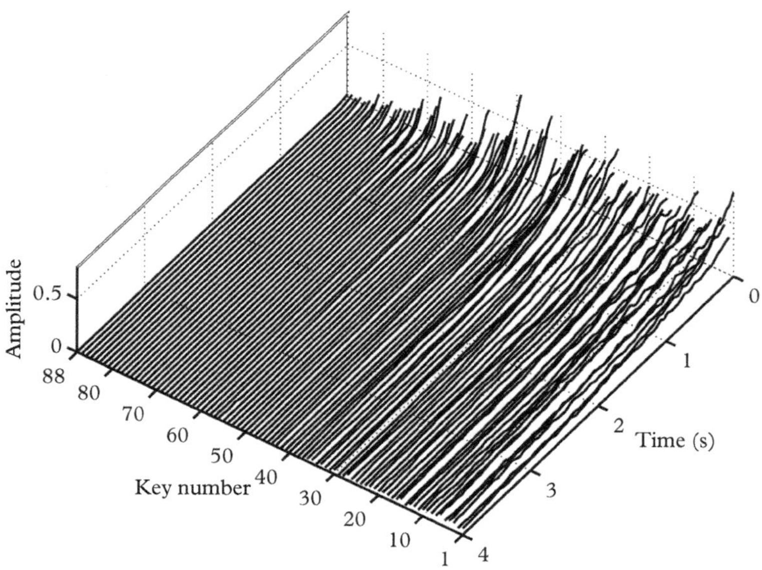

Figure 15.1.2(a) *Envelopes for the Knabe grand piano*

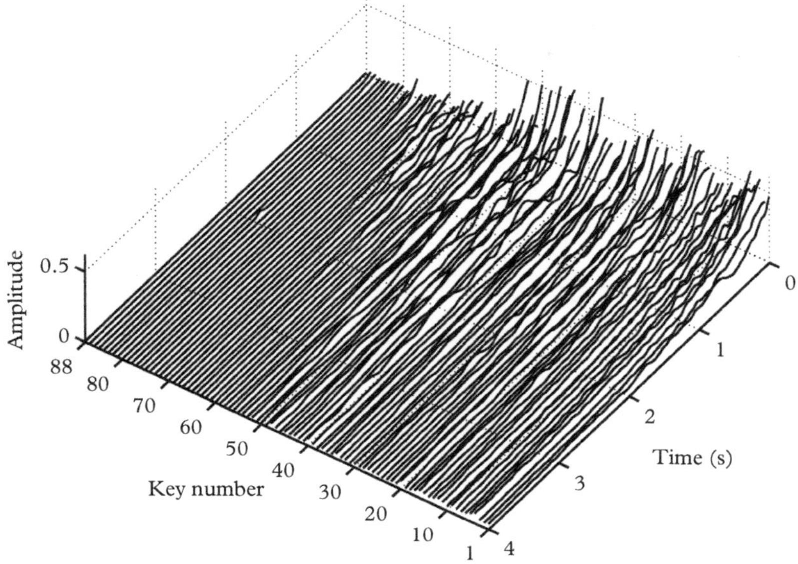

Figure 15.1.2(b) *Envelopes for the Yamaha console piano*

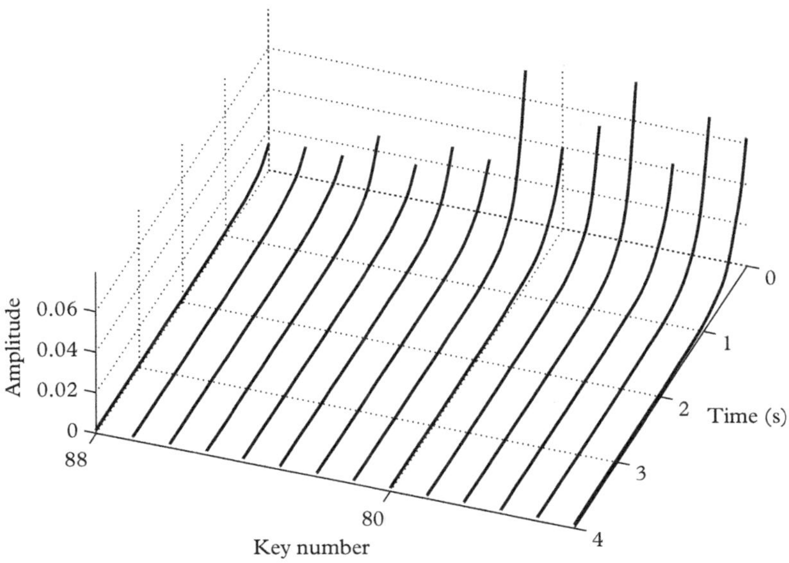

Figure 15.1.3 *Envelopes for B6 to C8 for Knabe grand*

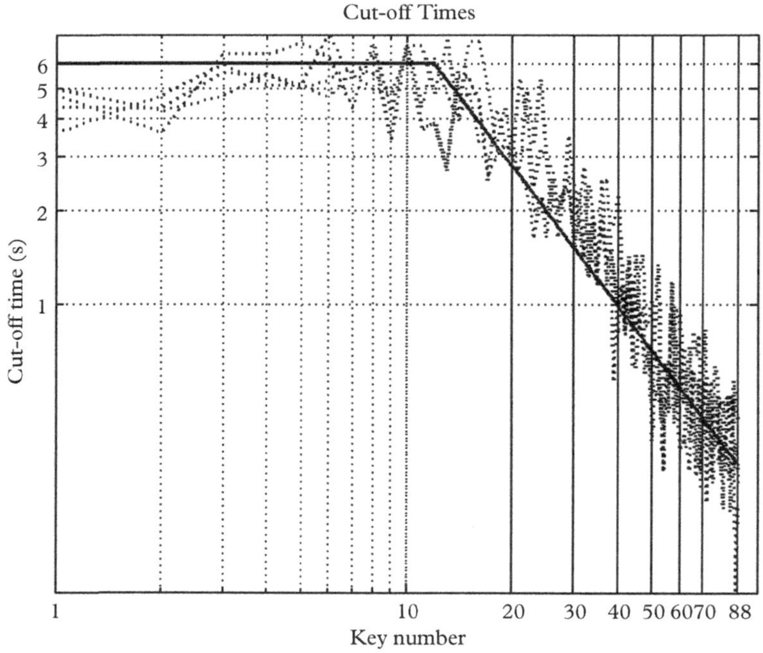

Figure 15.1.4 *Cut-off times for various pianos compared with the nominal cut-off time*

time. Figure 15.1.4 shows a plot of the cut-off times for several pianos as well as a plot of the so-called nominal cut-off time (the solid line) which appears to be a reasonable number for most pianos. Note that for the first 13 keys the nominal cut-off time is 6.0 s. For the 88th key the cut-off time is 0.3 s. In principle, this cut-off time should be applied to all of the key's waves. If, for example, all the waves were initially sampled over a period of 10 s, then, to make all the waves have the same length, the wave should be padded with zeros out to 6 s. Requiring all the waves to have the same length means that the resolution of the line and cumulative line spectra will be the same for each key. The reader can get more information on the benefits of padding from Appendix 1.

15.2 The Line Spectrum Map

Matthew Borland, in his master's thesis for the University of Waterloo, available at <http://www.uwspace.uwaterloo.ca/handle/10012/4302>, presents a plot of the line spectrum versus frequency versus key number. Figure 15.2.1 shows an alternative way to plot the same data for the Knabe piano where the line spectrum is plotted versus the key number and the logarithm of the frequency. Notice the diagonal ridges associated with the harmonics. Look at the ridge that starts at the 88th key and frequency 4186 Hz (look for point A) and runs across the page from right to left to the beginning of the first noticeable harmonic for the 15th key (this imaginary line flows from A to B). That line represents the ridge of the first harmonic (or the fundamental) frequency. The ridge for the second harmonic starts at the second harmonic of the C7 note (4186 Hz at key number 76 at

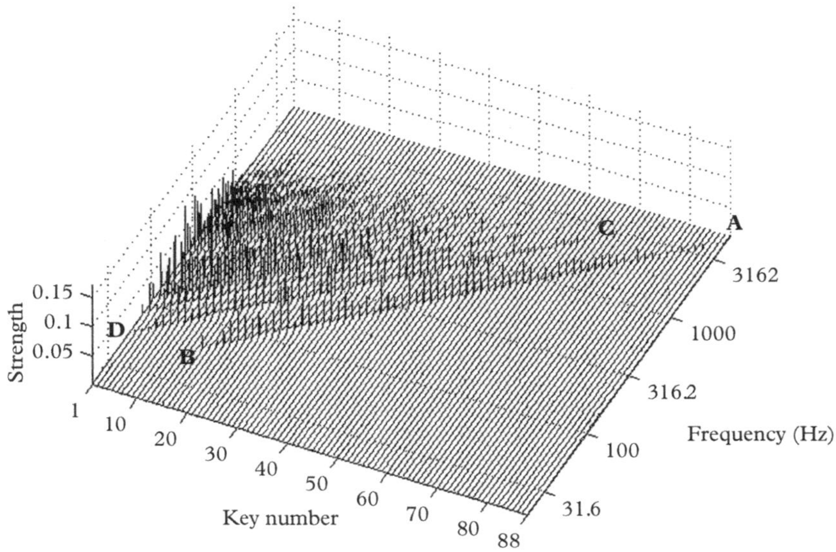

Figure 15.2.1 *Line spectrum map for the Knabe grand*

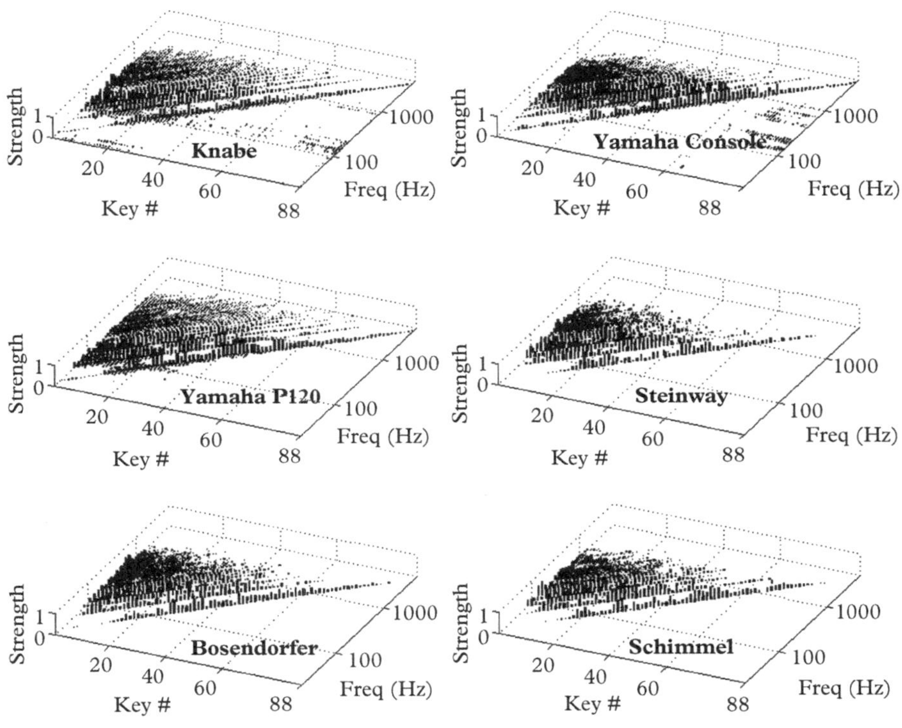

Figure 15.2.2 *Line spectrum maps for six pianos*

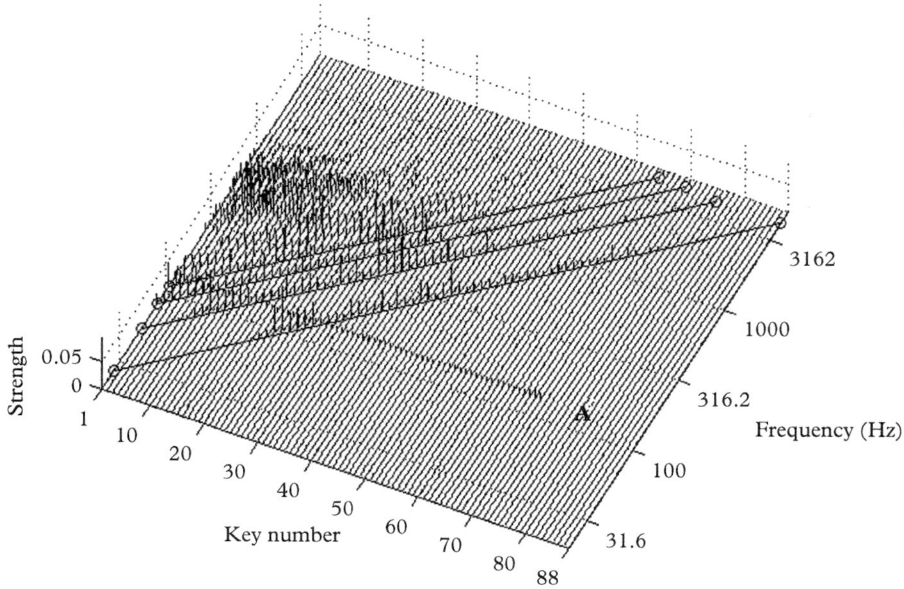

Figure 15.2.3 *Line spectrum map for a piano in a noisy environment (note the line at about 138 Hz denoted by A)*

point C) and runs from right to left to the second harmonic of the A0 note (55 Hz at key number 1 at point D). Although it is difficult to see, each harmonic has its own ridge. Several notes, especially the low-numbered ones, have no strength at the first harmonic. The straightness of the ridges is due to the logarithmic scale on the frequency axis.

The line spectrum maps for all six pianos are shown in Figure 15.2.2 where the Knabe, Yamaha console, and Yamaha P120 present a ridge for the first harmonic that extends across the whole key range while the corresponding ridge for the Steinway, Bösendorfer, and Schimmel is truncated at approximately the 11th key. Note that in all graphs the frequency is plotted logarithmically which allows for the straight lines.

The line spectrum map can sometimes be useful in warning the data analyst of troublesome background noise. Figure 15.2.3 shows the line spectrum map of a grand piano located in an office area in a manufacturing plant. One can see a strong component at approximately 130 Hz denoted by A.

In this case filtering like that done in Section 11.4 and Chapter 16 was required to remove the offending frequency before more analysis could be done.

15.3 Cumulative Line Spectrum Maps

The cumulative line spectrum map (CLSM) is to the line spectrum map as the cumulative line spectrum is to the line spectrum. For the reader who is uncomfortable with the idea of a cumulative line spectrum map here is an introduction.

An Introduction to CLSMs

Consider a hypothetical piano with just three notes: C1, G1, and A2. The time domain traces of the three notes are given in Figure 15.3.1.

The three line spectra are shown in Figure 15.3.2.

The three cumulative line spectra are shown in Figure 15.3.3.

The three curves in Figure 15.3.3 can also be plotted in a single three-dimensional graph as shown in Figure 15.3.4.

The reader has only to extend the hypothetical three-key piano to one with 88 keys and he/she will be able to understand the structure of the cumulative line spectrum piano map.

All Power in the Fundamental

Before we deal with the six pianos, consider two hypothetical pianos to get a feel for pathological extremes in the CLSM. The first hypothetical piano has all of its signal power at the fundamental frequency and has the CLSM shown in Figure 15.3.5.

Figure 15.3.5 is perhaps a strange-looking graph, but upon reflection it does make sense. The vertical axis gives the percentage of accumulated total power at each frequency. The axes in the plane of the page give the key numbers and the frequencies. Each key has all of its strength at its fundamental frequency, and all of the higher harmonics have no strength. Therefore, the cumulative line spectrum for each key is a single step of size 100% at the fundamental frequency for that

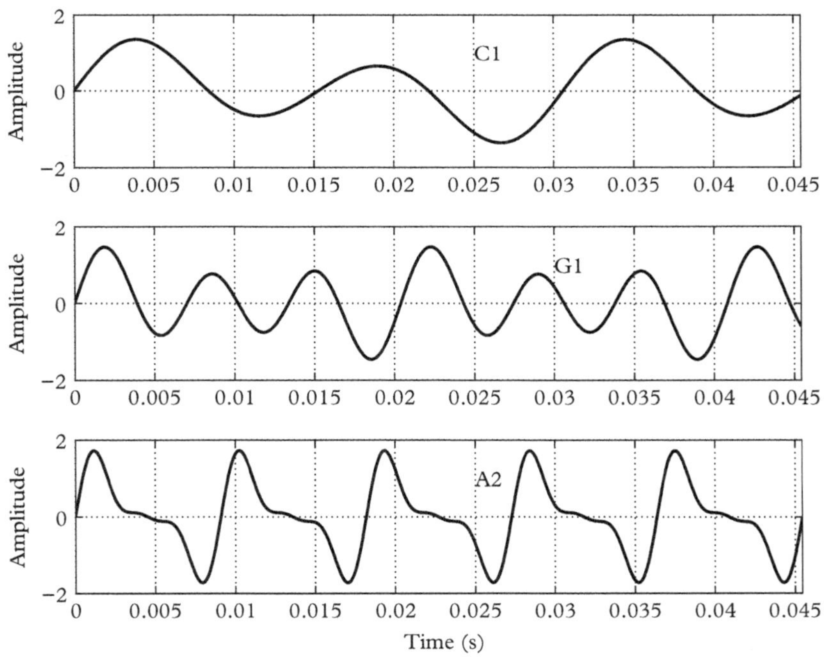

Figure 15.3.1 *Time domain plot of three hypothetical notes: C1, G1, and A2*

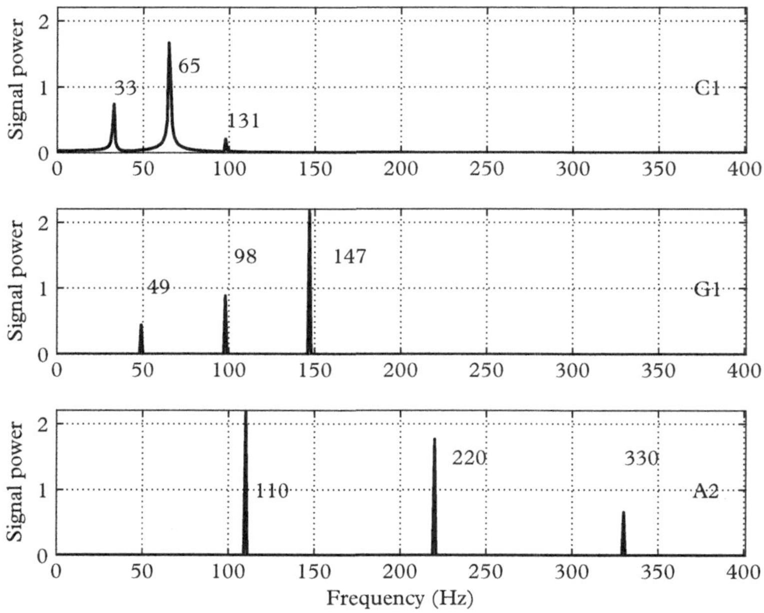

Figure 15.3.2 *Line spectra of three hypothetical notes: C1, G1, and A2*

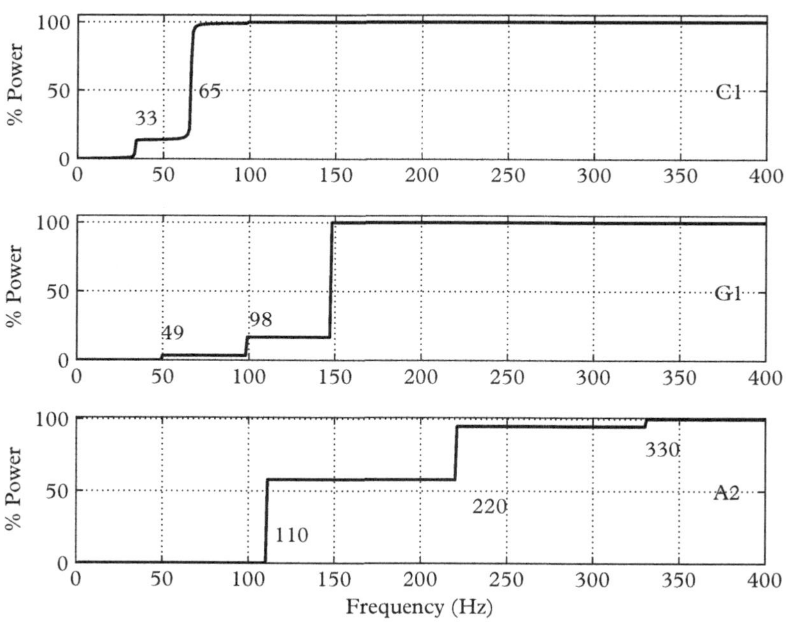

Figure 15.3.3 *Cumulative line spectra of three hypothetical notes: C1, G1, and A2*

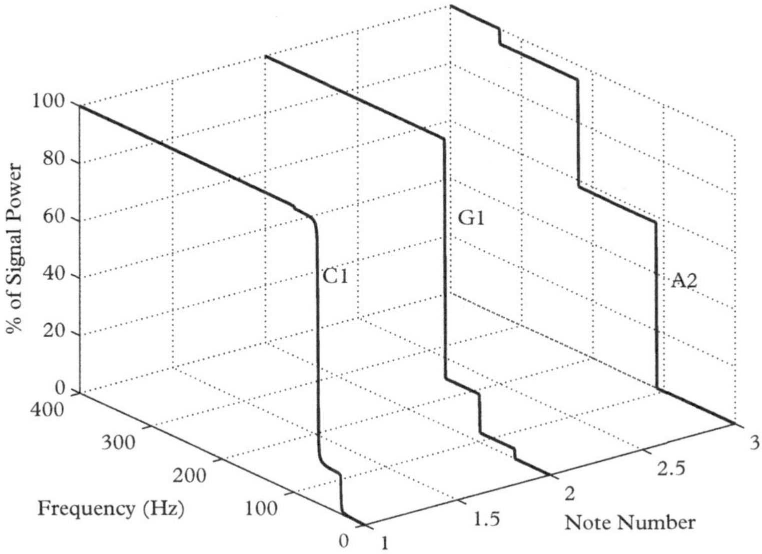

Figure 15.3.4 *Cumulative line spectrum map (CLSM) of the three hypothetical notes: C1, G1, and A2*

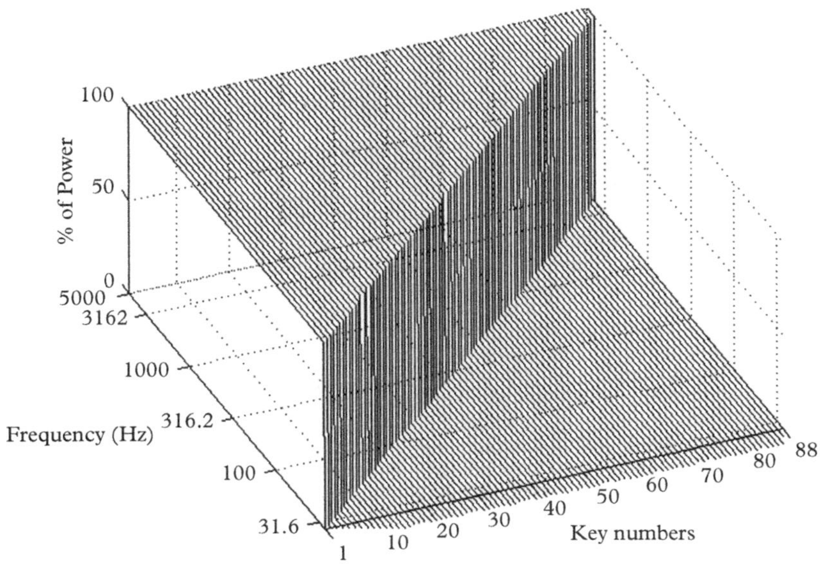

Figure 15.3.5 *CLSM of a hypothetical piano for which all keys have 100% of the power at the fundamental*

key. Because the fundamental frequencies of the keys increase exponentially I have chosen to plot the frequencies on a logarithmic scale, hence the straight line in the frequency–key number plane.

This piano would be described as one that emphasized the fundamental (to the extreme) over the higher harmonics (to put it mildly).

Dominant Power at the Fourth Harmonic

Consider the CLSM in Figure 15.3.6 where for each key 10% of the power resides at the fundamental and the second and third harmonics, while 70% resides at the fourth harmonic.

This piano would be described as one that emphasized the higher harmonics, especially the fourth. These two hypothetical pianos could serve as polar opposites and bring to mind the comments in Larry Fine's *The Piano Book* about how European pianos emphasize a different part of the spectrum compared to American pianos.

The Knabe Grand Piano

Figure 15.3.7 shows the cumulative line spectrum of all 88 notes for the Knabe grand. One axis in the plane of the page shows the 88 keys. The second axis in the plane of the page shows the frequencies plotted logarithmically. The vertical axis (out of the page) shows the accumulated signal power in percent. All values of the cumulative line spectrum that are less than 1.0% have been removed to enhance readability.

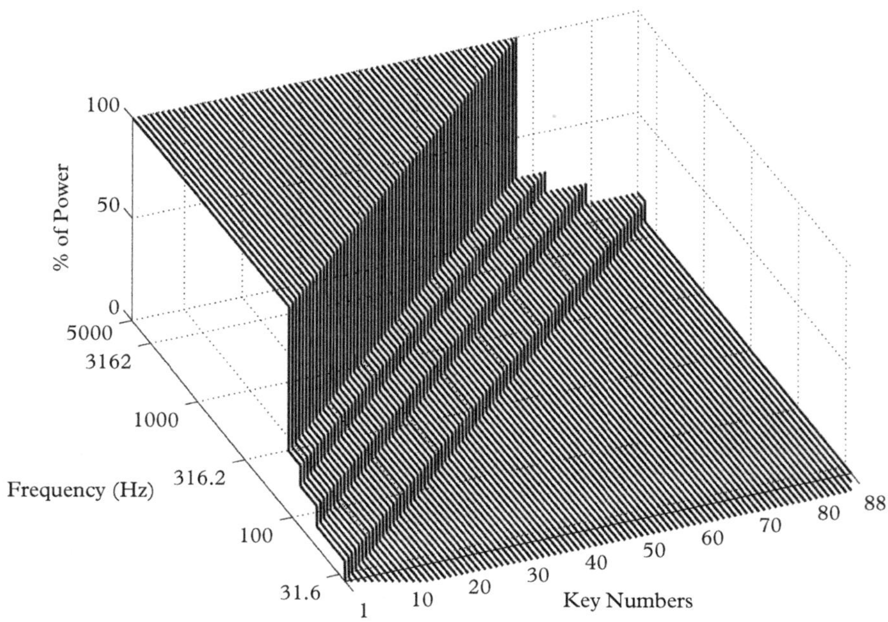

Figure 15.3.6 *CLSM of a hypothetical piano for which all keys have most of the power at the fourth harmonic*

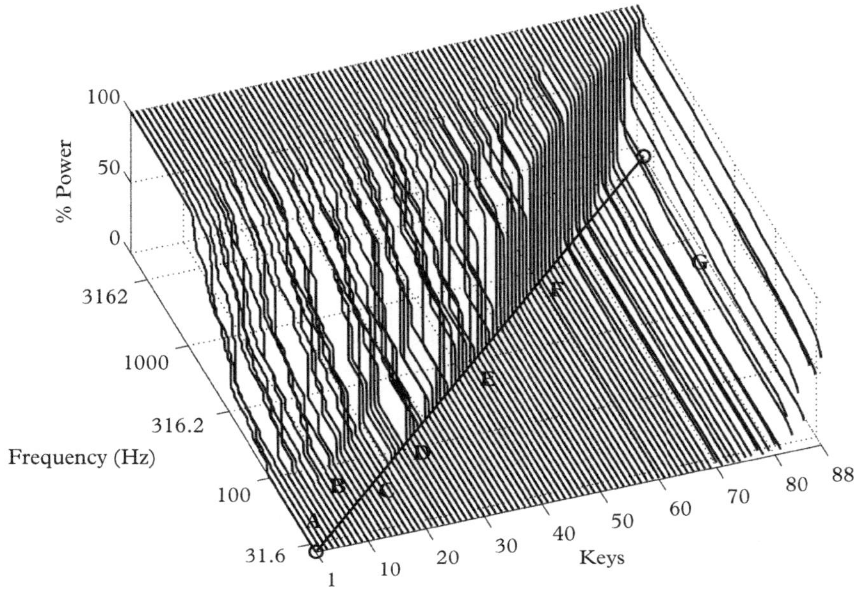

Figure 15.3.7 *CLSM of the Knabe grand piano*

The straight line in the plane of the page of Figure 15.3.7 with the small circles at each end represents the fundamental frequency for each key. For comparison purposes, Figure 15.3.5 showed that each key had 100% of its power appear at that frequency.

Returning to Figure 15.3.7, just above, and in parallel with, the fundamental frequency line is another straight line that is implied rather than drawn. It represents the second harmonic. Upon careful examination the reader can see a third straight line, also in parallel and also implied rather than drawn, representing the third harmonic. Upon even closer examination of the first three keys one can see the implied line for the fourth harmonic. Later on, Figure 15.3.8 will present a close-up of that portion and the lines will be drawn explicitly.

In Figure 15.3.7, the first three keys, A0, A0#, and B1, show no power until the fourth harmonic. The lowest eight keys denoted by A show no noticeable power until the third harmonic. The first eight keys have wrapped monochords. Starting at key nine, denoted by B, the strings are wrapped bichords. Some barely detectable power appears at the second harmonic at this point. At point C (key 17) in Figure 15.3.7 the strings are wrapped trichords (keys 17 to 24). Low but noticeable power appears at the fundamental frequency at this point.

At key 25 (point D) the bass section evolves into the tenor section. The stringing arrangement is different (no more wrapped strings) and there is a change in the nature of the graph at this point. The transition to the treble is denoted by point E (key 46). The upper treble section begins at point F (key 68). The keys beyond this point do not have dampers. There appears to be a slight change in the nature of the graph at this point. This transition point is consistent with that in the normalized spectral centroid plot of Figure 14.1.1.

The symbol G denotes the area where there is significant signal power below the fundamental. This portion of the signal power will be called "hammer knock" because it *likely* does not

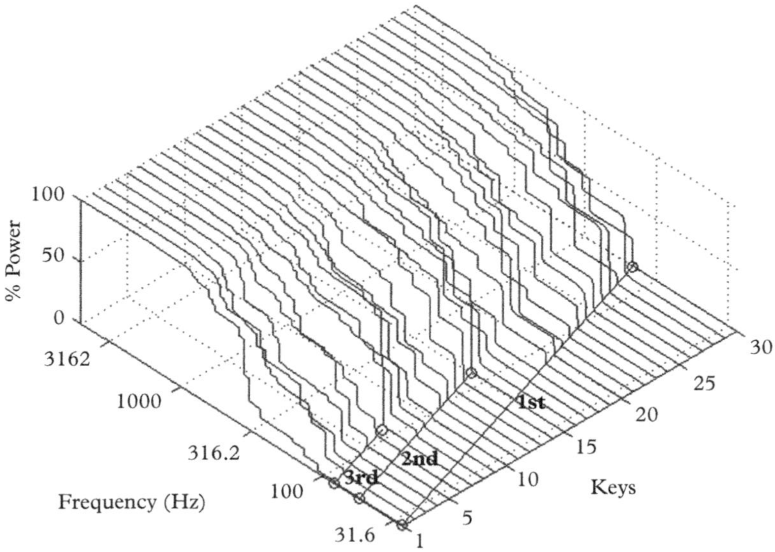

Figure 15.3.8 *CLSM of the Knabe grand piano's first 30 keys showing the absence of fundamental power in the lowest keys*

come from the hammer striking the strings but rather the key action mechanism. Note that the straight line, representing the fundamental frequency at each key, passes *under* the cumulative line spectrum traces although it may look as though it is on top of it in the figure.

One can therefore perceive a general trend: as the key number increases, more and more of the sound power moves toward the fundamental, and then beyond key 68 the power moves below the fundamental.

It is noteworthy that key 20 has noticeable signal power at a frequency lower than the keys numbered 1 through 8 because this group has no significant signal power at their fundamental frequency or even at the second harmonic frequency. That is, key 20, or E2, shows signal power at a frequency lower than the first frequency showing power for key 1, or A0.

Figure 15.3.8, which looks at only the first 30 keys, supports the comments of these last two paragraphs. It also shows the three parallel lines in the plane of the page representing the fundamental, the second harmonic, and the third harmonic frequencies.

That the "hammer knock" present in Figure 15.3.7 may be due to action noise is borne out by looking at the top two octaves of the Knabe using a piezoelectric soundboard transducer (Schatten EP-01) placed on the soundboard through one of the plate openings in the treble area instead of a Zoom H2 microphone. The transducer results are compared with the microphone results using the CLSM in Figure 15.3.9. Not only is the hammer knock greatly diminished but the spectral shapes are somewhat different. The latter part of that statement should not be too surprising, given that the spectral characteristics of the Zoom H2 microphone gathering the audible sounds is different from that using the Schatten transducer. The symbol A at the left-hand part of the graphs denotes the transition between the treble and upper treble sections.

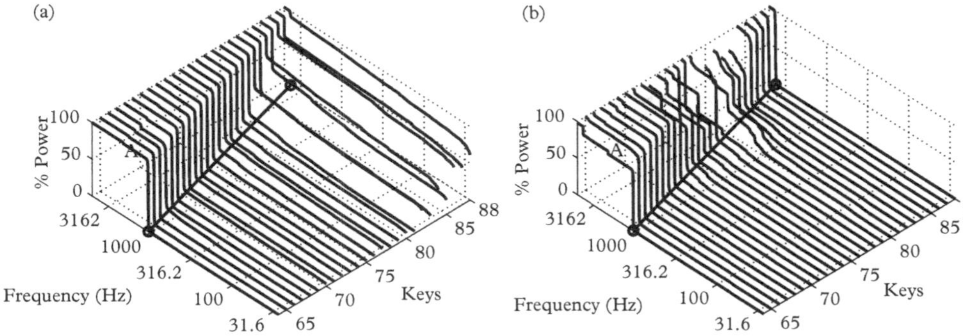

Figure 15.3.9 *CLSM of the Knabe's keys 64 to 88 using (a) an acoustic microphone and (b) a Schatten transducer*

CLSMs for the Six Pianos

The CLSM is shown for all six pianos in Figure 15.3.10.

Features common to most of the six pianos are:

1. Missing or relatively low power at the fundamental frequency for the low-numbered keys, usually those with wrapped strings.
2. Increasing contribution from the fundamental as the key number increases.
3. "Hammer knock" for the top octaves.

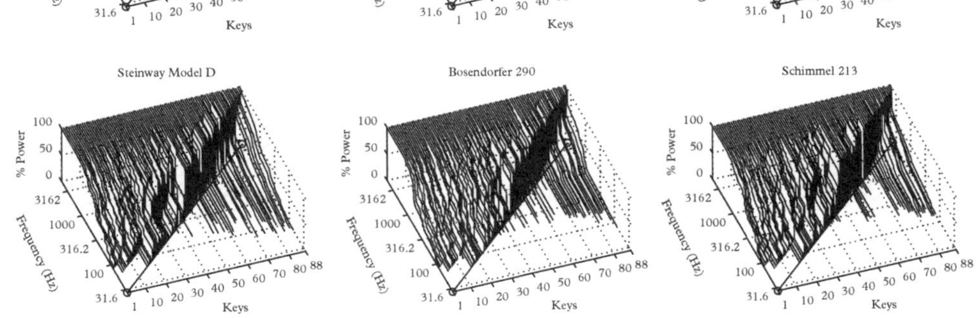

Figure 15.3.10 *CLSMs of six pianos*

The Yamaha console seems to have the most hammer knock while the Yamaha P120 and the Bosendörfer 280 appear to have less than the others.

A Closer Look at the Yamaha P120

Figure 15.3.11 presents the CLSM of the upper two octaves when using the Zoom H2 microphone and when using the line out connection on the P120 going directly to the computer.

When the Zoom H2 microphone is used the hammer knock is appreciable, perhaps even more than the Yamaha console. However, when using the line out connection it is less than any of the pianos. This suggests that hammer knock comes from the mechanical action of striking the key and the results are consistent with Figure 15.3.8 for the Knabe.

The Yamaha P120, being the only digital piano in the group, also has a different-looking CLSM for the lowest-numbered keys, as shown in Figure 15.3.12, which should be compared to Figure 15.3.8 for the Knabe. For the P120, the lowest five strings have a small but detectable power contribution at the second harmonic while the Knabe and the other pianos do not.

A Smoothed CLSM

Just for the sake of comparison, consider Figure 15.3.13, which shows a CLSM for a hypothetical piano based on the Knabe grand piano.

This "smoothed" CLSM has no hammer knock and the harmonic power is distributed from A0 to C8 in a manner somewhat similar to that for the Knabe, although significantly smoother. The sound of this "piano" compared to the Knabe is available in the exercises at the end of the chapter. One wonders what a real piano with these characteristics might sound like (and whether it could be made?).

Another View of the CLSM using Logarithmic Power Axes

Figure 15.3.7 shows the Knabe CLSM with linear axes for the power percentage. For this choice the low-numbered strings show no noticeable power in the first two partials. However, we know from line spectrum plots using decibels that there is power at those partials—they are just much

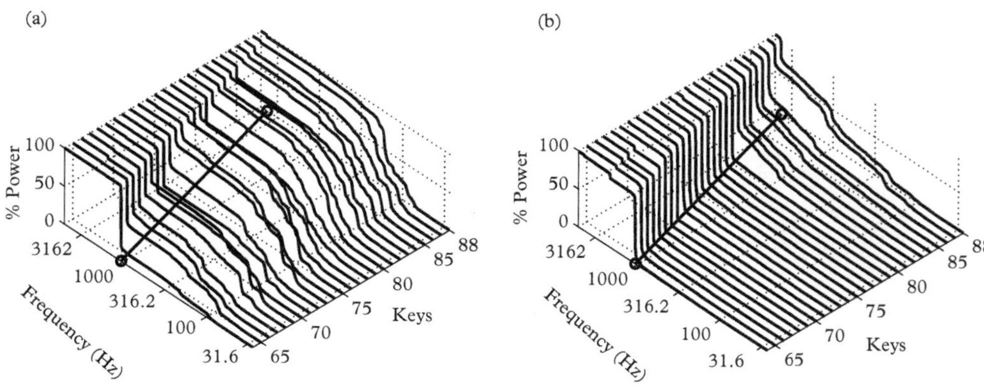

Figure 15.3.11 *CLSMs of the Yamaha P120 piano using (a) an acoustic microphone and (b) the "line out" connection on the piano to the "microphone in" connection on the computer*

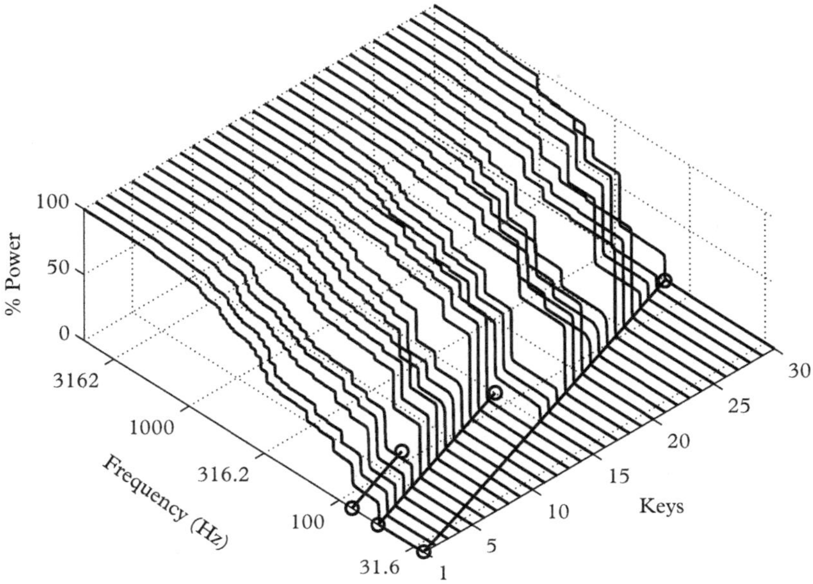

Figure 15.3.12 *CLSM of the Yamaha P120 piano keys 1 to 30 showing a slightly different distribution of power compared with the Knabe*

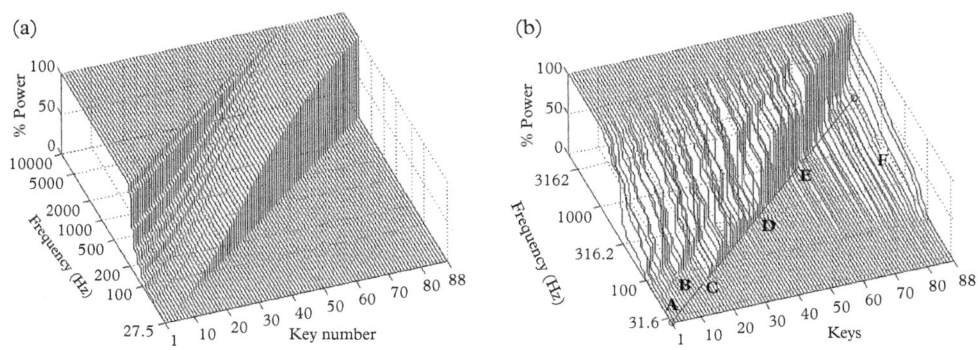

Figure 15.3.13 *(a) A hypothetical piano with a smoothed CLSM and (b) the Knabe CLSM*

lower than for the higher partials. If Figure 15.3.7 is re-plotted using a logarithmic axis for the power percentage, then the power for the first couple of partials can be seen, as shown in Figure 15.3.14. Now the transitions are not as distinct, striking, or dramatic. However, the logarithmic power scale shows that there is a gradual change in the amount of below-fundamental power as the key number increases. We will revisit this idea in Chapter 16 which deals extensively with hammer knock.

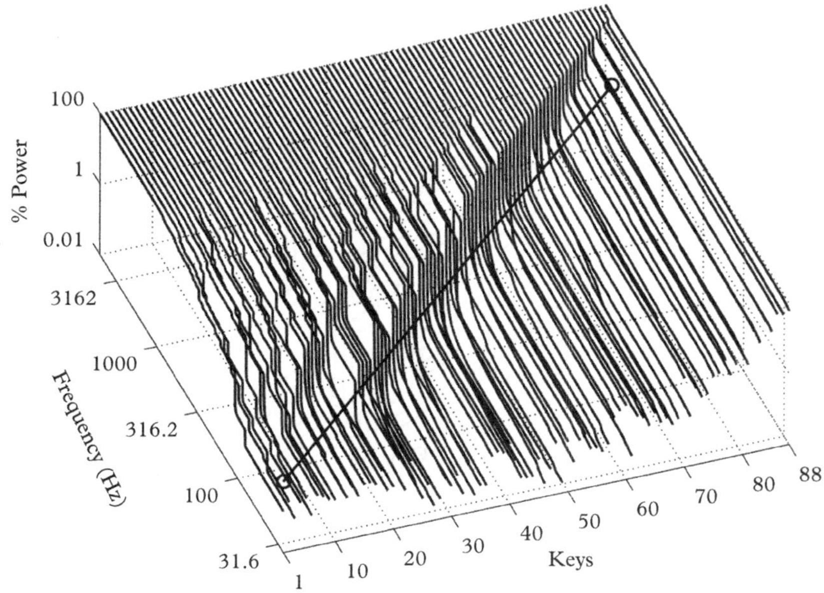

Figure 15.3.14 *CLSM of the Knabe piano with the power percentage plotted logarithmically to emphasize the low powers*

15.4 Summary

A third dimension was added to several of the graphs already discussed in the book. The first was the envelope map, which plotted the wave envelope for each key versus key number and time. This metric showed how the keys vary in response or in sustain time, how there are some transitions in the upper octaves, and that the highest octave has relatively little strength and sustain time—all in agreement with expectations.

The line spectrum map plotted the line spectrum versus the key number and frequency. It contained ridges associated with the harmonics and presented a measure of the relative strengths of the harmonics. A noise problem with one of the data gatherings was clearly revealed using this metric.

The cumulative line spectrum map (CLSM) was introduced and used to compare several pianos. The relative strengths of the harmonics were perhaps presented more clearly than in the line spectrum map. This metric suggested the presence of hammer knock and the weakness of the fundamental for the lowest-numbered keys. The CLSM was used to illustrate several hypothetical pianos. An alternative view where the percent accumulated power was plotted logarithmically was presented to show that the lowest keys do have some power at the fundamental.

15.5 Exercises

1. Using Matlab, run the script in the file PlayTheoAndKnabe.m and compare the sounds of the two pianos in Figure 15.3.13. The theoretical piano sounds artificial because it is using the Matlab *soundsc* function to generate the tones while the Knabe sounds a little stilted because it is using sampled tones rather than a recording of someone playing the tune on the Knabe. What could you do to make the theoretical piano sound more realistic?
2. Download Octave (free) and Audacity (free). Take a look at the YouTube video "Collecting Piano Data" (check the book's website for the link) and see if you can follow the instructions and collect an octave or two of piano notes starting at A0 and ending at B2. This will require you to download the files CollectPianoNotes.m and AnalyzePianoNotes.m from the book's website. If this is successful then you will have the notes in a directory specified by the script. Next, download the file ConstructCLSMOct.m and use it to construct a CLSM for the notes that you collected. Next, download the file StudyNote.m and use it to look at the time domain plot, the line spectrum, and the cumulative line spectrum for your choice of note. You may want to read Appendices 4 and 5 to see how some of the CLSMs in this book were done.

16

An Investigation into Hammer Knock and Key Striking

Hammer knock was discussed in Section 15.3 and there have been brief references to it in earlier parts of the book. The graphs presented so far suggest that the top octave from C7 to C8 exhibits the most hammer knock in that the cumulative line spectra show significant power below the fundamental.

The impact of the below-fundamental frequency power on the Knabe A7 key's sound will be studied by means of frequency domain filtering. Parts of the A7 spectrum will be removed and the sound of the remaining wave will be analyzed.

Since some of the data has been collected by dropping a felt-tipped brass rod on the keys, this method will be compared to that of pressing the finger down on the key. This complements the effect of strike force that was discussed in Section 14.3.

To determine if hammer knock is present in the lower octaves, the frequency and time domain characteristics of the C5 and A7 keys will be compared.

16.1 A Manual Keystroke with a Microphone as a Sensor

Figure 16.1.1 shows the line spectrum, cumulative line spectrum, and autocorrelation of the A7 note (3520 Hz fundamental frequency) resulting from a manual keystroke where the sound is collected using a Zoom H2 microphone mounted on the music desk. Note the splitting that occurs in the neighborhood of 7300 Hz. The sound of this wave can be heard in the file A7Unfiltered.mp3. Figure 16.1.1 (b) shows that approximately 40% of the wave's power occurs in the region below the first partial frequency at 3580 Hz. The first partial frequency of 3580 Hz is greater than the fundamental frequency because of the tuning designed to compensate for inharmonicity.

The conventional autocorrelation in Figure 16.1.1 (c) has a first peak at a time lag corresponding to 3675 Hz while the fine grid autocorrelation yields a peak at 3553 Hz, which is closer to the line spectrum peak at 3580 Hz.

Removing Power below the Fundamental

All of the power *below* 3300 Hz is removed, thereby keeping the first partial at about 3580 Hz and the higher partials near 7150 Hz and 7350 Hz. Therefore, we are removing the hammer knock part of the spectrum. The line spectrum is shown in Figure 16.1.2(a). The sound of this wave can

Spectral Analysis of Musical Sounds with Emphasis on the Piano. First Edition.
David M. Koenig, with comments by Delwin D. Fandrich.
Published in 2015 by Oxford University Press.

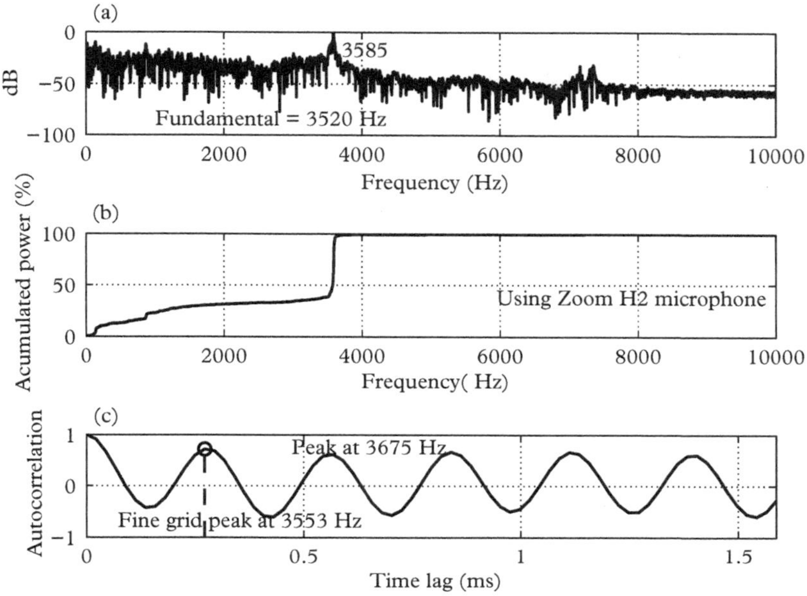

Figure 16.1.1 *Unfiltered Knabe A7 (using acoustic microphone): (a) line spectrum, (b) cumulative line spectrum, and (c) autocorrelation*

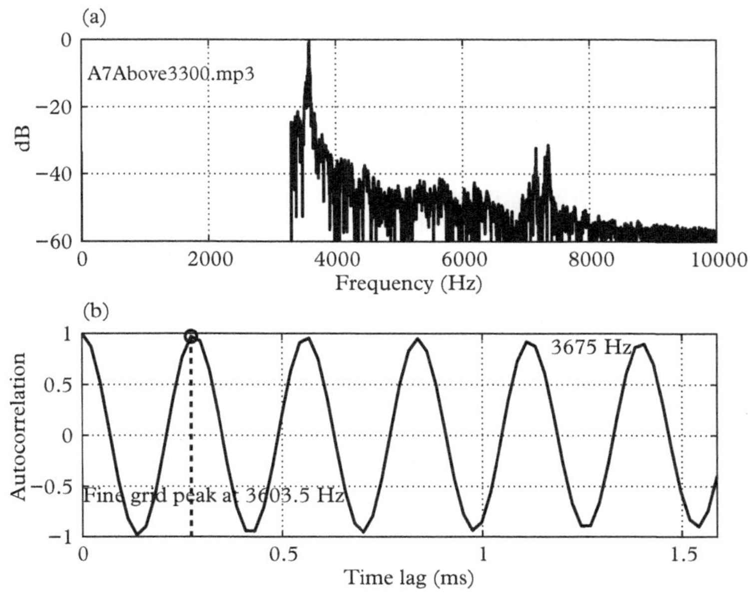

Figure 16.1.2 *Knabe A7 with power below 3300 Hz (hammer knock power) removed: (a) line spectrum, and (b) autocorrelation*

be heard in the file A7Above3000.mp3 and it is rather weak and strange, although according to the conventional autocorrelation in Figure 16.1.2(b) it does have the same pitch (3675 Hz) as the unfiltered wave. The fine grid autocorrelation indicates that the pitch is 3603 Hz.

Removing Power above the First Partial

All of the power *above* 4000 Hz is removed, leaving the first partial at 3580 Hz and the hammer knock power. Therefore, effectively, only the higher partials near 7150 Hz and 7350 Hz are being removed. The line spectrum is shown in Figure 16.1.3(a). The sound of this wave can be heard in the file A7Below4000.mp3, and it is effectively the same as the unfiltered A7 in A7Unfiltered.mp3, suggesting that the higher partials near 7150 Hz and 7350 Hz have little impact on the sound of the note. This is not too surprising since those peaks are approximately 35 dB down. The conventional autocorrelation in Figure 16.1.3(b) supports this contention, showing a peak at a time lag corresponding to 3675 Hz. However, the fine grid AC estimates the pitch at 3504 Hz.

Keeping Only the Hammer Knock Power

Finally, all of the power *above* 3300 Hz is removed, leaving just the below-fundamental hammer knock power. The line spectrum is shown in Figure 16.1.4(a). This sound can be heard in the file A7Below3000.mp3 and it is like a strike of a muted key, with little musical character. The autocorrelation in Figure 16.1.4(b) shows that the pitch of this sound is basically undefined, with a weak peak at a time lag corresponding to 1025 Hz, which agrees with the peak found using the fine grid autocorrelation.

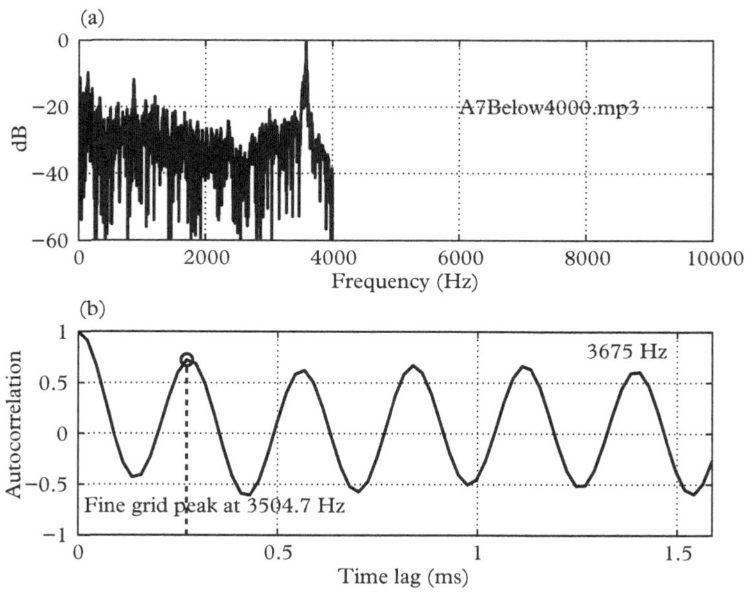

Figure 16.1.3 *Knabe A7 with power above 4000 Hz removed: (a) line spectrum, and (b) autocorrelation*

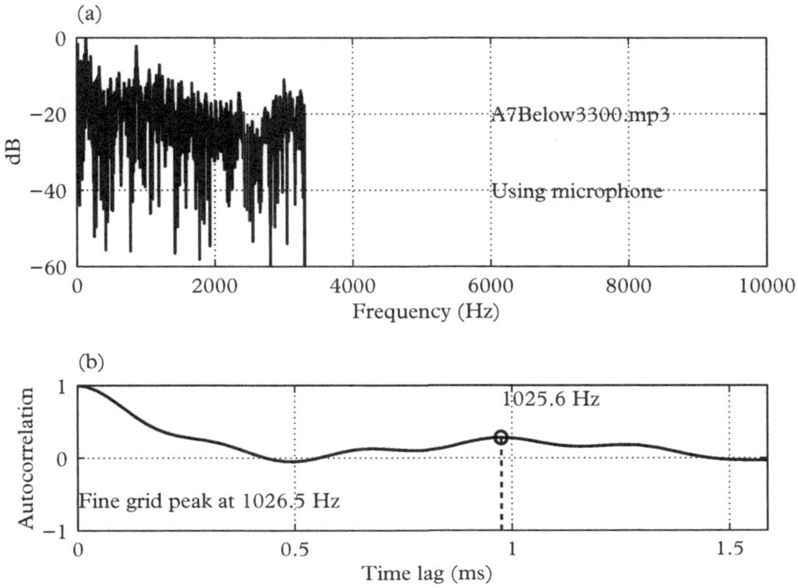

Figure 16.1.4 *Knabe A7 with power above 3300 Hz removed: (a) line spectrum, and (b) autocorrelation*

These spectra and autocorrelations, along with the mp3 files, suggest that the sound of the keys with the significant hammer knock, i.e., with significant power below the first partial, is dependent on that below-fundamental power. The pitches estimated by the autocorrelation are consistent with the mp3 files in that the first two experiments generate waves that have the same pitch. The below-fundamental or hammer knock wave has a pitch of 1025 Hz, although the autocorrelation "peak" is not particularly strong and therefore the pitch estimate is suspect. In fact, the line spectrum in Figure 16.1.4(a) looks quite flat, as in the case of band-limited white noise, so the lack of a tone is not surprising.

16.2 A Manual Keystroke with a Korg Contact Microphone as a Sensor

The above experiment is repeated, replacing the Zoom H2 microphone with a Korg CM-100L contact microphone attached to a soundboard rib in the treble area. The output of the Korg was fed to the "line in" port of the Zoom H2. The use of this pickup device was proposed by Ralph J. Onesti in *Piano Technician's Journal* (June 2013). The spectra for the A7 are shown in Figure 16.2.1(a) and (b). When compared to Figure 16.1.1 it appears that there is significantly more hammer knock power. When the Zoom H2 microphone is used, 40% of the power was attributed to hammer knock. When the Korg pickup is used almost 80% of the power is below the first partial. The autocorrelation in Figure 16.2.1(c) shows that the first peak points to a pitch of 3746 Hz but it is not the strongest peak, just the first one.

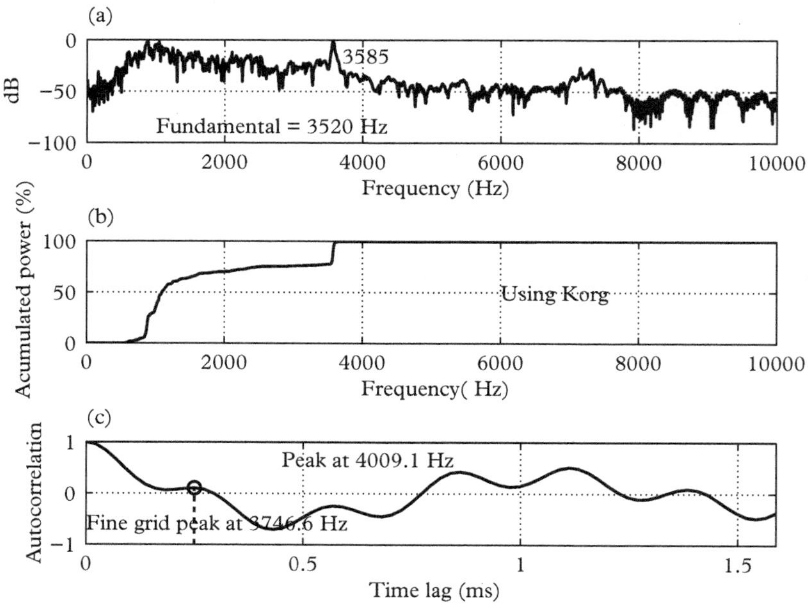

Figure 16.2.1 *Unfiltered Knabe A7 (using Korg transducer): (a) line spectrum, (b) cumulative line spectrum, and (c) autocorrelation*

16.3 A Manual Keystroke with the Schatten Soundboard Transducer

The same transducer (Schatten EP-01) referred to in Figure 15.3.9 is used to gather the sound of A7. The spectra are shown in Figure 16.3.1.

As was shown in Figure 15.3.9, Figure 16.3.1 exhibits effectively no hammer knock. The Schatten and Korg transducers are basically located at the same spot on the soundboard except that the Schatten transducer is on top of the soundboard surface and the Korg transducer is clipped to a rib on the underside of the soundboard. Why they produce such different spectra is a conundrum.

16.4 Looking at the Early Part of the Wave

There are opinions that the early part of the sound wave contains the influence of the hammer knock and also exhibits "chaotic behavior." We have already looked at the early part of the wave when the subject of longitudinal vibrations in wrapped bass strings was studied in Chapter 5, Section 5.11. In several chapters we used the spectrogram to look at the time dependence of the line spectrum. Here we take a slightly different approach and trim the A7 wave such that the early part of the wave before 20 ms is removed. Figure 16.4.1 shows time domain traces of the A7 key. Figure 16.4.1(a) shows the wave before any cutting, 16.4.1(b) shows the portion beyond the cut point at 0.020 s, and 16.4.1(c) shows the portion before the cut point. Figure 16.4.2 shows the line spectra for these three portions and supports the contention that there is no significant difference.

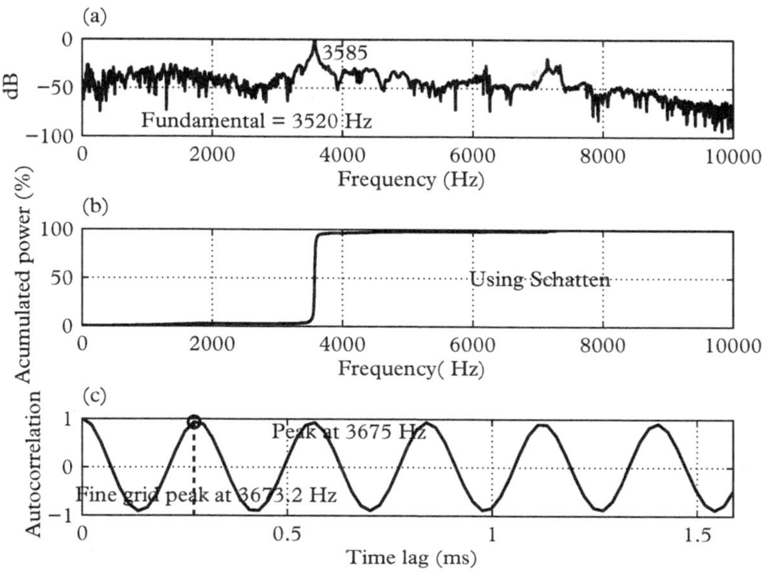

Figure 16.3.1 *Unfiltered Knabe A7 (using Schatten transducer): (a) line spectrum, (b) cumulative line spectrum, and (c) autocorrelation*

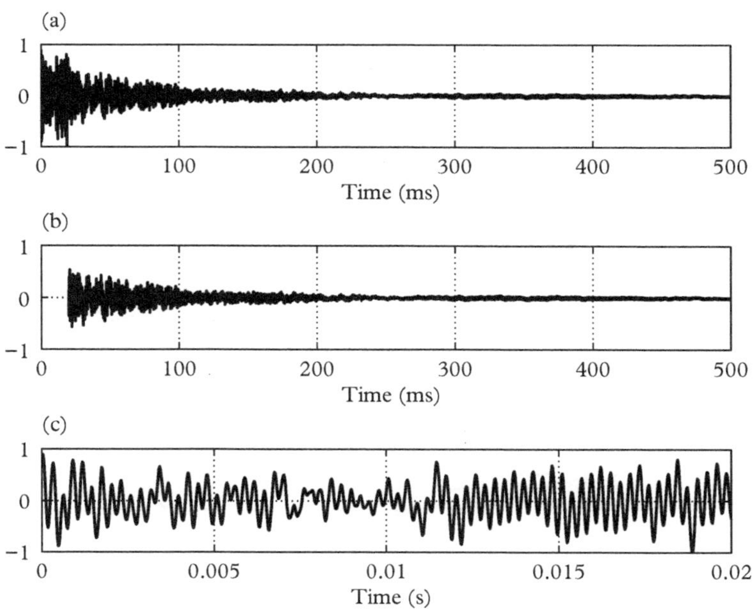

Figure 16.4.1 *Knabe A7 time domain plot: (a) before trimming, (b) after removing the early portion, and (c) the early part of the wave*

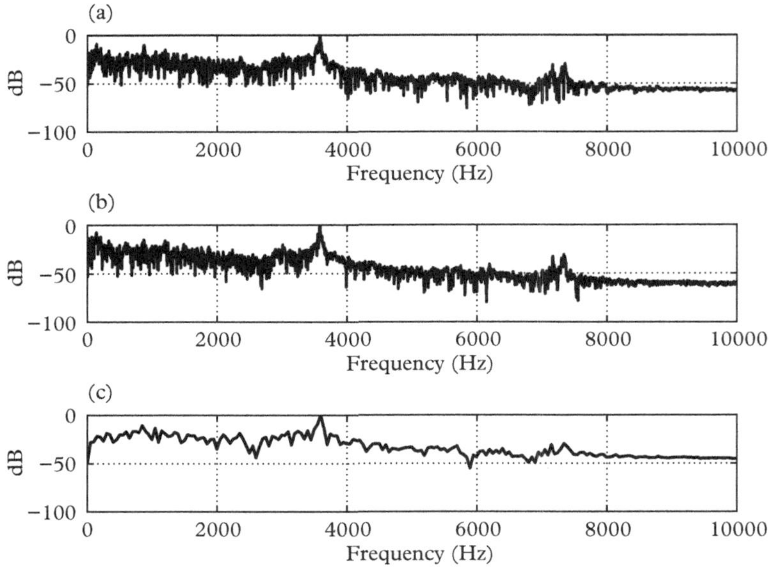

Figure 16.4.2 *Knabe A7 line spectrum: (a) before trimming, (b) after removing the early portion, and (c) the early part of the wave*

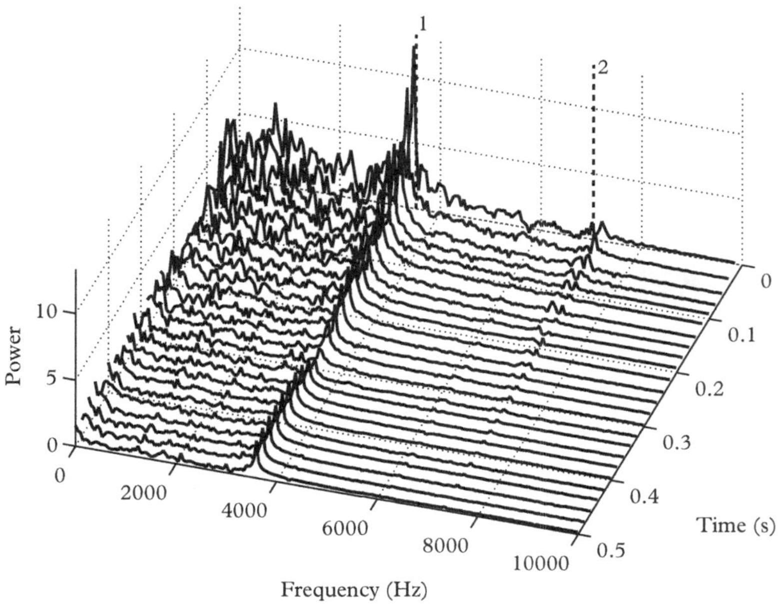

Figure 16.4.3 *Snapshots of Knabe A7 line spectrum with "1" and "2" denoting the location of the first and second partials*

The results in Figure 16.4.2 are consistent with the series of snapshots of the line spectrum of A7 shown in Figure 16.4.3, where "1" and "2" denote the first two partials. The ridge for the first partial near 3520 Hz appears to diminish in strength over the 0.5 s time sample period. The ridge for the second partial near the window at 7040 Hz starts significantly down from that of the first partial and loses considerable strength to the point where it is not differentiable from the floor of noise. Note that the two peaks in the neighborhood of 7040 Hz that show up in Figure 16.4.2 dissipate quickly into one peak. At frequencies less than the first partial there is considerable strength from the hammer knock which also loses strength quickly. However, in spite of all this, a major change, as in the presence of "early chaos," in the character of the tone does not appear to take place over 0.5 s.

16.5 Manual Striking versus the Dropping Rod Strike

Appendix 4 discusses the methods for gathering the data from the pianos. The simplest method consists of pressing the key with the finger such that a suitable tone results. To make the strike more repeatable a felt-tipped brass 0.5″ outer diameter rod is dropped 20 mm onto the key. The rod passes through a 67 mm long hole that has been drilled through a board that is positioned over each key. The brass rod is 258 mm long. Before the strike, the rod is all the way through the hole, poised 20 mm above the key surface.

Figure 16.5.1 shows the results for the Knabe A7 key. There is no significant precursor when the key is struck by pressing a finger down. With the dropping rod, Figure 16.5.1(c) shows that

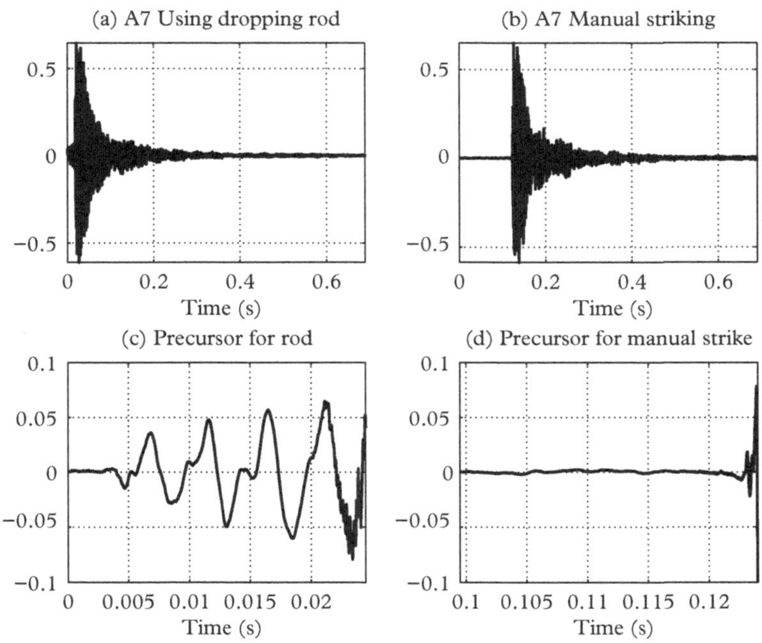

Figure 16.5.1 *Knabe A7 from dropping rod: (a) whole wave, and (c) precursor. Knabe A7 from manual strike: (b) whole wave, and (d) precursor*

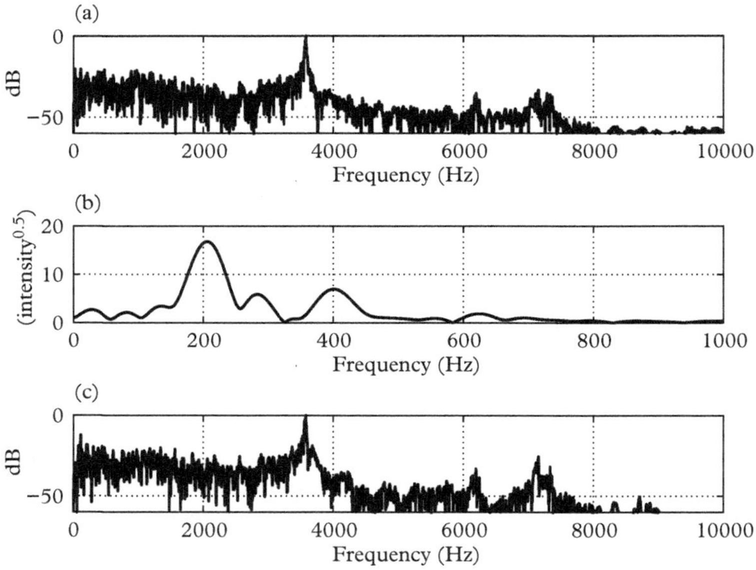

Figure 16.5.2 *Knabe A7 line spectra: (a) main wave from rod drop, (b) precursor from rod drop, and (c) main wave from manual strike*

there is a precursor, approximately 20 ms in duration, having low frequency and low amplitude. Notice how the amplitude of the precursor grows until the main wave starts.

If one assumes that the dropping of the rod is without friction and that the drop distance is 20 mm then, using a formula from high school physics, one gets 0.064 s (64 ms) for the duration of the drop:

$$t = \sqrt{\frac{2S}{g}} = \sqrt{\frac{2 \times 20\,\text{mm}}{9800\,\text{mm s}^{-2}}} = 0.064\text{ s}.$$

This drop time is about triple the duration of the precursor, at least as detectable from Figure 16.5.1(c).

The line spectra are shown in Figure 16.5.2. The two spectra for the overall waves in Figures 16.5.2(a) and (c) are similar and the power is plotted in dB. The line spectrum for the precursor in Figure 16.5.2(b) is plotted in linear units (actually $\sqrt{\text{intensity}}$) over a narrower frequency range and shows strength at approximately 195 Hz and some strength at what might be the second harmonic at 390 Hz. The precursor duration is about 20 ms, so a 195 Hz wave could have about four cycles during that period of time. Therefore, it appears that the movement of the rod generates a tone with power near 195 Hz. Do not be misled by the units on the y-axis.

16.6 Analysis of Hammer Knock for Knabe C5 and A7

We know that the keys in the top octave exhibit hammer knock or below-fundamental-frequency power. We have not seen hammer knock in the lower octaves—but could it be there? We will start by characterizing the hammer knock and background noise for the A7 key which we think has hammer knock and then repeat the analysis for the C5 key, which we think does not.

Figure 16.6.1 shows the time domain plots, the line spectra, and the cumulative line spectra for the two keys. In Figure 16.6.1(a) and (d), notice the difference in the amplitude peaks of the two waves. The spectral analyses were done on the portion of each wave between the beginning and the cut-off point so that the background noise would not dominate. The two cumulative line spectra in Figure 16.6.1(c) and (f) suggest that there is no hammer knock for C5 and that up to 40% of the A7's power comes from hammer knock.

The snapshots of the line spectra shown in Figure 16.6.2 also support the argument of less hammer knock for the C5 key. However, compare the power axes in Figure 16.6.2(a) and (b), which are plotted in $\sqrt{\text{intensity}}$. The C5 peak at 522 Hz is on the order of 30 while that for A7 at 3569 Hz is 6. If there were below-fundamental-frequency power in the C5 wave with strength equal to that of A7, it would be completely washed out by the strength of the power at the first partial.

Figure 16.6.3 shows another set of line spectrum snapshots with a frequency range that does not include the first partial. Note that in both snapshots the power plotted, again in $\sqrt{\text{intensity}}$, has the same span. Now, one can see that C5 indeed has below-fundamental-frequency power that, using the power axis as a measure, is actually stronger than that for A7. It is not just background noise because Figure 16.6.3(a) shows that it dies away within 0.5 s. The C5 below-fundamental-frequency power does not show as strongly in Figure 16.6.2(a) and Figure 16.6.1(c) because the C5 first partial's strength is so dominant compared to that for A7. Consequently, we tentatively suggest that there is hammer knock on all the keys and that for the lower-numbered ones it is washed out by the strength of the harmonics.

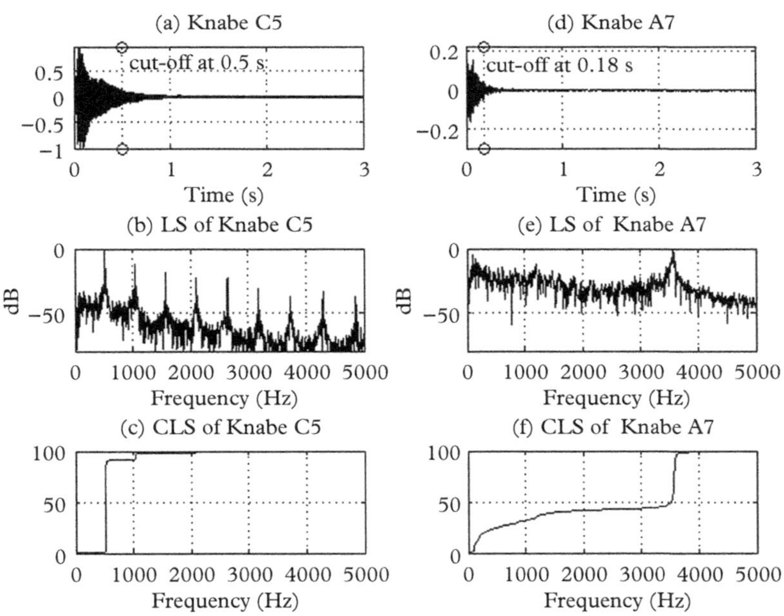

Figure 16.6.1 *Knabe C5: (a) time domain (TD), (b) line spectrum (LS), and (c) cumulative line spectrum (CLS). Knabe A7: (d) TD, (e) LS, and (f) CLS*

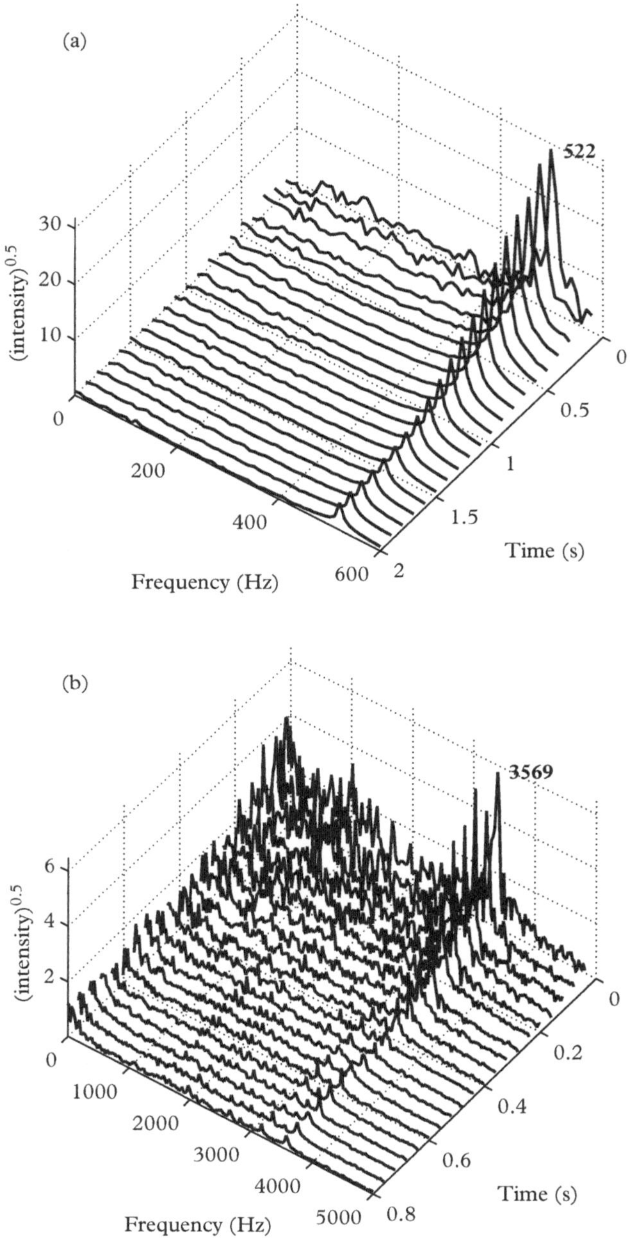

Figure 16.6.2 *Snapshots of the line spectrum for (a) Knabe C5, and (b) Knabe A7*

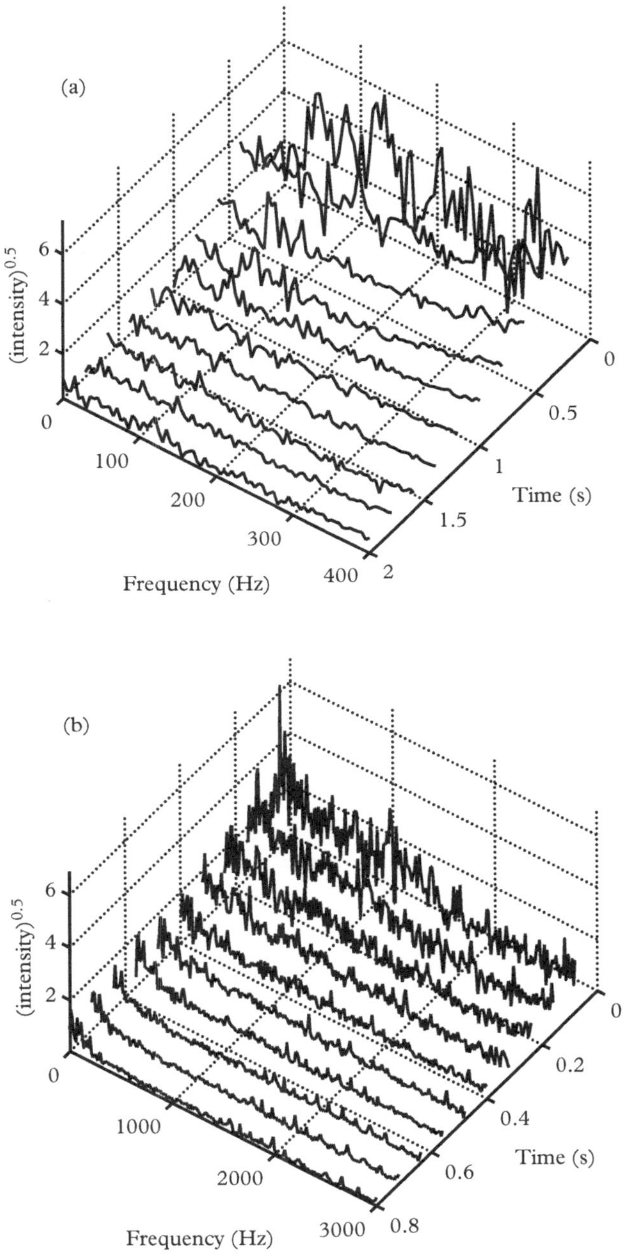

Figure 16.6.3 *Snapshots of the low frequency region of the line spectrum for (a) Knabe C5, and (b) Knabe A7*

16.7 Summary

The impact of the below-fundamental-frequency power on the Knabe A7 key's sound was studied by means of frequency domain filtering. Parts of the A7 spectrum were removed and the sound of the remaining wave was analyzed. It appears that the below-fundamental power makes a noticeable contribution to the sound of the A7 note.

Since some of the data has been collected by dropping a felt-tipped rod on the keys, this method can be compared to that of pressing the finger down on the key. The rod-dropping method appears to generate a precursor wave with power in the neighborhood of 200 Hz. Otherwise, the rod-dropping results match those of the finger-pressing method.

To determine if hammer knock is present in the lower octaves, the frequency and time domain characteristics of the C5 and A7 keys were compared. It appears that both keys have below-fundamental power but the amplitude of the C5 is so much stronger than that of the A7 that hammer knock does not appear in the overall spectra.

The reader is encouraged to check the book's web page where post-publication results based on a time-dependent spectral centroid show how a note's power center shifts with time.

17 Evaluation of a Wapin Bridge Conversion

On conventional piano bridges there are two pins that are staggered and inserted at angles to give the strings side bearing and to hold them tightly so they will contact the bridge and transmit the strings' vibrations to the soundboard. The Wapin procedure uses three pins. The middle and back pins are slanted while the front pin is vertical. There are many different claims as to why this arrangement improves the sound of a piano but I will not repeat them here, mostly because they appear to me to be more subjective and speculative than objective. Instead, I will use the spectral centroid and the cumulative line spectrum map to compare the pre-Wapin conversion and post-Wapin conversion characteristics of a Steinway Model B piano located in the Ithaca College School of Music.

17.1 Spectral Centroid Method

Figure 17.1.1 shows the spectral centroids before and after the conversion. The curve for the fundamental frequencies of each key is shown as small circles. The solid line shows the spectral centroid before the conversion and the dotted line shows it after the conversion. It appears that the dotted line (post-conversion) lies above the solid line on the average for the first twenty or so keys. This suggests that the Wapin conversion moved power into the higher harmonics for those keys.

Figure 17.1.2 shows the cumulative difference in Hertz between the two spectral centroids. Figure 17.1.2(a) is plotted in Hertz while Figure 17.1.2(b) is plotted in cents. The cumulative difference rises for the first 19 keys and then decreases for the remaining keys. The early rise in the curve suggests that *on average* the post-Wapin notes have a higher spectral centroid. That is, there is more power in the higher harmonics for keys 1 through 19 *after* the Wapin conversion. The curve decreases for keys 20 through 88 indicating that the fundamental is emphasized more after the conversion. This decrease is less pronounced in the graph using cents, Figure 17.1.2(b), because the number of Hertz percent increases geometrically with key number.

Spectral Analysis of Musical Sounds with Emphasis on the Piano. First Edition.
David M. Koenig, with comments by Delwin D. Fandrich.
Published in 2015 by Oxford University Press.

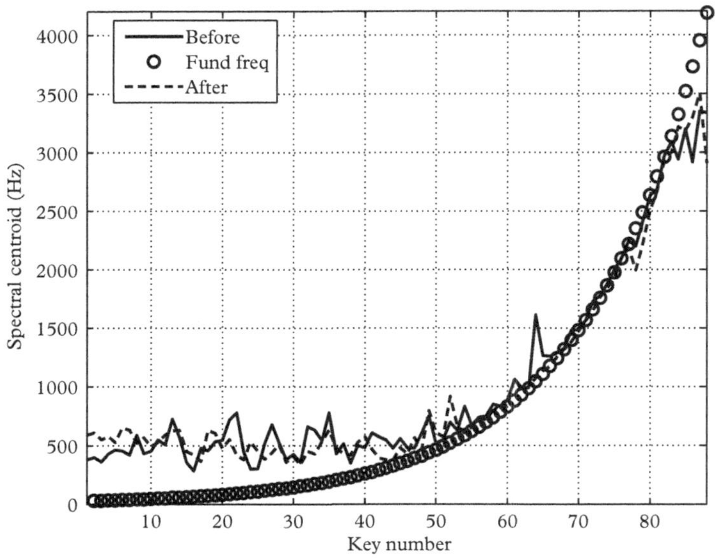

Figure 17.1.1 *Spectral centroids before and after the Wapin conversion of a Steinway Model B*

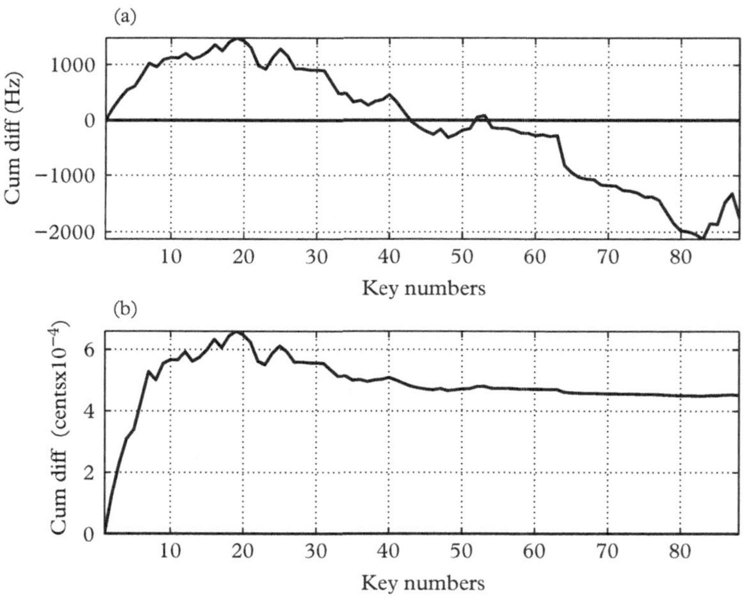

Figure 17.1.2 *Cumulative difference of spectral centroids for Wapin conversion: (a) plotted in Hertz, and (b) plotted in cents × 10,000*

17.2 Cumulative Line Spectrum Map Method

Figure 17.2.1 shows the cumulative line spectrum map of the before and after conditions. There are three lines in the frequency–key number plane representing the first, second, and third harmonics. We have shown these lines earlier when first presenting the CLSM. The termination of these lines is denoted by small circles.

There are six lines in the frequency–key number plane representing the transitions for the Steinway Model B. They run parallel with the frequency axis. The keys in the first zone, numbered from one to eight (A0 to E1), have a wrapped monochord string. The keys in the second zone, from 9 to 20 (F1 to E2), have wrapped bichord strings. These two zones are sometimes called the bass zone. The keys in the third zone, from 21 to 51 (F2 to B4), have unwrapped trichords

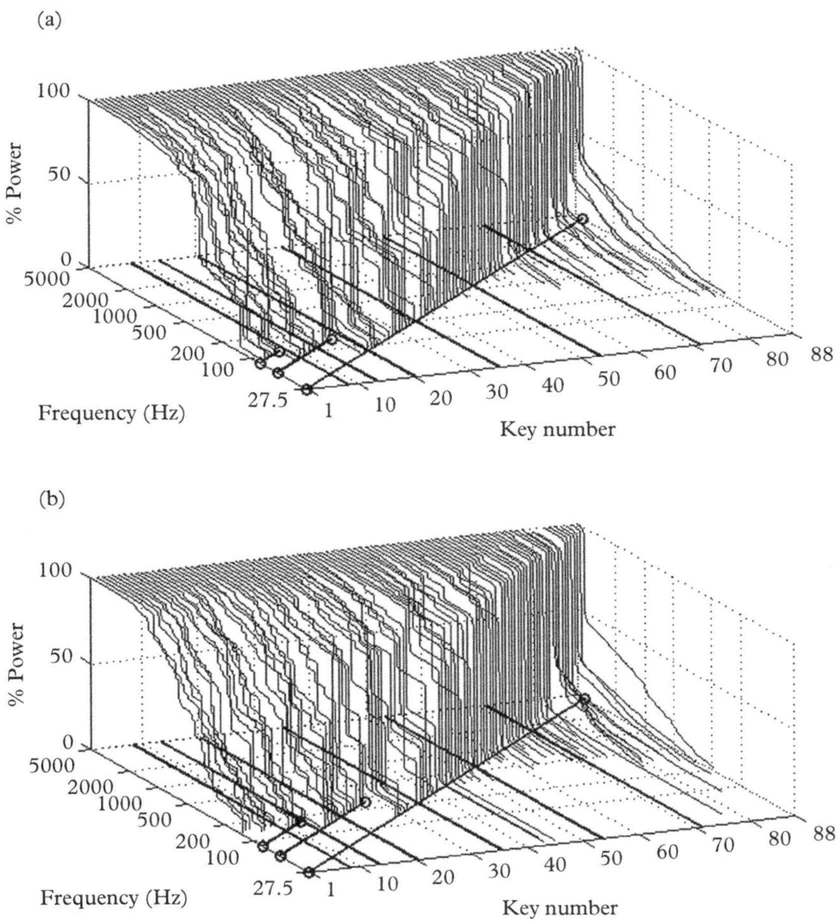

Figure 17.2.1 *CLSM for Wapin conversion of Steinway Model B (all 88 keys): (a) before conversion, and (b) after conversion*

and are sometimes referred to as making up the tenor zone. The keys 52 to 68 make up the lower treble zone. Keys 69 to 88 make up the upper treble zone and do not have dampers.

For the pre-Wapin condition, Figure 17.2.1(a), the lines denoting the above zones and the lines denoting the harmonics appear to correlate with the changes in power. For the post-Wapin condition, Figure 17.2.1(b), there is a different relationship (you have to study the graphs fairly carefully to see the differences).

The CLSMs for the keys in the treble section look similar for the two cases but the CLSMs for bass and lower tenor section are quite different, as shown in Figure 17.2.2. After the Wapin conversion, the first 13 keys show negligible power at the first two harmonics, and in general for the first 30 keys the power appears at higher partials than before the conversion. This is consistent with the results of applying the spectral centroid method. At the transition point after key 20 there is a clear change in the structure of the CLSM for the pre-Wapin condition. However, after the Wapin conversion there is not a clear change in the CLSM structure at that point.

Look for point A in Figure 17.2.2(b) for the post-Wapin case. The corresponding location in Figure 17.2.2(a) for the pre-Wapin case is different: the post-Wapin graph displays no power in this section while there is significant power in the pre-Wapin graph. Point B in the post-Wapin

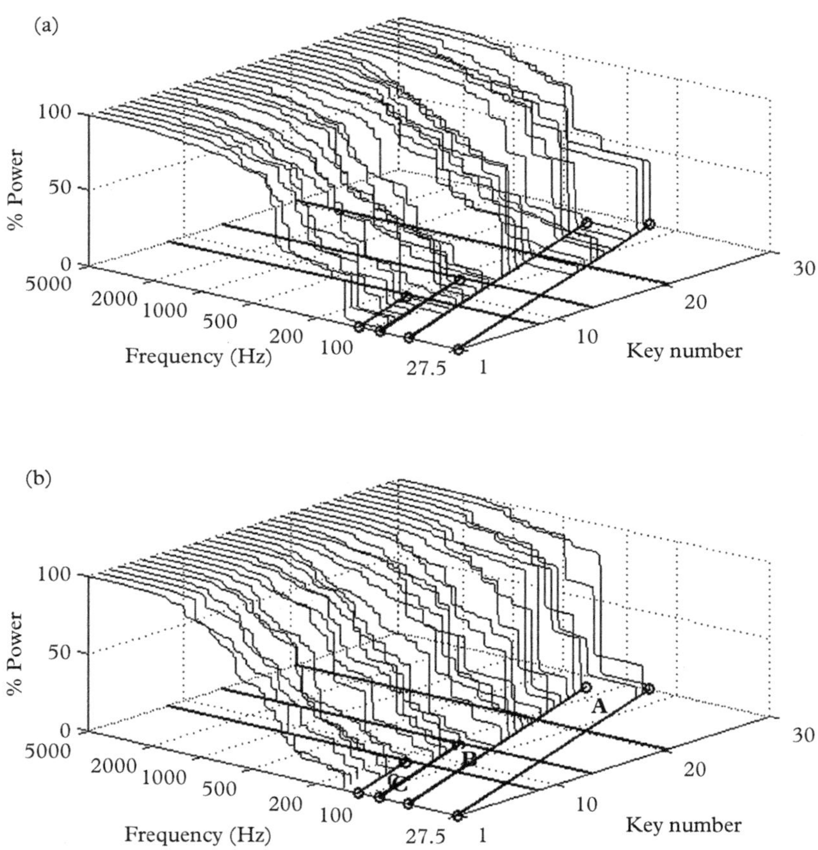

Figure 17.2.2 *CLSM for Wapin conversion of Steinway Model B (keys 1 to 30): (a) before conversion, and (b) after conversion*

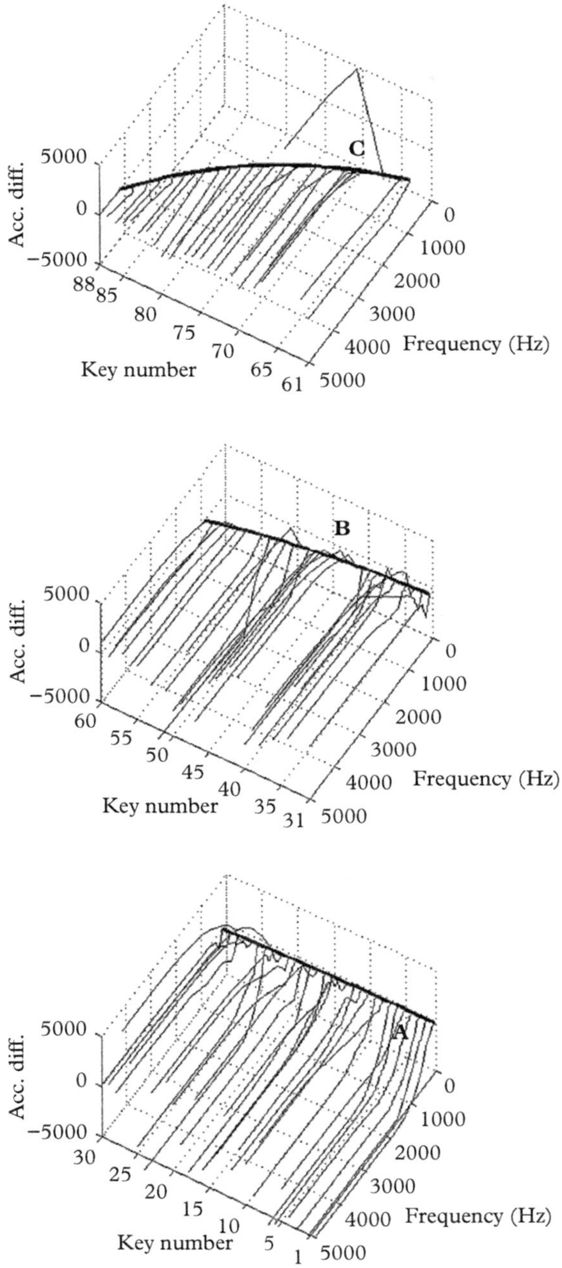

Figure 17.2.3 *The cumulative difference (post–pre) of the Wapin conversion CLSMs for three sections of the piano's compass*

graph also designates an area where there is no power but there is noticeable power in the same area in the pre-Wapin graph. The same analysis applies to the zone denoted by C.

Therefore, based on both the spectral centroid method and the CLSM method, one might expect that the lower keys would sound different after the Wapin conversion, there being more power in the higher partials for the lower notes after the conversion.

A three-dimensional plot of the difference of the CLSMs might also provide insight. However, the simple difference of the CLSMs at every frequency sample for each key produces a chaotic, unruly picture, with little potential for interpretation or insight. Previously, the cumulative sum has been used to address situations where the data has an erratic appearance when dealing with line spectra and differences between spectral centroids. The same approach can be applied to the difference between the CLSM for the post- and pre-Wapin conversion cases. Figure 17.2.3 shows the result. There are three plots in Figure 17.2.3, each for a third of the piano's compass starting with the lowest numbered keys on the far right. In each of the three plots, the cumulative difference for each key is plotted versus the frequency starting at the fundamental frequency, which is denoted in each graph by the thick line. Notice how the curve representing the fundamental frequency curves in the plot for the upper third of the compass. This is because frequency is plotted linearly rather than logarithmically.

For keys 1 through 5, point A shows a downward trend in the cumulative difference. Since the difference is post-Wapin minus pre-Wapin, a downward trend means that the post-Wapin power is weaker at the lower key numbers or that there is more power in the upper harmonics. At other ranges of keys the situation varies. For example, point B in the neighborhood of key 45 shows an upward trend signifying that there is more power near the fundamental for the post-Wapin condition. Aside from the one outlier at key 63, the upper third of the compass shows little difference.

17.3 Summary

Can one infer any general conclusions from this chapter? Probably not because, again, the analyses are carried out on a sample of size one. Had there been 20 pianos of the same brand and model that underwent Wapin bridge conversions, then the spectral centroid and the CLSM might be useful for arriving at firm conclusions. If I accumulate more data on Wapin conversions, the results will be placed on the website mentioned in the preface.

18

Similarities between Pianos and Repeatability of Data-Gathering Methods

Two Steinway Model D pianos from Ithaca College will be compared using the spectral centroid and the cumulative line spectrum map. The repeatability of the methods will be tested using the Knabe grand piano. Appendix 4 shows the equipment used to collect the data analyzed in this section.

18.1 Comparing Two Steinway Model D's

Two Steinway Model D's from the Ithaca College School of Music are compared. The first was manufactured in 1997 and the second in 2008.

Envelopes

Figure 18.1.1 shows the envelopes of the two Model D's. The overall structure of the two envelopes is similar and Figure 18.1.2 shows a closer look at the keys between C3 and C5. The latter plot shows similar structure in the neighborhoods around E3 and A3#.

Root Mean Squares

The RMS (root mean square) plot in Figure 18.1.3(a) shows that there are differences between the two pianos. For keys 1 to 35 the first Steinway has consistently higher values, suggesting that it has a bit more sound power. The area from key 35 to 50 shows a similar structure for both pianos. This difference is borne out in Figure 18.1.3(b), which shows the cumulative difference of the RMS values. The cumulative difference rises until about key 35, suggesting that the RMS value of the first Steinway is consistently greater than that of the second. Beyond key 35 the cumulative difference is constant, suggesting that the RMS values are the same for both pianos. Of all the metrics, the RMS may be the one most prone to microphone placement. I attempted to place the microphone in identical locations but there may have been some error.

Spectral Analysis of Musical Sounds with Emphasis on the Piano. First Edition.
David M. Koenig, with comments by Delwin D. Fandrich.
Published in 2015 by Oxford University Press.

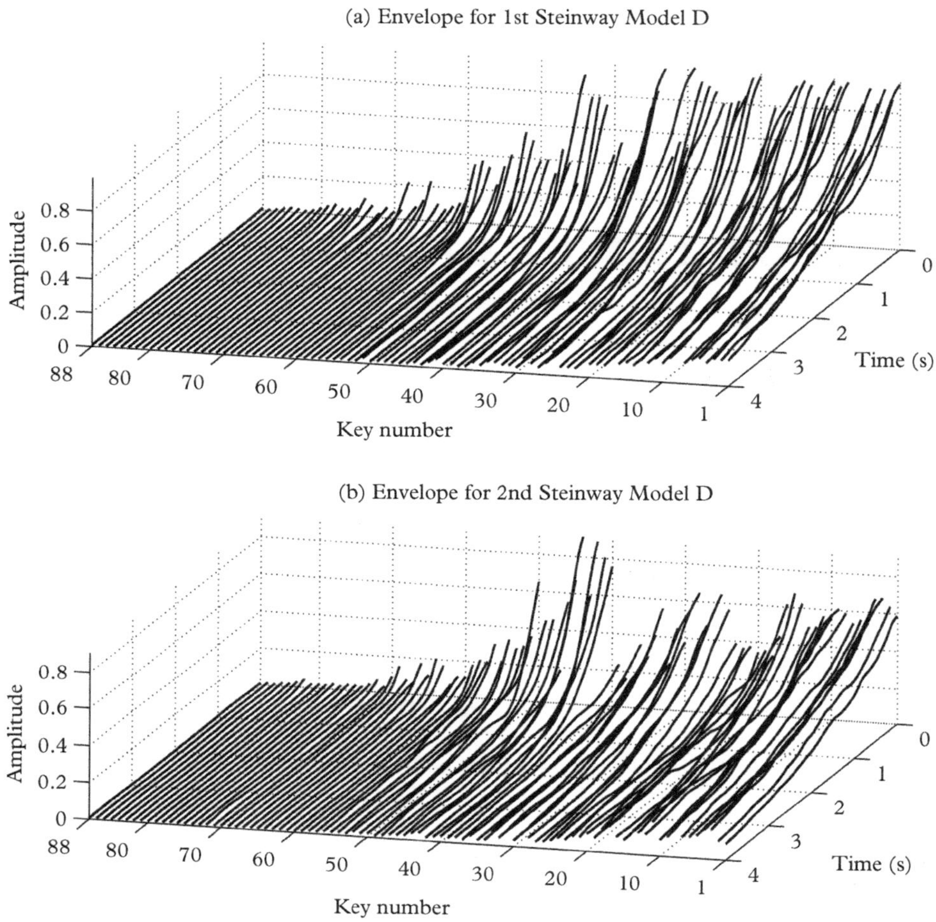

Figure 18.1.1 *Envelopes for two Steinway Model D's for all 88 keys*

Spectral Centroids

The spectral centroids are shown in Figure 18.1.4. There seem to be three areas where the centroid is slightly different: around keys 38, 46, and 60.

An alternative visualization of the spectral centroids of the two Steinways is shown in Figure 18.1.5. The two spectral centroids (with the fundamental frequency subtracted) are plotted side by side in Figure 18.1.5(a) in a manner similar to Figure 18.1.4 except that cents is used instead of Hertz. (In the upper octaves the same number of cents covers a much wider range of Hertz than in the lower octaves.) The difference in the spectral centroids is shown in Figure 18.1.5(b) and the cumulative difference is shown in Figure 18.1.5(c). As expected, the range of the cumulative difference is significantly less than that for the comparisons of the Knabe and Bosendorfer and for the Knabe and NY Steinway in Section 14.2; however, there is a small, non-trivial difference.

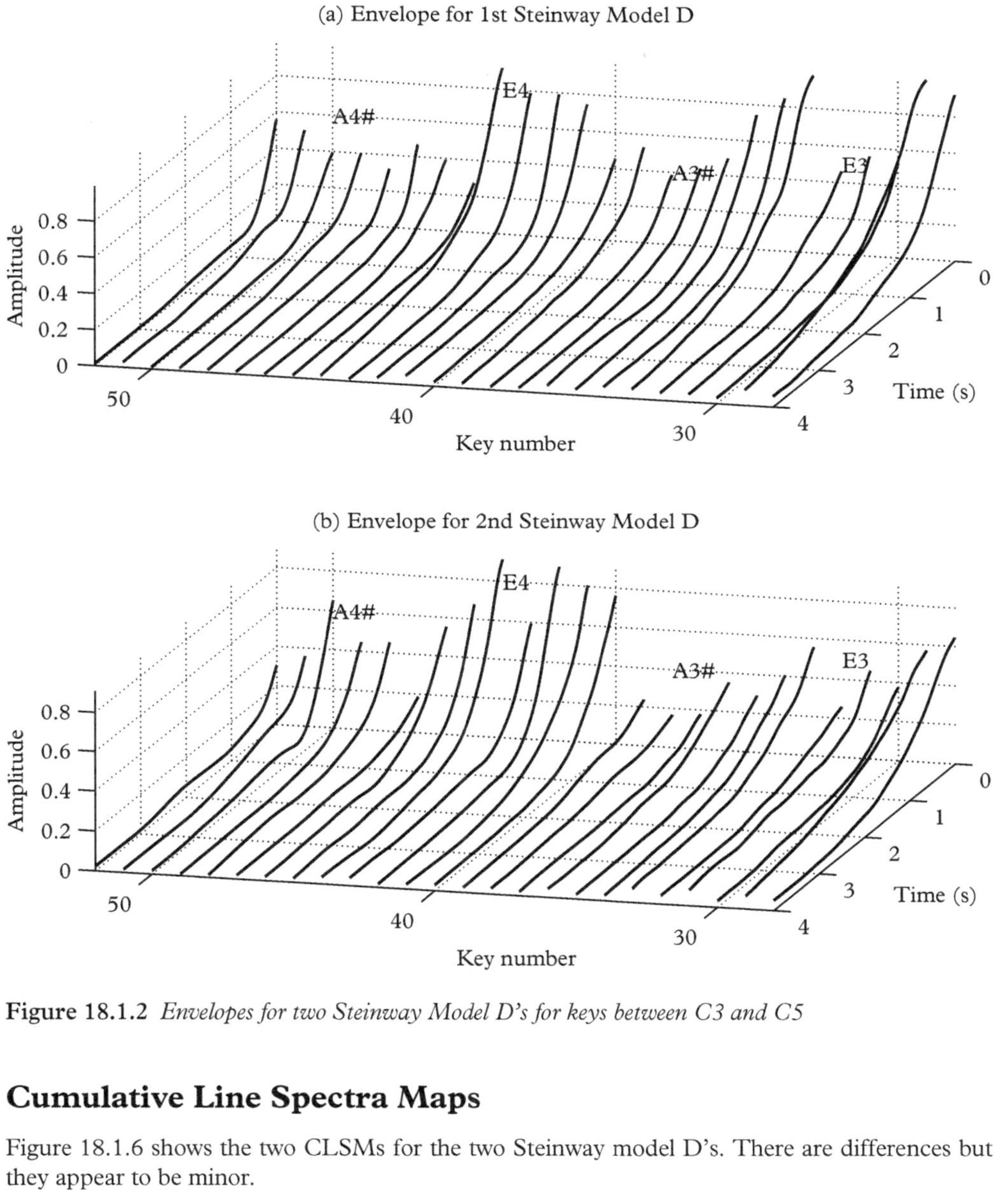

Figure 18.1.2 *Envelopes for two Steinway Model D's for keys between C3 and C5*

Cumulative Line Spectra Maps

Figure 18.1.6 shows the two CLSMs for the two Steinway model D's. There are differences but they appear to be minor.

18.2 Repeatability of Data-Gathering Method

To test this concern I gathered data for the Knabe grand and then repeated the data-gathering process on the same day using an Earthworks microphone with a Roland Duo Ex interface.

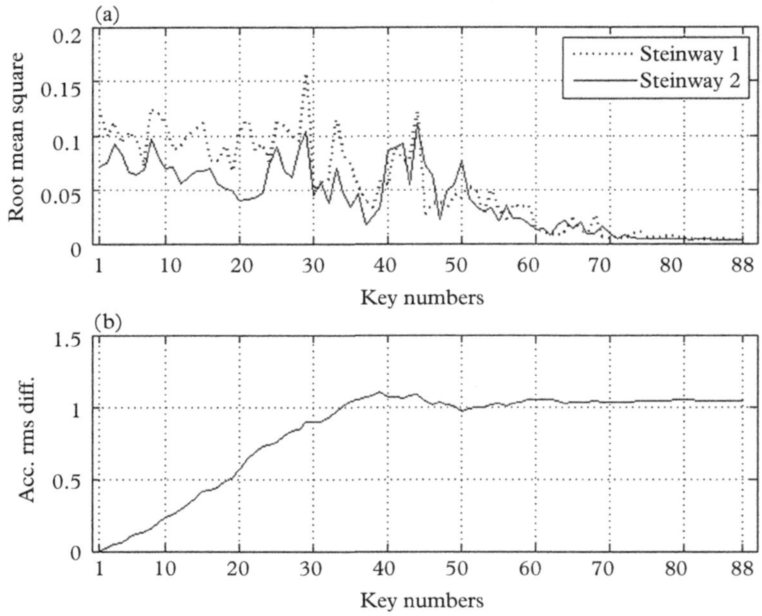

Figure 18.1.3 *RMS values for the two Steinway Model D's: (a) separate traces for each piano, and (b) the cumulative difference between the RMS values for the two pianos*

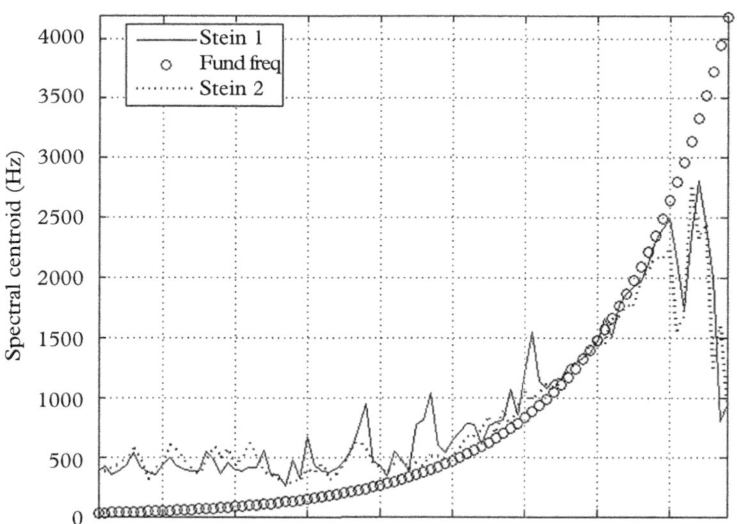

Figure 18.1.4 *Spectral centroids of the two Steinway Model D's along with the fundamental frequency*

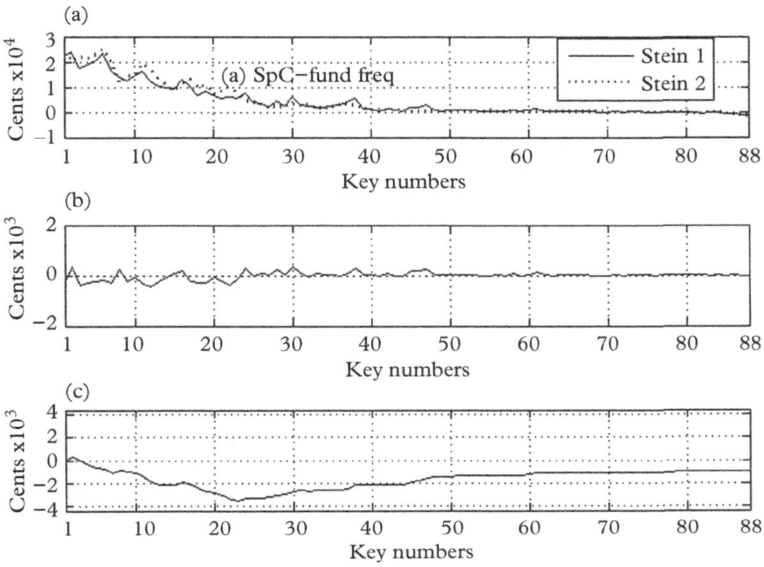

Figure 18.1.5 *Spectral centroids of the two Steinway Model D's: (a) spectral centroid minus the fundamental frequency, (b) the difference of the two spectral centroids, and (c) the cumulative difference of the centroids*

Envelopes

Figure 18.2.1 shows the two envelopes. They are not identical but there are enough overall similarities to justify the use of envelopes to illustrate how the nature of the tone changes as one moves from A0 to C8.

Root Mean Squares

The overall trend of decreasing RMS with increasing key is the same for both runs in Figure 18.2.2. A couple of the keys that show extreme strength or weakness on one run do so for the other run but there are also some significant outliers. The cumulative difference in Figure 18.2.2(b) when compared to that in Figure 18.1.3(b) (note the *y*-axis scales on the two graphs) suggests that the differences between the two Knabe experiments are significantly less than the differences between the two Steinways, as one would expect.

Cumulative Line Spectrum Maps

Figure 18.2.3 shows the cumulative line spectrum maps for the two runs. Once again the two maps are similar in overall structure but one can find a few differences, especially in the amount of hammer knock for the upper octave.

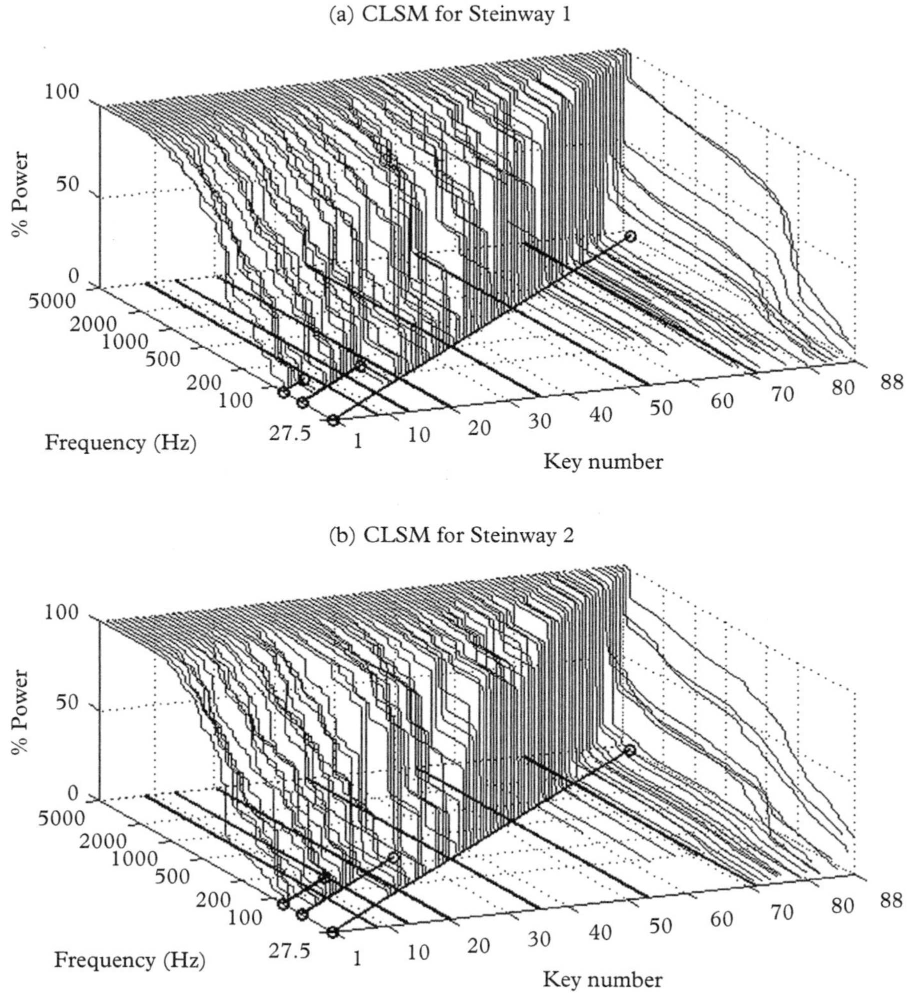

Figure 18.1.6 *CLSMs of the two Steinway Model D's*

Spectral Centroids

Figure 18.2.4 shows the spectral centroids for the two runs. Here the two runs are quite similar except for the uppermost octave. The curve traced out with small circles represents the fundamental frequency.

Finally, we apply the cumulative sum approach to the spectral centroids from the two data collection exercises. Figure 18.2.5(a) shows the difference between the spectral centroid and the fundamental frequency for all 88 keys for both collection exercises. The difference between the

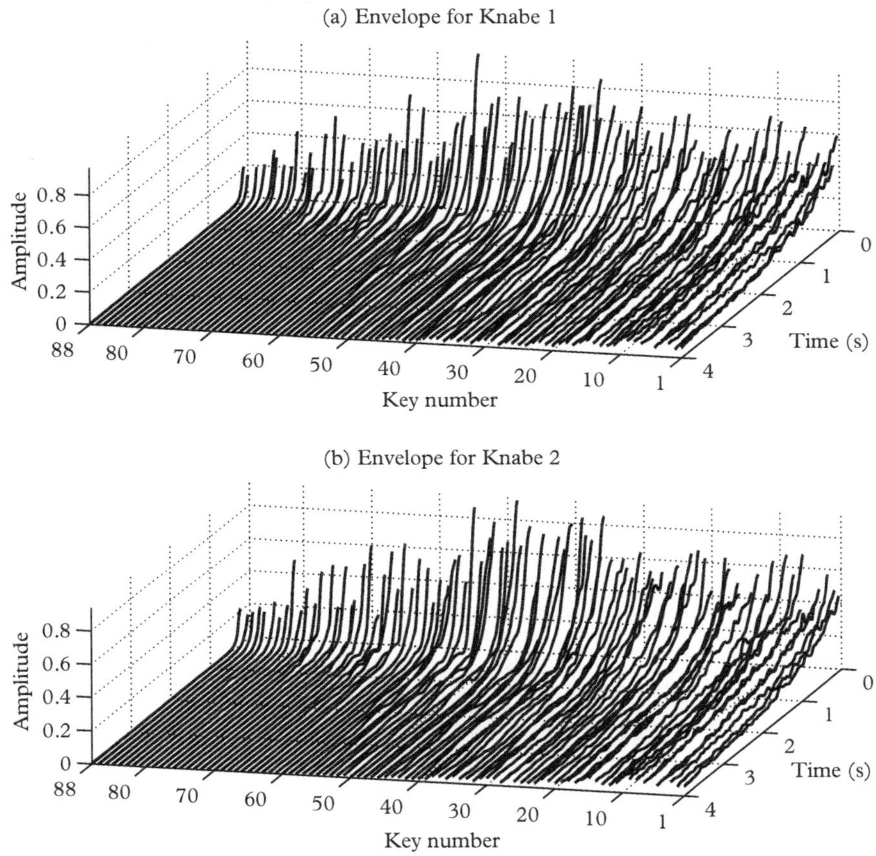

Figure 18.2.1 *Envelopes for two different data collections on the Knabe grand piano*

two spectral centroids is shown in Figure 18.2.5(b), followed by the cumulative difference in Figure 18.2.5(c).

The reader should juxtapose Figure 18.2.5(c) with those for the other comparisons between (1) the Knabe and the Bosendorfer, Figure 14.2.1(c), (2) the Knabe and the New York Steinway, Figure 14.2.2(c), in Chapter 14, Section 14.2, and (3) the two Steinways, Figure 18.1.5(c). For the two Knabe grand collection exercises, Figure 18.2.5(c) shows that the spectral centroid cumulative difference stays well within ±1000 cents. In contrast, for the comparison between the Knabe and the Bosendorfer, the spectral centroid cumulative difference varied between zero and 13,000 cents. This data collection exercise suggests to me that, given the background noise, the equipment, and my technique, the cumulative difference of the spectral centroid will not vary more than about ±1000 cents for nearly identical conditions. Were these methods used in a quality control effort, the necessary improvements in equipment, technique, and environment would probably attenuate the variance of the cumulative difference considerably.

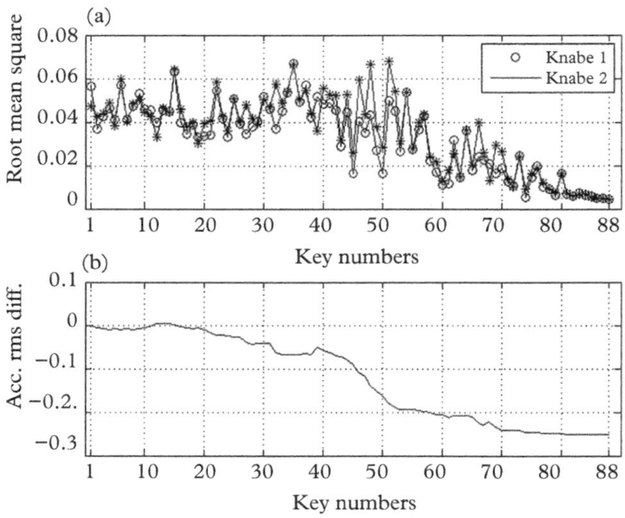

Figure 18.2.2 *RMS values for the two data collections on the Knabe grand piano: (a) separate traces for each piano, and (b) the cumulative difference between the RMS values for the two pianos*

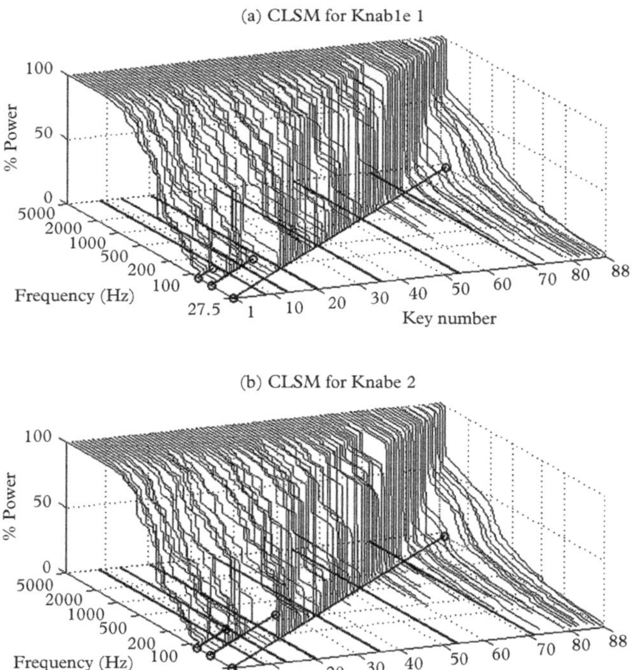

Figure 18.2.3 *CLSMs for the two data collections on the Knabe grand piano*

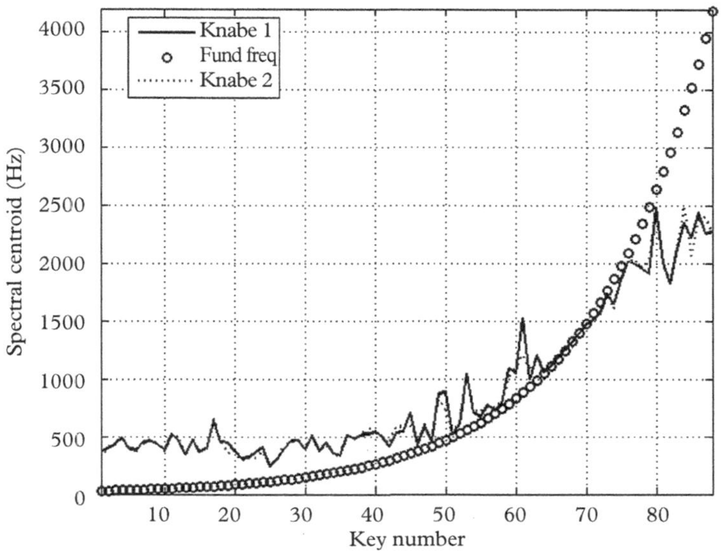

Figure 18.2.4 *Spectral centroids for the two data collections on the Knabe grand piano along with the fundamental frequency*

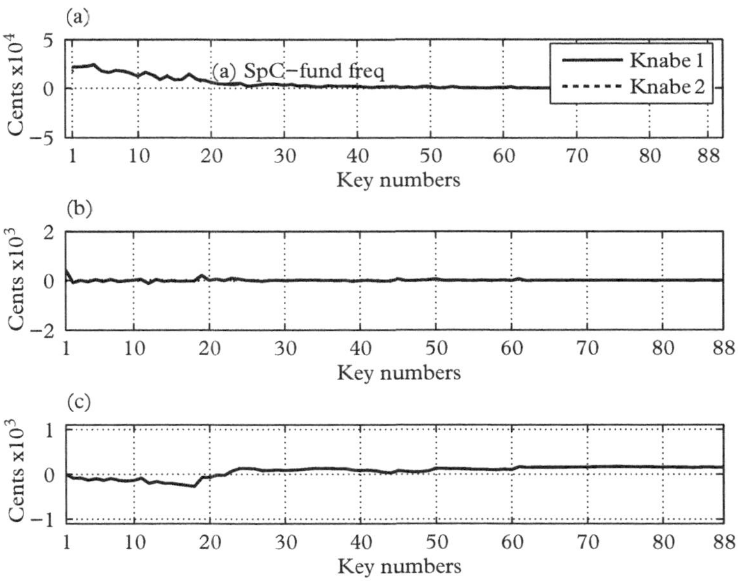

Figure 18.2.5 *Spectral centroids of the two data collections on the Knabe grand piano: (a) spectral centroid minus the fundamental frequency, (b) the difference of the two spectral centroids, and (c) the cumulative difference of the centroids*

18.3 Summary

Two Steinway Model D pianos from Ithaca College were compared using the spectral centroid and the cumulative lines spectrum map. Basically, the two Steinways were similar with a few small differences. The repeatability of the envelopes map, root mean squares, spectral centroid, and cumulative line spectrum map methods were tested using the Knabe grand piano. The results suggested that there was some experimental error either due to technique or to humidity changes or both. In any case, not enough, I feel, to make the methods suspect.

19

Two Metrics Applied to Other Instruments

Now that the reader is familiar with the concept of a cumulative line spectrum map for a piano, we will try the same thing for other musical instruments mentioned in earlier parts of the book. Most of the musical instruments have a traditional range of perhaps two to three octaves. For each of the instruments there will also be a graph of the normalized spectral centroid.

19.1 Trombone

Figure 19.1.1 shows a cumulative line spectrum and the spectral centroid for E2 to A4# played by the music student mentioned in Chapter 1. The axis labeled "note numbers" gives the number of the trombone tone played: 20 refers to E2 and 51 to A4#. In the CLSM plot, Figure 19.1.1(a), there are lines representing the fundamental frequency and the second harmonic. One sees no significant power at the actual fundamental frequencies for the nine lowest notes. For example, E2 shows nothing until the second harmonic around 160 Hz. The ninth note, B2 (note number 27), is the first to show noticeable power at the fundamental frequency. As the fundamental frequencies increase, more of the power occurs at this frequency rather than at higher harmonics. At the 29th note, G4, there appears to be a significant shift in the amount of power in the fundamental compared to the higher harmonics. This suggests that the trombone, like the piano, exhibits a lack of power at the fundamental frequency at the lowest notes. Likewise, the normalized spectral centroid plot, Figure 19.1.1(b), shows that there is considerable signal power in the higher harmonics for the lowest notes.

19.2 Cornet

The reasonable range for me on my cornet is from G3# below middle C to C6, two octaves above middle C. The key of the cornet is B♭, so playing G3 on the cornet sounds the same as F3 on the piano. Figure 19.2.1 shows the cornet map with straight dotted lines in the frequency–keys plane representing the fundamental frequency and the second harmonic for each note from G3# to C6. The note numbers refer to the corresponding notes on the piano.

As with the piano and the trombone, the lower notes are not represented strongly at the fundamental frequency. Only at note 55 (D5# on the piano but F5 on the cornet) does the fundamental

Spectral Analysis of Musical Sounds with Emphasis on the Piano. First Edition.
David M. Koenig, with comments by Delwin D. Fandrich.
Published in 2015 by Oxford University Press.

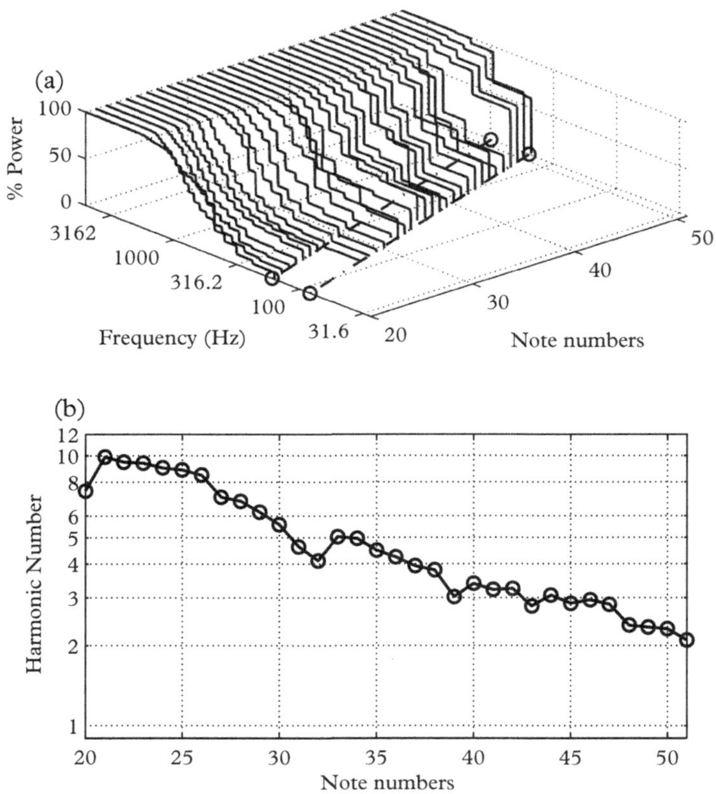

Figure 19.1.1 *Trombone: (a) CLSM, and (b) spectral centroid divided by the fundamental frequency of each note*

start to dominate. Unlike the trombone, the cornet does not have significant power above the seventh harmonic.

19.3 Bassoon

The bassoon was played from A1# to E4 and the CLSM is shown in Figure 19.3.1. Not until note 41 (C4#) does the fundamental frequency have a significant contribution and it never dominates. Likewise, not until note 25 (A2) is there a significant contribution from the third harmonic. Even the highest note E4 has a spectral centroid that is located at the second harmonic.

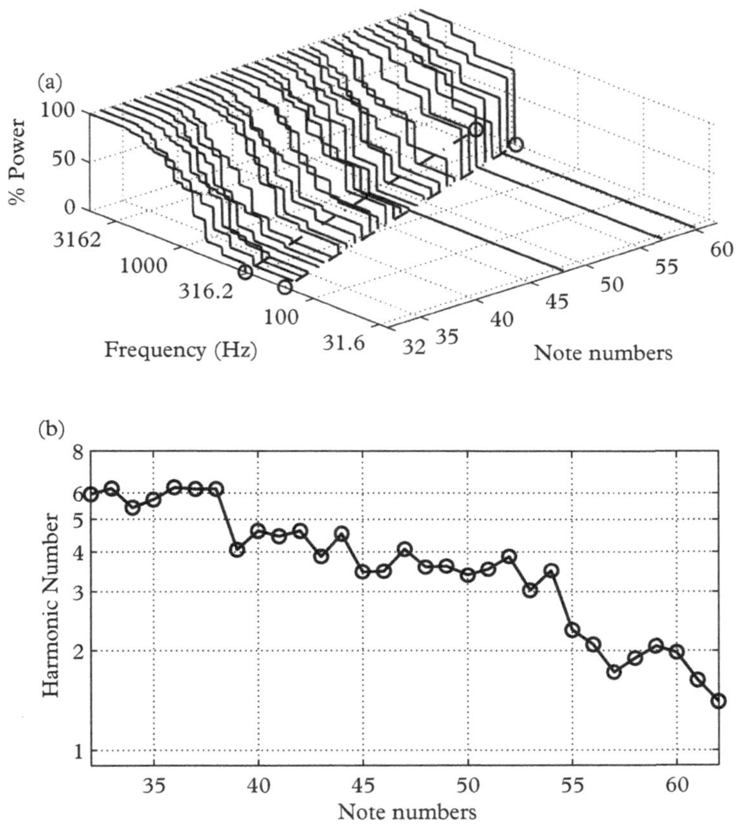

Figure 19.2.1 *Cornet: (a) CLSM, and (b) spectral centroid divided by the fundamental frequency of each note*

19.4 Flute

A local flute teacher was kind enough to let me record a chromatic sequence from B3 to D7. The map for this can be seen in Figure 19.4.1.

Unlike any of the other instruments studied so far, the lowest note, B3 (or note number 39), shows considerable strength at the fundamental frequency with a normalized spectral centroid of less than 2.0. At notes 65 and above (C6# and higher) the signal is almost completely composed of the fundamental, with little contribution from the higher harmonics (the harmonic number for the normalized spectral centroid is effectively unity).

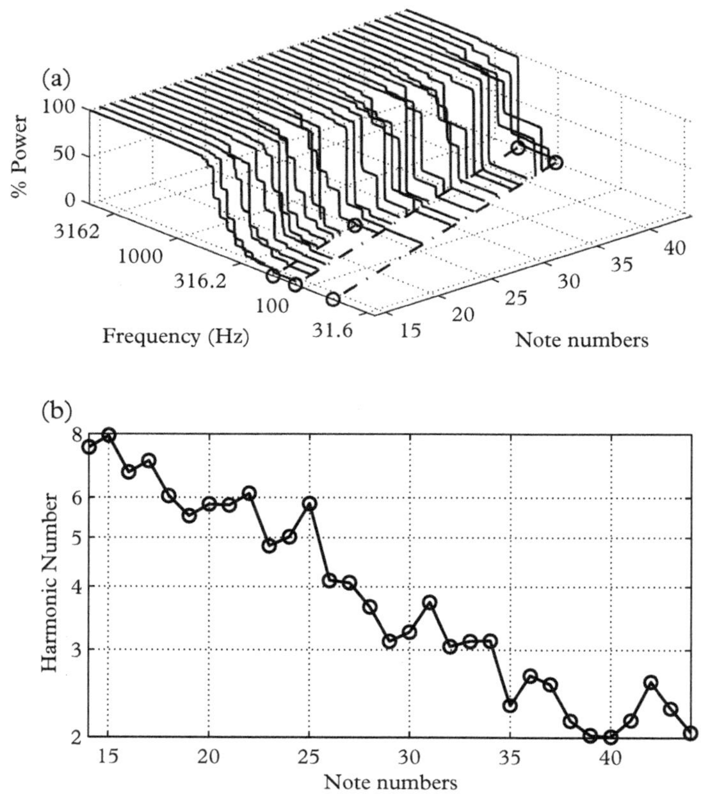

Figure 19.3.1 *Bassoon: (a) CLSM, and (b) spectral centroid divided by the fundamental frequency of each note*

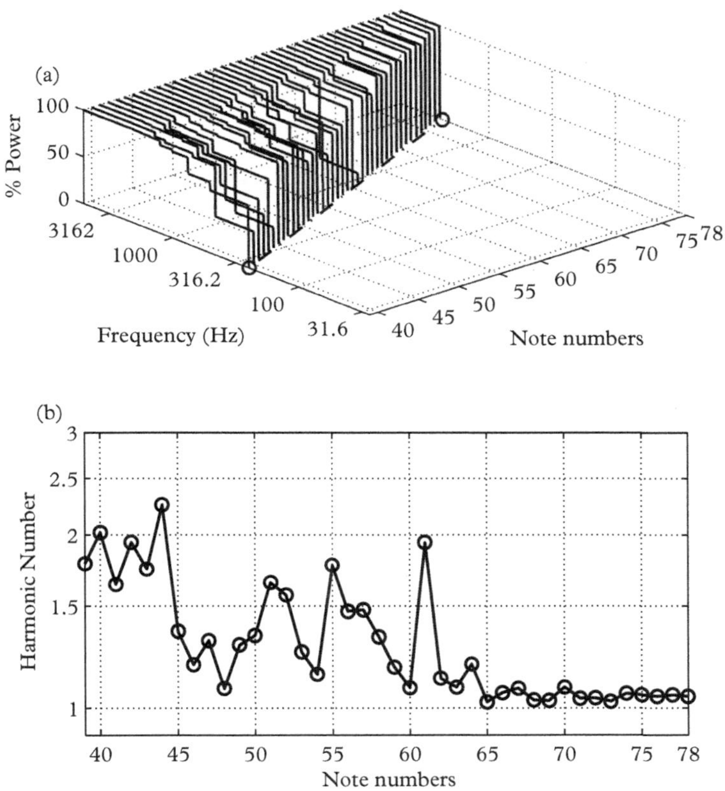

Figure 19.4.1 *Flute: (a) CLSM, and (b) spectral centroid divided by the fundamental frequency of each note*

19.5 Summary

The cumulative line spectrum map and the spectral centroid map were constructed for a trombone, cornet, flute, and bassoon. All instruments but the flute exhibited a weak fundamental in the low notes. Also, all instruments, including the flute, showed decreasing spectral centroid with increasing note number.

20
Use of the Metrics in Production and Development

I've tried to cover (or perhaps just uncover) spectral analysis of musical sounds with emphasis on the piano. Now it is time to ask a question. Can these metrics be useful?

20.1 Limit Samples

In Section 15.1 I mentioned using the envelope maps for quality control. In fact, I think *several* of the metrics presented in Part 3 could be used along with *limit samples*. To develop limit samples a panel of experienced piano technicians and performers would do the following:

1. Agree on a small number of tone qualifiers such as mellowness, brightness, and so forth. Let's say that the panel members could agree on four different categories. This, at least to me, would be difficult because I have trouble deciding if a piano's tone is bright or mellow or harsh or rich in higher harmonics, etc., but if the posts on Frank Baxter's Piano World are any indication, many individuals are able to discern these qualities.
2. Search for pianos that the panel members can agree are exemplars of these qualities. This would also be difficult and would take a considerable amount of time. The search would have to be comprehensive. Let's say that the panel members could find three pianos that satisfied the conditions of each category. So, there would be 12 pianos chosen and each would be a member of just one group that strongly exhibited one of the four characteristics. Let's designate the groups as A, B, C, and D.
3. Apply the array of metrics presented in this book to the 12 pianos and see if one could see any correspondence between, say, a cumulative line spectrum map having a unique characteristic with each category. For example, the CLSMs, the spectral centroid maps, or the envelope maps for the mellow pianos might have a characteristic that (a) did not appear for the pianos in the other groups and (b) did appear consistently among the pianos in its group.

If this could be done, and admittedly it's a big "if," then perhaps the CLSM along with other metrics like the spectral centroid map and the envelope map could be used to evaluate pianos. At this point a set of limit samples would be available for reference.

Spectral Analysis of Musical Sounds with Emphasis on the Piano. First Edition.
David M. Koenig, with comments by Delwin D. Fandrich.
Published in 2015 by Oxford University Press.

The next step, equally difficult, would be to see if there was a consistent way of modifying the pianos from group A to take on the characteristics of those in category B, and use the metrics to help decide if the modification was a success. If successful, these results would provide sets of criteria for evaluating modifications.

Finally, a search would commence for pianos with a certain defect. These pianos would be analyzed using the metric array and one would attempt to correlate the defect with a characteristic in one or more of the metrics. If successful, the metrics could be used to detect/confirm flaws in pianos.

The concept of a limit sample (or limit samples) is widely applied in manufacturing technology. Applying it in the manufacture of pianos would not be easy. However, it might be beneficial in decreasing the variability of the pianos coming out of the factory.

20.2 Grandfathers or Exemplars in Quality Control

The cumulative sum of the differences of the spectral centroid for two pianos could be used for quality control in the following way. Among the pianos being manufactured, an exemplar, sometimes called a "grandfather," could be chosen. The samples from this exemplar piano would be stored. Each new piano coming off the production line would be sampled. The cumulative sum of the spectral centroid difference between the newly produced piano and the grandfather could be used to evaluate the degree to which that piano differed from the exemplar. With experience one could develop cumulative sum criteria for deciding if the piano was worthy of leaving the plant or was in need of adjustments, and what adjustments should be made. Should there be a piano that was superior in some way to the grandfather, its samples would replace those of the exemplar.

The cumulative sum of the differences of the cumulative line spectrum maps, mentioned briefly in Section 17.2, has, I think, potential as a metric for comparing a production line piano with an exemplar but, as the book goes to print, I have not studied it enough to recommend it.

20.3 Some Soapbox Comments

Subjective Evaluation and Blindfold Tests

I have read many comments by experts in Larry Fine's *Piano Book*, his *Piano Buyer*, and Frank Baxter's Piano World Internet forum about certain pianos that

a. "emphasize the higher harmonics"
b. "have a strong fundamental"
c. "have an American sound"
d. "have a European sound"
e. "have that Baldwin sound"
f. "a richer and warmer tone"
g. "a huge range of color"
h. "a clearer bass"
i. "more complex."

These attributes are lifted from just one Piano World post. Other characteristics have been mentioned that only a wine connoisseur could appreciate.

This idea of making subjective evaluations cries out for blindfold tests where experts would be challenged to name the brand or country of origin of an unknown piano. Could they identify the presence or absence of strength in higher or lower harmonics and could that evaluation be supported by the shape of, say, a CLSM or a spectral centroid map? Could any of the metrics in this book be used to identify and/or quantify any of these characteristics and attributes? Could Steinway Model B pianos be characterized such that they could be identified and differentiated from a Yamaha C5? Carrying out such a test might only be feasible in a large school of music having a variety of pianos available for practice and performance.

20.4 Summary of Part Three

The interaction between two notes via the bridge was illustrated in the time and frequency domains: striking a note could stimulate those harmonics of another un-struck note which the two notes share. The weakness of the fundamental in the low-numbered notes was studied and it was found that many sounds do not need strength in the fundamental if there is a strong representation in the higher harmonics. A native characteristic of pianos, inharmonicity, was demonstrated and quantified. Line spectra were used to follow an amateurish attempt to tune a key. They were also used to follow an intervallic approach to tuning a key using partial coincidences.

Several novel metrics were introduced to aid in the characterization of pianos. The spectral centroid map was presented as a metric that could be used as an alternative and compact way to show how the power of various notes were distributed amongst its harmonics (or partials). The cumulative difference of the elements of a spectral centroid map was used to compare two pianos. The time envelope of all the 88 keys of a piano unsurprisingly showed that the sustain of a key decreased markedly with key number. The envelope plot showed how dramatically the sound strength changes character at the treble–upper treble transition.

In the line spectrum map, ridges showed the presence of the various harmonics (and of troublesome background noise). The chosen six pianos demonstrated similarities and differences in the maps. The cumulative line spectrum map was shown to be a companion of the line spectrum map. It uniquely exposed the absence of power at the low-numbered keys and the presence of hammer knock in the top octave.

Hammer knock was examined for the A7 note on the Knabe grand. The analysis suggested that the portion of the spectrum below the fundamental plays a significant part in the sound and pitch of the wave. The early part of the A7 wave was compared to the whole wave and found to have similar characteristics. On comparing the C5 note and the A7 note, the data suggested that there may be hammer knock in all the keys to some extent but that it is washed out by the stronger tone as one progresses down from the upper octaves. The line spectrum was used to demonstrate the impact of using the dropped rod to generate test samples as compared to manually striking the keys.

The CLSM and the spectral centroid map were used to evaluate the effects of a Wapin bridge conversion. For the one case presented here, it appears that the Wapin conversion moved power to the higher harmonics for the lowest keys. The cumulative difference of the CLSMs was used to illustrate the effects of the conversion.

The repeatability between two pianos of the same model was studied via the CLSM and the spectral centroid map. In this case, two model D Steinways showed that there are many more similarities than differences. A similar study looked at two different data gatherings from the same piano and showed that there is some variability in the methods (or could the variation be the result of slightly different conditions on the two days of the experiment?).

Finally, there were several suggestions as to how the metrics presented in this book might be used in piano manufacture.

Several metrics, such as the cumulative line spectrum map, the spectral centroid map, and especially the cumulative difference of CLSMs, are probably new to the reader. It is hoped that these metrics can be useful to piano technologists.

21 Defining and Understanding Piano Tone

Contributed by Delwin D Fandrich

In a 1964 landmark decision that attempted to separate art from obscenity, Supreme Court Justice Potter Stewart famously wrote,

> I shall not today attempt further to define the kinds of material I understand to be embraced within that shorthand description [i.e., obscenity]; and perhaps I could never succeed in intelligibly doing so. But I know it when I see it . . .

So it is with music. Noise to one listener is exciting music to another. And a piano perceived as "harsh and strident" by one is "bright and articulate" to another. One person's "warm and rich" piano will be "dead" or "dull" to another. All the various words used to describe piano tone do so in a subjective way and are based on the experience and preferences of the individual doing the describing; there have been no definitive rules to be followed. Although none of us has been able to define piano tone in words that will stand up to mathematical analysis, we all think we "know good piano tone when we hear it." And for hundreds of years we have continued to attempt to express piano tone using terms like "warm" or "bright" or "soft" or "harsh."

In this book, author and researcher David Koenig attempts to bring order to chaos. As a piano designer, I've been asked to comment on this book as it relates to my own work. Developing a desired piano tone is a complex process that has long been guided primarily by trial and error. This book gives us a box full of analytical tools that provide fresh ways of quantifying piano tone. Maybe one day soon we'll be able to define tone in terms that go beyond "I know it when I see it."

Because of my interest in piano design, manufacturing and rebuilding, I have a vested interest in seeing this succeed and, in this chapter, I'll attempt to articulate how and why various parts of Koenig's work excite me.

21.1 A Designer's Perspective

I have approached this book as a piano designer, not as a mathematician. In that capacity I am continually looking for ways to better view and quantify what is happening in the overall piano system—from the action and hammers to the strings to the soundboard and, ultimately, to the air

Spectral Analysis of Musical Sounds with Emphasis on the Piano. First Edition.
David M. Koenig, with comments by Delwin D. Fandrich.
Published in 2015 by Oxford University Press.

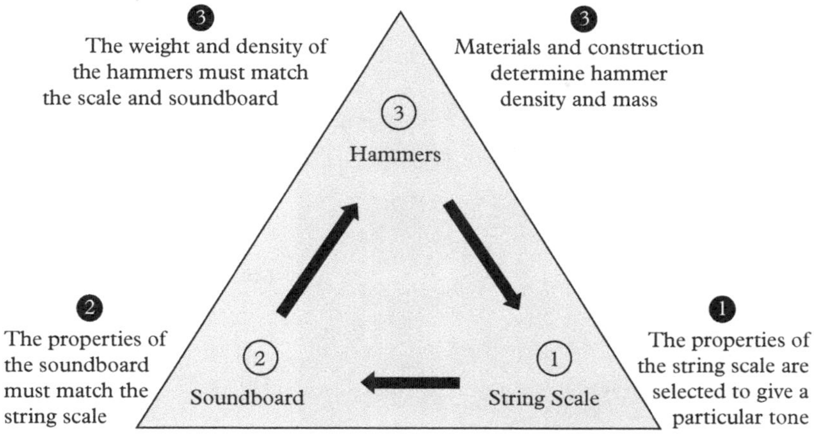

Figure 21.1.1 *The whole system is interactive. A change in any one of the three will have an effect on the others and on the overall performance of the piano*

surrounding the soundboard. It is a continuous and interrelated cycle. In my lectures on the basic concepts of piano design, one of the illustrations I use is something I call the "design triad," as shown in Figure 21.1.1.

The whole system is interactive and we have no way to predetermine with accuracy just what result we will get when we make specific changes. We can, and do, make educated and experienced guesses about these interactions, and the more experience and knowledge we accumulate the closer we come to our desired results.

In the end, we try it and listen or try it and measure. I make it a point to read every study and thesis and research paper that I can find on the piano and its function but, while I find most of them very interesting, few of them are really helpful when it comes to designing a piano. Koenig's work has introduced me to several analytical tools I've not seen before and it is my hope that these tools will help to take some of that guesswork out of the process. At the least they will help us to better visualize and quantify these interactions.

21.2 The Value of Signal Analysis

Scientists and researchers have been attempting to measure and visualize sound for hundreds of years. Some of these attempts were quite ingenious, such as the example from c.1915 shown in Figure 21.2.1.

Signal analysis has come a long way since it depended on the motion of a flame in response to a musical note. But we still use it for much the same reason; we want to know more about whatever tone-producing mechanism we are studying.

I have found signal analysis useful and generally use it for one of two reasons: (1) I want a visual representation of how two or more components of a piano are working together, and (2) I am comparing the wave envelope—usually of the same note—on two or more pianos (or before and after some physical change). In both cases, I am usually trying to evaluate what, if any, change in tone might have been brought about by some physical change made to one or more parts of my design triad.

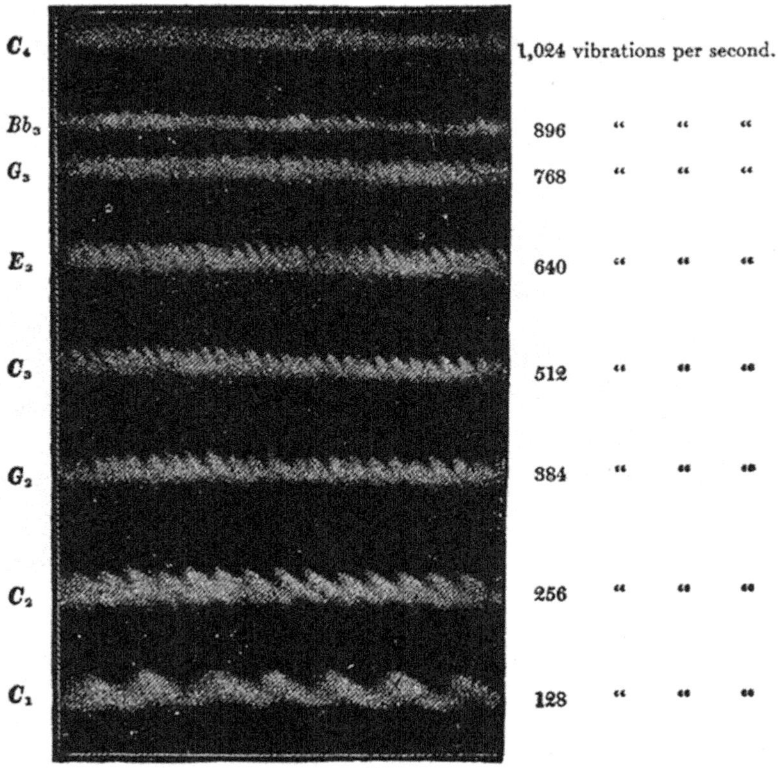

Fig. 1
Photograph of the motion of the flames while singing the vowel *ä* as in father.

The lower line is the fundamental, and the others are the 1st, 2nd, 3rd, etc., overtones in the order of their pitch. One wave of the fundamental corresponds to two in the first overtone, three in the second, four in the third, and so on.

Figure 21.2.1 *Visualizing wave harmonics using flames*

By looking at different visual representations of a piano's tone, we can more clearly understand how its various parts are working together (or not). They can help us picture and understand things in ways that even critical listening simply cannot match.

The Time Domain

The first visual picture of a piano tone that most of us see will be the simple time domain plot as shown elsewhere in this book and repeated below.

This plot is popular because it is so easy to get—any sound editing program such as Audacity (free) or Sony's somewhat easier to use Sound Forge (approximately 60) will work—and they give us a quick overall picture of the amount of sound power in the initial attack, and then they show how rapidly (or slowly) the vibrating energy in the strings is transferred to the soundboard system

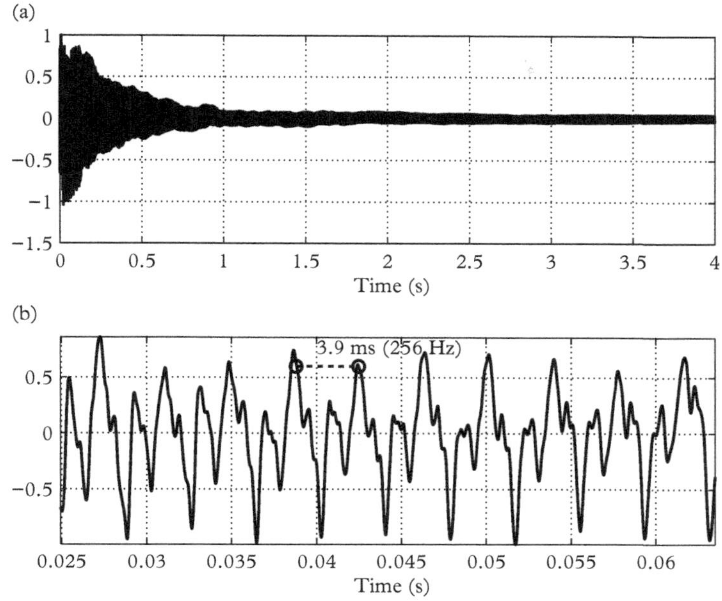

Figure 10.1.4 *Time domain plot of Knabe C4: (a) long view, and (b) close-up view showing a period of 3.9 ms or 256 Hz*

and converted into sound energy. It gives us a visible record of a note's rate of decay that we can easily save for later study.

We can compare the time domain plot from one string (or set of strings) with others from pianos that have been modified in some way to see what effect those changes have made in those two parameters. For example, if we have one region of the piano that has a good, strong attack but a wave envelope that decays too rapidly we may want to increase the stiffness of the soundboard assembly in that area. We might do this in several ways but let's assume we've increased the height or width of a rib. We could get at least a rough idea of the effect that change has had on the rate of energy transfer from the strings to the soundboard assembly. By carefully controlling the number and types of changes we make to the various elements of the test piano(s) we can use these plots to show us—in broad terms, at least—what effect those changes have made to the initial attack sound and the following decay rate.

The time domain plot is not very subtle, however. It's a bit like looking at a black and white photograph of a beautiful flower. We can see the shape of the flower and its size but there is a lot of information—color and motion—missing. The time domain plot tells us nothing about the frequency spectrum or about how specific harmonics within that spectrum decay or interact with each other. For that we have to find some other way of looking at the sound envelope.

The Line Spectrum

With the advent of digital signal processing, piano designers have had access to FFT (fast Fourier transform) spectrum analyzers to generate simple *line spectrum* plots such as the one shown in

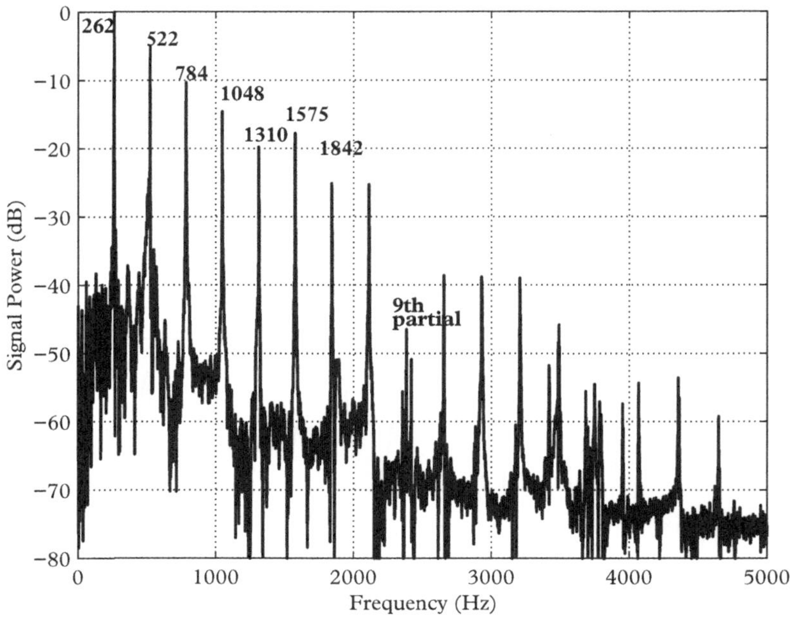

Figure 10.1.5 *Knabe C4: (a) line spectrum*

Figure 10.1.5a, first shown in Section 10.1. I don't try to analyze the mathematics behind the FFT, I'm just glad I can turn the thing on and get useful information from it.

The line spectrum plot gives us a quite a lot more information; we can see the relative strengths—amplitudes—of the various harmonics of a sound envelope. We can tell whether or not we're getting any sound energy at the fundamental (always an important consideration but especially so with shorter pianos). And we can detect things like longitudinal waves.

The line spectrum is useful in studies such as quantifying the effect of changes to that rib. Did the strength of the lower harmonics change? Or the higher harmonics? The answer will tell us if changing the height of the rib was the right choice. Perhaps we should have left the height alone and changed the width. Or both.

While this information is certainly useful, it is still limited; it is rather like watching a movie in black and white. It's a lot better than viewing a series of still photographs of a moving event but it is still leaving out a lot to be desired. What we can't tell with the line spectrum is what happens to the sound envelope over time.

We've now added a bit of color to the photograph. But it is still a brief snapshot taken of a single point in time.

The Spectrogram (or "Waterfall" Display)

The piano's tone package is unlike that of, say, the pipe organ, which produces a relatively constant mix of harmonics at a relatively constant volume as long as the key is depressed. In contrast, the piano produces an instantaneous—percussive—sound that is somewhat chaotic in nature but which very quickly settles into a waveform based on its fundamental frequency (whether there

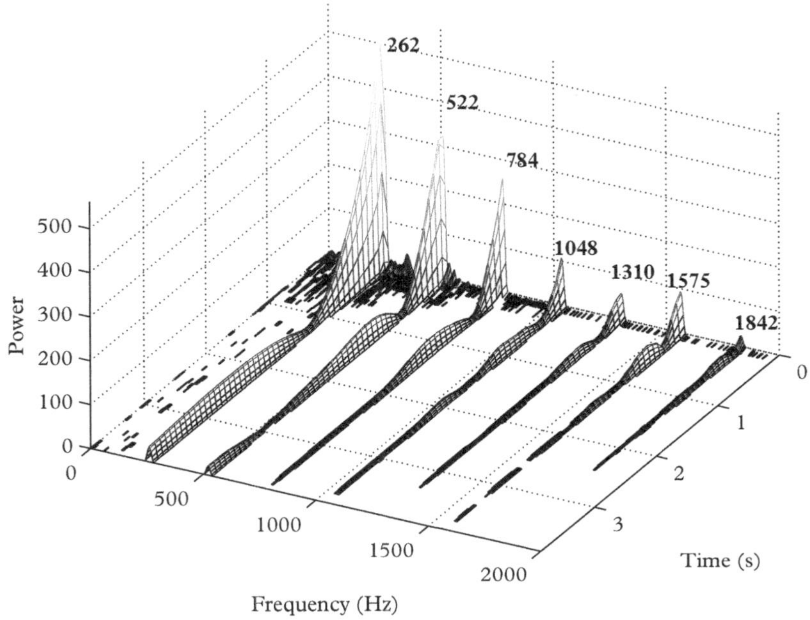

Figure 10.1.6 *Spectrogram of Knabe C4*

is any energy there or not) and which immediately begins to decay at some rate that varies significantly not just from piano to piano but from note to note.

This rate of decay should vary when the same note is played at different intensities. The sound envelope changes from note to note and from piano to piano. It also changes on the same note when the force of the hammer blow changes against the strings. Or at least it should. Most skilled piano tuner/technicians and experienced pianists can readily hear and recognize this change (commonly called the *dynamic* response of the piano). It is this dynamic change that makes possible the piano's ability to produce *forte* and *pianissimo* sound rather than just loud and less loud sound.

Studying and quantifying this dynamic change is challenging and requires more than a line spectrum taken at a single instant in time. To view this we need to introduce the time element into our picture of the waveform in some way. This is done using a *spectrogram*—also called a *waterfall* plot—such as that in Figure 10.1.6. This picture combines the information presented in the simple time domain plot and the line spectrum plot and, with a little experience in reading them, it can be extremely useful. Several spectrograms of the same note taken with different hammer velocities can tell us quite a lot about the interaction between the hammer, the strings, and the soundboard assembly. Finally, at least for a given note played with a given amount of force, we are able to watch a movie in color.

Using the Spectrogram to Improve Factory Processes

In the factory, if we're testing different methods of piano hammer-making, we can press a few sets using one process, change the felt or readjust things and press a few more sets. By installing these hammers on two different actions fitted to the same piano and recording the sound produced by each, we can create a visual picture of the tone they produce. With some experience at analyzing the resultant spectrograms, we can learn quite a lot about how hammers made using different processes actually perform in practice.

The argument can, of course, be made that, given a certain amount of voicing experience and ear training, a skilled technician can hear these differences with more subtlety than can be provided by any amount of signal analysis. And this is at least partly true. I can usually tell with fair accuracy how a certain hammer is going to perform just by feeling it, and most piano technicians who have worked with replacing and voicing hammers for a few years can also do this.

But the dirty little secret is that most workers in modern piano factories have little, if any, voicing experience or ear training. Indeed, I've encountered altogether too many workers in hammer departments—and voicing departments, for that matter—who have never heard a piano of any manufacture actually played in concert. Given this reality, signal analysis takes on a stronger and more critical role in the modern factory. Once standards are established, the factory has a way to evaluate tone performance even without golden ears.

Multiple Uses for the Spectrogram

Outside of the factory, I regularly use the spectrogram when comparison testing different string configurations on my string test frame. This is a research tool on which I can install two different strings—or groups of strings—that differ in only one parameter, and then I can compare the waveforms produced by each. For example, I can compare two wrapped strings having the same speaking lengths, the same core and wrap wire but with different backscale lengths. Aside from being able to hear a difference between the two strings, with the spectrogram it is possible to visually observe with a fairly high degree of certainty precisely what is happening as a direct result of just the one variable.

They can also prove a point to a skeptical management. Several years ago, for example, I used spectrograms to demonstrate the benefits of removing bass bridge cantilevers. This required shortening the speaking lengths of already very short strings and it went against more than a century of conventional wisdom. Even though the difference in tone between the two test samples was significant and could easily be heard, it was comparing the spectrogram of each sample that carried the day. Combining the information presented in both the time domain plot and the line spectrum plot, the resulting spectrogram is much more useful than either studied on its own.

These three tools—the time domain plot, the line spectrum plot, and the spectrogram—have been readily available to anyone even moderately interested in researching pianos and piano tone. It takes little to get started—a measurement microphone of reasonable quality, a modern computer with a built-in soundcard, and some readily available FFT software. Since I have to travel with a reasonably powerful laptop anyway the only extra gear I have to carry with me are a microphone (along with the requisite pre-amp and phantom power supply) and a microphone cable. It's all quite incredible when compared with the large, bulky, heavy, and frightfully expensive standalone signal analyzer I used back in the 1980s.

21.3 The Cumulative Line Spectrum Map

Up to this point in this book, I was on fairly familiar ground. But I became excited when I saw demonstrated a couple of new—at least new to me—ways of looking at piano tone. The first of these, and possibly the most powerful and useful, is the cumulative line spectrum map (CLSM). Koenig writes, "The cumulative line spectrum is a running sum of the line spectrum" (see Section 2.10) and "the cumulative line spectrum map is to the line spectrum map as the cumulative line spectrum is to the line spectrum" (see Section 15.3).

All of the displays I've mentioned above help us study what is happening with individual notes but they don't tell us much about the tone qualities of the instrument as a whole. The cumulative line spectrum changes all that; it displays the relative power in the most significant harmonics for each note—the voice of the whole piano—all on one concise, three-dimensional map.

Looking at a Piano's Voice

The Knabe Grand (from Section 9.2)

The Knabe semi-concert grand is 7′ 8″ long and appears to have been built in 1927. It was recently rebuilt by Beethoven Pianos in Manhattan. Keys 1 through 8 are single strung (or monochords) and wrapped. Keys 9 through 16 are double strung (bichords) and wrapped. Keys 17 through 24 are wrapped trichords. Keys 25 to 88 are triple strung (trichords) and not wrapped. Keys 1 through 24 make up the bass section (see Figure 9.5.1). This piano has the bass–tenor break between keys 24 and 25. The tenor zone contains keys 25 through 45. The tenor–treble break is between keys 45 and 46. The treble–upper treble break (beyond which there are no dampers) is between keys 66 and 67.

In Figures 15.3.7 and 15.3.8 we can see at a glance which harmonics of each note contribute the most amount of sound energy to that note, and we can see trends and easily spot inconsistencies. In the CLSM for the Knabe grand, it is clear that the lowest notes depend heavily on energy

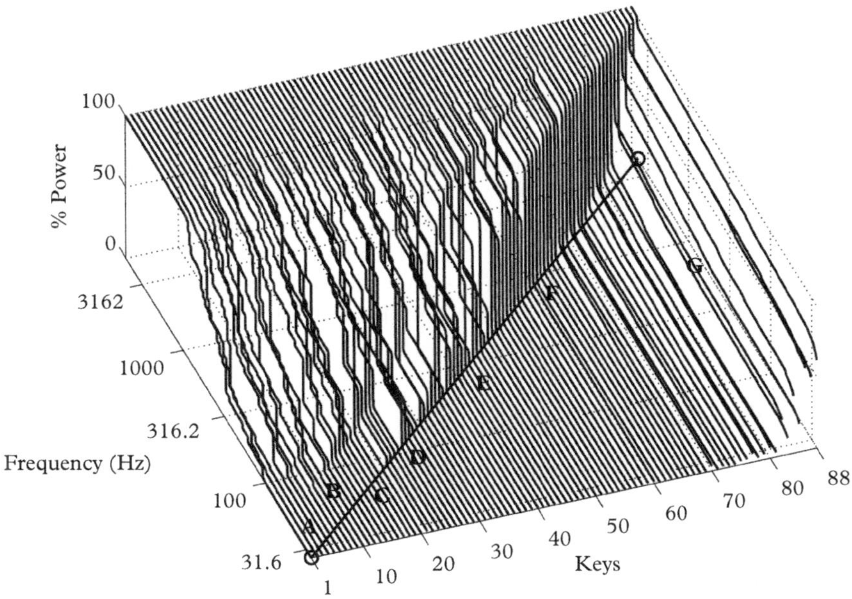

Figure 15.3.7 *CLSM of the Knabe grand piano*

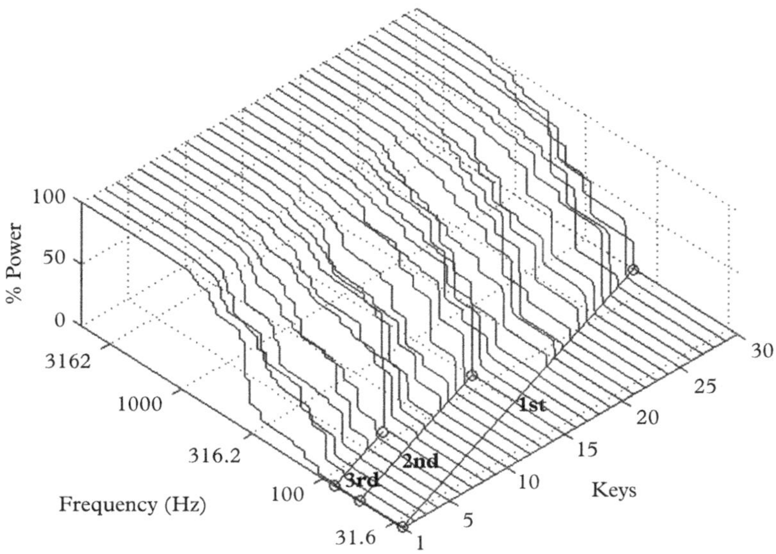

Figure 15.3.8 *CLSM of the Knabe grand piano's first 30 keys showing the absence of fundamental power in the lowest keys*

in the higher harmonics. In fact, the lowest 18 notes show little or no discernible power in the fundamental, and I find this somewhat surprising. The piano is long enough that I would expect to measure at least some power in the fundamental. If this were a new production piano I'd be studying the fundamental design to find out why that fundamental is missing.

There are noticeable variations in the note-to-note spectrums that could be attributable to either measuring inconsistencies, room resonances, uneven hammer voicing, or more than likely a combination of all three.

In the CLSM for the whole piano, it is obvious that the note-to-note audible sound power gradually shifts from the higher harmonics of the struck string(s) to the fundamentals, and after about note 70 the higher harmonics add very little energy to the sound envelope.

It would be hard to overstate the potential of the CLSM to graphically demonstrate the tone qualities of a piano. Assuming that proper measurement techniques—microphone standards and locations, hammer strike forces, and so forth—and analysis protocols can be established, it is the first analysis technique I've seen that has the potential to reasonably compare the sound of one piano with that of another.

Hammer Knock

One of the components of piano tone about which little has been written is the phenomenon known as *hammer knock*. It can be clearly heard, of course, but it has been difficult to study. It shows up on line spectrum plots and waterfall plots but it is hard to visualize and study. It shows up very clearly on the CLSM. Hammer knock—the term is something of a misnomer since very little of the knocking sound seems to come from the hammer—is an inherent part of every piano tone. It is part of the percussive attack sound that defines the piano as both a percussion and a string instrument. The piano would not sound like a piano without its presence. Still, enough is enough!

Through most of the scale, hammer knock contributes almost subtly to the sound envelope but does not overwhelm it. The CLSM, however, clearly shows not only the presence of hammer knock but its strength as it becomes an increasing percentage of the overall sound envelope from approximately note 70 upwards. The amount and character of hammer knock varies from one piano to the next—until it becomes the predominate sound heard in the top few notes.

There are several things we think we know about hammer knock:

- Hammer mass—other factors being equal, pianos with heavier hammers in the treble section will create more hammer knock.
- Strike point ratio—hammer strike points closer to the front termination generate more hammer knock. Most pianos have a strike point ratio of something between $1/12$ th and $1/18$ th of the speaking length at C-88. Assuming a speaking length of 54.0 mm (older designs would have been closer to 48.0 mm) this places the hammer strike at 3.8 mm from the V-bar. Given the stiffness of three taut piano strings, it should be no surprise that much of the hammer's impact force is transferred directly to the V-bar/capo tastro bar.
- Hammer shape—hammers with a rounded or flattened striking point will create less signal and more hammer knock. Even with the hammer strike point nominally 3.8 mm from the V-bar/capo tastro bar, an effective hammer strike width of 2.0 to 3.0 mm means that some part of the hammer will be 2.8 mm (or less) from the V-bar.
- Design of V-bar/capo tastro bar—V-bars with a more rounded shape will increase hammer knock, as will thinner capo tastro bars.

We can form some generalities from the above (abbreviated) list, but more research is clearly needed. While some hammer knock is critical to defining the tone of the piano, too much can overwhelm the desired tone of strings in the mid-to-high treble. The CLSM will make research into hammer knock easier and more productive. It has the potential of becoming a significant aid in understanding and quantifying the phenomenon.

21.4 Autocorrelation

In the piano the concept of pitch—or the confusion thereof—generally only comes up when we're discussing the low-to-mid bass. Especially in short pianos there is little, if any, sound power in the fundamental or (sometimes) in the second and third harmonics. Most of the acoustical power is contained in the higher partials, and sometimes much higher partials. As discussed earlier in this book, some of this acoustical information gets filled in through various means and we do perceive the "pitch" of the lowest notes on most pianos—see Chapter 11.

The greater problem is that the higher partials of these strings are inharmonic; that is, they are not precise mathematical multiples of the fundamental. Due to inharmonicity—see Section 5.10 and Chapter 12—these harmonics have a frequency somewhat higher than would be expected if the strings were perfectly flexible. This can become particularly problematic in shorter pianos and/or pianos with very large and stiff core wire in (particularly) the monochord wrapped strings. Here the frequency of those high harmonics can actually be closer to the next note up in the scale than they are to the fundamental frequency of the note being played. At some point we begin to have difficulty discerning the actual pitch of the note being played.

A study of autocorrelation—Section 2.11 and Appendix 1—can help us better understand why we have this difficulty, and with that understanding comes the clues toward alleviating or at least lessening the problem.

21.5 Conclusion

I confess that I have not yet fully understood the entire theoretical basis for the different ways of looking at piano tone provided in this book. Thankfully, we do not need to know every detail of how our tools work in order for us to use them effectively.

The piano industry is undergoing a major change as this is being written. Many take the position that the piano "has already been invented and there is nothing left to change or improve." And it is certainly true that there are many exceptional pianos available today. Since its conception during the late 1800s, however, what has come to be called the "modern" piano industry has been divided into two separate camps—the hand-crafted instruments that offer exceptional performance albeit at a very high price, and the mass-produced instruments that sacrifice performance for a significantly lower price.

The introduction of computer numerically controlled (CNC) machinery has narrowed and blurred this performance gap. The performance of mass-produced pianos has been steadily improving as manufacturers have figured out how to optimize their designs and their production processes to better complement each other. This optimization has been aided by the tools that digital signal analysis have made practical. Some of these tools have long been available to the piano designer and rebuilder; others, such as the cumulative line spectrum map, are new, at least to me. I am hopeful that this book will lead to the continued development and application of these tools. I believe the design evolution of the piano still has a way to go, and this book will help us along that way.

Appendix 1
Mathematical Basis for Part One

The line spectrum is derived from the Fourier transform (FT). We start with the continuous Fourier transform and move to the discrete Fourier transform. There is a short discussion of the fast Fourier transform, the fine grid line spectrum, and an alternative to the fine grid discrete Fourier transform via padding. The impact of a window on spectral shapes is discussed with the aid of the convolution operation. We spend some time presenting the Fourier series and conclude the appendix with a detailed discussion of the spectral centroid. As we have mentioned in the preface, one can profitably read this book without needing to understand the contents of this appendix.

The Continuous Fourier Transform (FT)

By definition, the FT of a *continuous* time domain sound wave, $y(t), -\infty < t < -\infty$, is

$$Y(f) = \int_{-\infty}^{\infty} y(t)\, e^{-j2\pi f t} dt = CFT\{y(t)\}, \tag{A1.1}$$

where

$$\exp(-j2\pi ft) = \cos(2\pi ft) - j\sin(2\pi ft). \tag{A1.2}$$

Equation A1.1 can be considered as a weighted sum of the sound wave $y(t)$ over all time with the weighting factor of $e^{-j2\pi ft}$. The FT moves $y(t)$ from the time domain to $y(f)$ in the frequency domain for $-\infty < f < \infty$. Note that $CFT\{*\}$ represents the operation of making a continuous Fourier transform out of whatever is inside the brackets.

The inverse FT of $Y(f)$ is defined as

$$y(t) = \frac{1}{2\pi} \int_{-\infty}^{\infty} Y(f)\, e^{j2\pi f t} dt = ICFT\,[Y(f)]. \tag{A1.3}$$

The inverse FT is a weighted sum of the transformed quantity over all frequencies and $ICFT\,[*]$ is a symbol representing the operation of a time-domain function on whatever is inside the brackets. Combining Eqs. A1.2 and A1.3 gives

$$Y(f) = CFT\,\{ICFT[Y(f)]\}. \tag{A1.4}$$

Likewise, one could write

$$y(t) = ICFT\{CFT[y(t)]\}.$$

Convolution

One of the staples of signal processing is the convolution operation where two quantities that multiply in one domain convolve in the other domain and vice versa. For example, if two time domain signals, $y(t)$ and $w(t)$, multiply to form $z(t)$, as in

$$z(t) = y(t)\, w(t), \tag{A1.5}$$

then in the frequency domain the FTs of the time domain product convolve, as in

$$CFT\ \{y(t)\, w(t)\} = Z(f) = \int_{-\infty}^{\infty} Y(f-u)\, W(u)\, du. \tag{A1.6}$$

Although it may not be obvious, Eq. A1.6 can be considered as a weighted sum of $W(u)$ with a weighting function $Y(f-u)$ for every f in the frequency domain. Equation A1.6 can also be written as

$$Z(f) = \int_{-\infty}^{\infty} W(f-u)\, Y(u)\, du. \tag{A1.7}$$

Since the variable of integration is u, the negative sign in the argument $f-u$ means that the weighting function moves in the opposite direction to u as the integration proceeds left to right. Perhaps the reader can see why the adjective "convoluted" means "complicated, intricately involved."

The inverse relationship is also true. If one can write the product of the FTs in the frequency domain,

$$P(f) = W(f)\, Y(f), \tag{A1.8}$$

then in the time domain one can write

$$ICFT\ [W(f)\, Y(f)] = p(t) = \int_{-\infty}^{\infty} w(t-u)\, y(u)\, du = \int_{-\infty}^{\infty} w(t)\, y(t-u)\, du, \tag{A1.9}$$

where, again, the integration operation can be considered as a weighted sum.

In Eq. A1.9 it is sometimes not immediately obvious how the weighting function $w(t-u)$ should be applied. The reader is invited to look at the Wikipedia entry for convolution where the weighting operation is animated.

The Discrete Fourier Transform

We deal with sampled values in the discrete time domain, as in $y_1, y_2, \ldots, y_N$, where N is the number of samples and is usually chosen to be even. The samples occur at times

$$t_1, t_2, \ldots, t_N \text{ or } 0, h, 2h, \ldots, (N-1)h,$$

where $t_k = t_{k-1} + h$, $k = 2, \ldots N$ and $t_1 = 0$. The sampling interval h is related to the sampling frequency F_s by

$$F_s = 1/h. \tag{A1.10}$$

The Nyquist or folding frequency is

$$f_{Ny} = \frac{F_s}{2} = \frac{1}{2h}. \tag{A1.11}$$

The discrete Fourier transform (DFT) is defined as

$$\begin{aligned} Y_n &= \sum_{k=1}^{N} y_k e^{-j2\pi f_n t_k} \qquad n = 1,\ 2,\ \ldots,\ N \\ &= DFT\left\{y_k\right\}, \end{aligned} \tag{A1.12}$$

where $t_k = 0,\ h,\ 2h,\ \ldots,\ (N-1)h$. Equation A1.12 can be considered as a weighted sum of the N sound wave samples $y(t_k)$, $k = 1, \ldots, N$ with the weighting factor of $e^{-j2\pi f_n t_k}$. The sum in Eq. A1.12 might be considered as a discrete time approximation to the integral in Eq. A1.1; however, it is also a starting point for the DFT definition.

The N values of Y_n occur at the N frequencies, $f_1, f_2, \ldots, f_N$ where

$$f_n = \frac{n-1}{Nh},$$

or at the frequencies

$$0, \frac{1}{Nh}, \frac{2}{Nh}, \ldots, \frac{\frac{N}{2}}{Nh}, \ldots, \frac{N-1}{Nh},$$

or at the frequencies

$$0, \frac{1}{Nh}, \frac{2}{Nh}, \ldots, \frac{1}{2h}, \ldots, \frac{N-1}{Nh}.$$

The interior frequency $1/(2h)$ is the Nyquist or folding frequency f_{Ny}, and the duration of the samples is $T_d = (N-1)\,h$. The separation in the frequency grid (or the frequency resolution) is $\frac{1}{Nh} = \frac{1}{T_d}$, hence the longer the duration of the signal, the better the resolution.

The inverse DFT is

$$\begin{aligned} y_k &= \frac{1}{N}\sum_{n=1}^{N} Y_n e^{j2\pi f_n t_k} \qquad n = 1,\ 2,\ \ldots,\ N \\ &= IDFT\left[Y_n\right]. \end{aligned} \tag{A1.13}$$

Equation A1.13 can be considered as a weighted sum of the DFT $Y(f_n)$, $n = 1, \ldots, N$, with the weighting factor being $e^{j2\pi f_n t_k}$.

The sequence $y_1, y_2, \ldots, y_N$ can be complex, but for all of our work the sequence is real since it represents samples of a sound wave in the time domain. The sequence $Y_1, Y_2, \ldots, Y_N$, representing the DFT of the sound wave samples, is, in general, complex. That is, an element Y_k can have a magnitude $|Y_k|$ and a phase θ, as in $Y_k = |Y_k| \exp(j\theta)$ where $\theta = \tan^{-1}\left[\frac{\text{Im}\{Y_k\}}{\text{Re}\{Y_k\}}\right]$. The magnitudes of the DFT components, $|Y_1|, |Y_2|, \ldots, |Y_N|$, make up the components of the line spectrum. These magnitudes are symmetrical about the Nyquist or folding frequency f_{Ny} Consequently, for plotting purposes only the first $N/2$ magnitudes are used. These magnitudes are also the amplitudes mentioned in Section 1.2. Note that by taking just the magnitudes and discarding the phases in constructing the line spectrum, there is a loss of information.

Keep Eqs. A1.12 and A1.13 in mind because we will see similar equations later on when we talk about the Fourier series.

As with the continuous FT, we can write

$$y_k = IDFT\,[DFT\{y_k\}]. \tag{A1.14}$$

The Fast Fourier Transform

The fast Fourier transform or FFT is a method of efficiently computing Eqs. A1.12 and A1.13. The values of the DFT at the frequencies $0, \frac{1}{Nh}, \frac{2}{Nh}, \ldots, \frac{\frac{N}{2}}{Nh}, \ldots, \frac{N-1}{Nh}$ with the spacing of $1/(Nh)$ are often sufficient to gain the needed insight. Maximum efficiency is obtained if N is a power of 2.

The Fine Grid DFT or Fine Grid Line Spectrum

Sometimes one wants to look at a finer frequency grid. In this case, we resort to calculating

$$Y(f_m) = \sum_{k=1}^{N} y_k\, e^{-j2\pi f_m t_k}, \tag{A1.15}$$

where $f_m = f_1, f_2, \ldots, f_M$ (note that M is used, not N) and where the spacing between the frequencies is dependent on the problem at hand. For example, if the frequency range of interest is $[f_L, f_R]]$ then the frequency spacing could be chosen as

$$\Delta f = \frac{f_R - f_L}{M}, \tag{A1.16}$$

where the analyst chooses M. This allows one to specify any desired frequency spacing at the cost of increased computation time since the FFT cannot be used. The use of the Matlab® function *ndgrid* can ameliorate this problem, especially if there are repeated calculations over the same frequency and time grids. The following snippet of Matlab code

```
>t=(0:N-1)*h;              % a vector of time samples
>delF=(fR-fL)/M;           % the desired spacing between the frequencies
>f=fL:delF:fR;             % a vector of frequencies where we wish to
                             calc the DFT
>[T F]=ndgrid(t,f);        % a 2-D time x frequency grid
>A=exp(-1i*2*pi*F.*T);     % the matrix factor in the calculation
                             of the fine grid DFT
```

illustrates how the *ndgrid* function can be used. The fourth line,

```
>[T F]=ndgrid(t,f); % a 2-D time x frequency grid
```

generates an $N \times M$ grid in the timefrequency plane. The fifth line,

```
>A=exp(-1i*2*pi*F.*T);   % the matrix factor in the calculation of the
                            fine grid DFT
```

calculates the value of A at each point in the grid. The fine grid DFT symbolized by Q is calculated from

```
>Q=abs(y*A);
```

where y is a $(1, N)$ vector of sound wave samples and A is an (N, M) matrix. The calculation of A=exp(-1i*2*pi*F.*T) takes a relatively long time but it can be reused if there are subsequent samples of $y(t)$ that have the same sampling interval and the same length. Also, the built-in function *ngrid* speeds up the calculation of A by orders of magnitude over a calculation using two nested for/next loops.

Often for fine tuning on the piano I choose a narrow interval of, say, 20 cents about the target frequency (please note that I am not an experienced or competent piano tuner). As is explained in Appendix 3, there are 100 cents per semitone and the relationship between cents and frequency is given by Eq. A.1.17:

$$r = 2^{\frac{1}{12}}$$

$$n = \text{floor}\left(49 + \frac{\log_{10}\left(\frac{\mathrm{f}}{440}\right)}{\log_{10}(\mathrm{r})}\right) \tag{A.1.17}$$

$$HpC = \frac{f(n)}{200}\frac{(r^2 - 1)}{r},$$

where f is the frequency in Hertz and HpC is the conversion factor in Hz/cent. (This equation is derived in Appendix 3.) Therefore 20 cents at 440 Hz is equivalent to 76.4 Hz.

For example, if there are 44,100 samples from a sequence sampled at 22,050 Hz (a sampling interval of $4.5351 \times 10{-}^5$ s), if the frequency range of interest is [430, 450] Hz, and if M is chosen to be 1000 then the theoretical resolution in the fine DFT would be

$$\frac{20}{1000} \cong .02\,\text{Hz}.$$

The FFT resolution for the same problem is

$$\frac{1}{Nh} = \frac{1}{44100 \times (4.5351 \times 10^{-5})} = 0.5\ \text{Hz}.$$

Padding

As an alternative to using Eq. A1.15, one can pad the time domain samples with zeros (append zeros to the end of the sequence), thereby increasing the length of the sample. Since the separation in the frequency grid is $1/(N\,h)$, using padding appears to effectively increase the resolution but it is really allowing interpolation between the points on the original frequency grid. The FFT can be applied to the padded series. In some cases, this leads to greatly decreased computation times, especially if the analyst pads the time domain data to a length equal to a power of 2.

The following example compares padding with the fine grid approach. A 220 Hz sine wave was sampled at 10,000 Hz, producing 256 samples of a wave over a period of 0.0255 s. The line

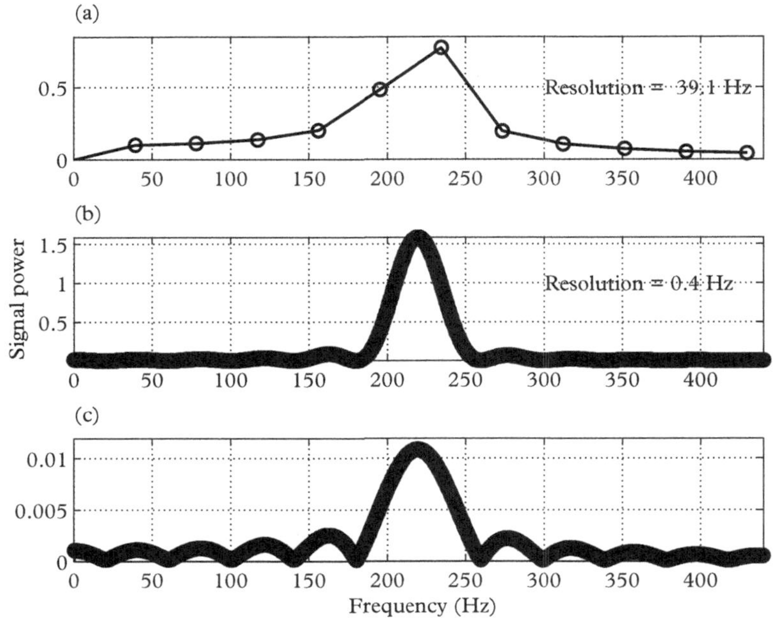

Figure A1.1 *Effect of padding on the line spectrum: (a) line spectrum directly from FFT calculation without padding, (b) fine grid line spectrum using 1024 points over a span of 440 Hz, and (c) line spectrum from FFT with padding out to 23,256 points*

spectrum using the FFT was calculated and is shown in Figure A1.1(a). The calculation time varied on my computer around 0.2 ms and the resolution was 39.1 Hz. The fine grid DFT, using 1000 points in the frequency domain over the range [0, 440] Hz, is shown in Figure A1.1(b). The calculation time varied around 10 ms and the resolution was 0.4 Hz. Finally, the original 256 samples were padded with zeros to 23,256 points and the line spectrum was calculated using the FFT—Figure A1.1(c). The calculation time varied around 3 ms and the resolution was 0.4 Hz. This example suggests that the penalty for using the fine grid DFT over padding is not prohibitive.

In some cases, I collect the result of striking a piano key over a period of 10 s. For the upper octaves the sustain time is on the order of 0.5 s so using the whole 10 s sequence means that one is analyzing mostly background noise. To address this, only the early part of the wave where the signal is strong is kept. The balance of the 10 s is padded with zeros.

Interpretation via Convolution and Windows

To sample a wave one effectively multiplies a continuous wave of infinite extent $y(t)$, $-\infty < t < \infty$, by a boxcar (or rectangular) window and then samples the windowed continuous wave of finite extent. The boxcar window is defined as

$$
\begin{aligned}
b(t) &= 1 \quad \text{for } 0 \leq t \leq T_{\text{d}} \\
b(t) &= 0 \quad \text{for } t < 0 \text{ and } t > T_{\text{d}},
\end{aligned}
$$

where T_d is the length of time over which the sampling takes place. Therefore, we have multiplied in the time domain, as in, $y_w(t) = b(t)\,y(t)$, then sampled the result of the multiplication, obtaining $y_{w1}, y_{w2}, \ldots, y_{wN}$. This multiplication in the time domain is equivalent to convolution in the frequency domain, as in

$$Y_w(f) = \int_{-\infty}^{\infty} B(f-u)\,Y(u)\,du, \tag{A1.18}$$

where

$$Y(f) = \int_{-\infty}^{\infty} y(t)\,e^{-j2\pi f t}dt \tag{A1.19}$$

and where $B(f)$ is the FT of $b(t)$, $Y(f)$ is the FT of $y(t)$, and $Y_w(f)$ is the FT of the windowed (but still continuous) sample. Therefore, the act of picking out a section of a wave having finite duration and taking the FT of the result means that, instead of the true FT, $Y(f)$, one gets $\tilde{Y}(f)$, which is a weighted-average version of $Y(f)$ where the weight is $B(f)$, the FT of the boxcar window.

The above discussion further suggest that we should be interested in the shape of $|B(f)|$, which is shown in Figure A1.2, plotted in dB units for samples of size 44,100 and 88,200. As the sample size increases the central peak get sharper and the side lobes are more attenuated.

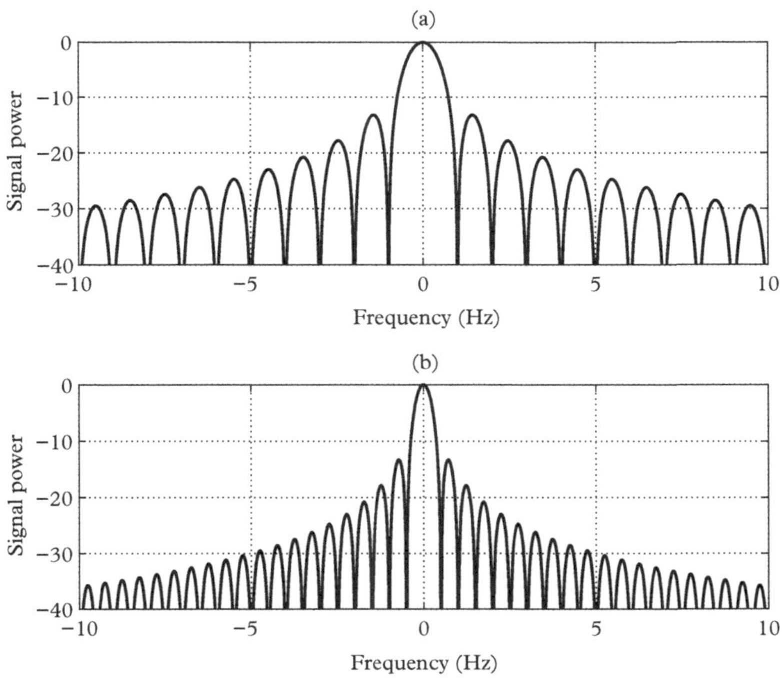

Figure A1.2 *Fine grid line spectrum of the rectangular window with sample sizes of (a) 44,100, and (b) 88,200*

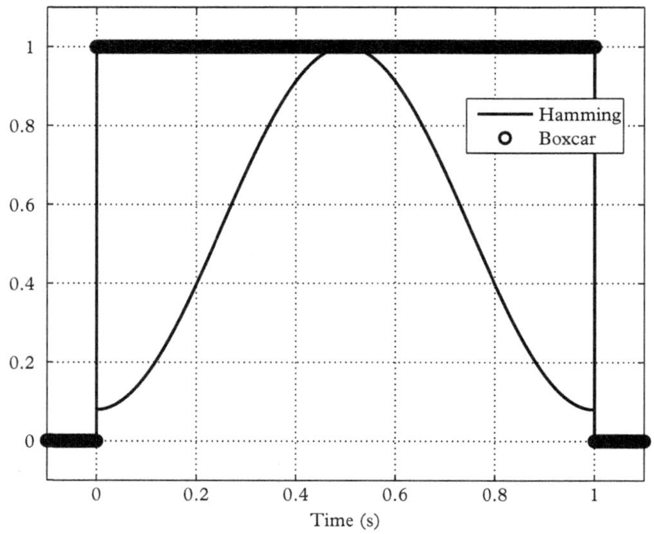

Figure A1.3 *Comparison of Hamming and boxcar windows in the time domain*

The central peak in the weighting function $B(f)$ tends to pick out the desired value of $Y(u)$ as the weighted averaging in Eq. A1.18 proceeds, somewhat like a delta function might. However, there are lobes on either side of the peak that tend to pick up nearby values of $Y(u)$, thereby *smearing* the average somewhat. Therefore, $Y(f)$ might be looked upon as a smeared version of $Y(u)$ with less smearing as the sample size increases.

Some signal processing analysts like to use different windows to address the smearing. The Hamming window is popular and has the shape shown in Figure A1.3, which also shows the boxcar window.

Consider the case of the signal $y(t)$ having a duration of 1.0 s and being sampled at 44,100 Hz. In effect, it would be the result of multiplying a wave of infinite extent by the Hamming window function $w(t)$ shown in Figure A1.3. The Hamming window takes the place of the boxcar window used previously. Figure A1.3 shows that the beginning and ending portions of the signal are multiplied by values slightly less than 0.1. Halfway through the duration the window has risen to its full value. The spectrum of the Hamming window, $W(f)$, is shown in Figure A1.4.

The main peak for the Hamming window is not as narrow as that for the boxcar window. On the other hand, the side lobes are down more than 40 dB for the Hamming window as compared to 15 to 30 dB for the boxcar window. This means that the ability to pick out and resolve individual peaks might be a little better for the boxcar window but the influence of neighboring peaks might smear the peak of interest.

As an example, consider an attempt to analyze two sinusoids:

$$y(t) = \sin(2\pi f_1 t) + \sin(2\pi f_2 t),$$

where $f_1 = 440$ Hz and $f_2 = 441.75$ Hz. Figure A1.5 shows the signal in the time domain, which is sampled at 44,100 Hz for 1 s (so there are 44,100 samples).

Note the beating due to the closeness of the two frequencies.

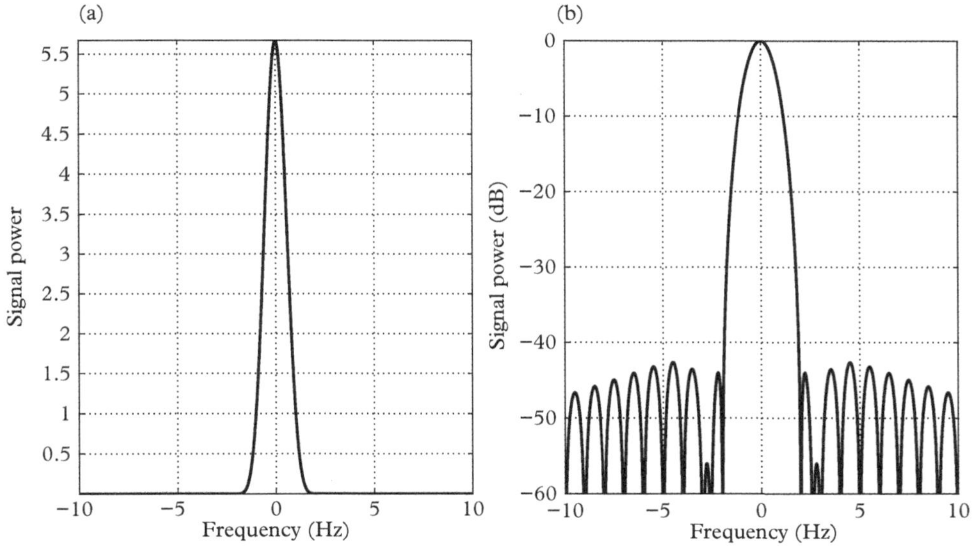

Figure A1.4 *Fine grid line spectrum of the Hamming window using (a) linear units, and (b) using dB units*

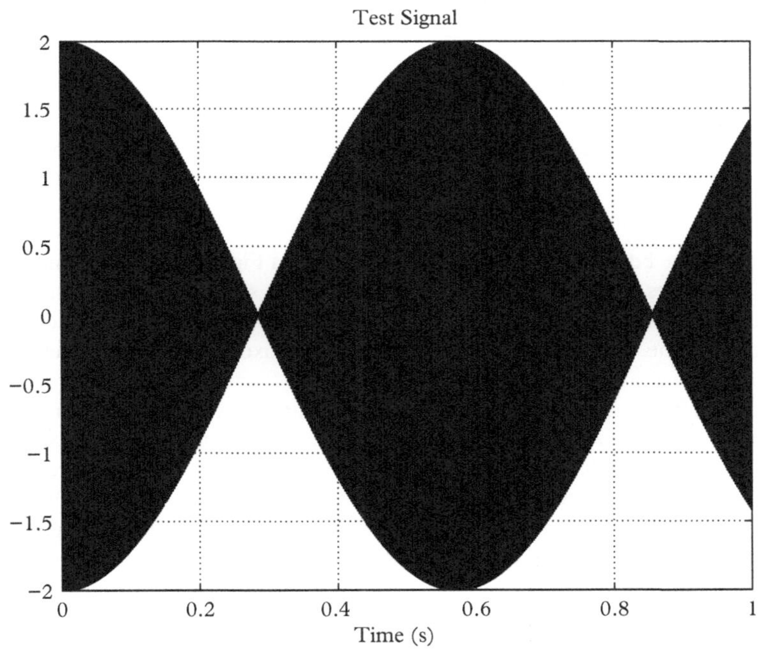

Figure A1.5 *Test signal: the sum of two sine waves of frequencies 440 Hz and 441.75 Hz*

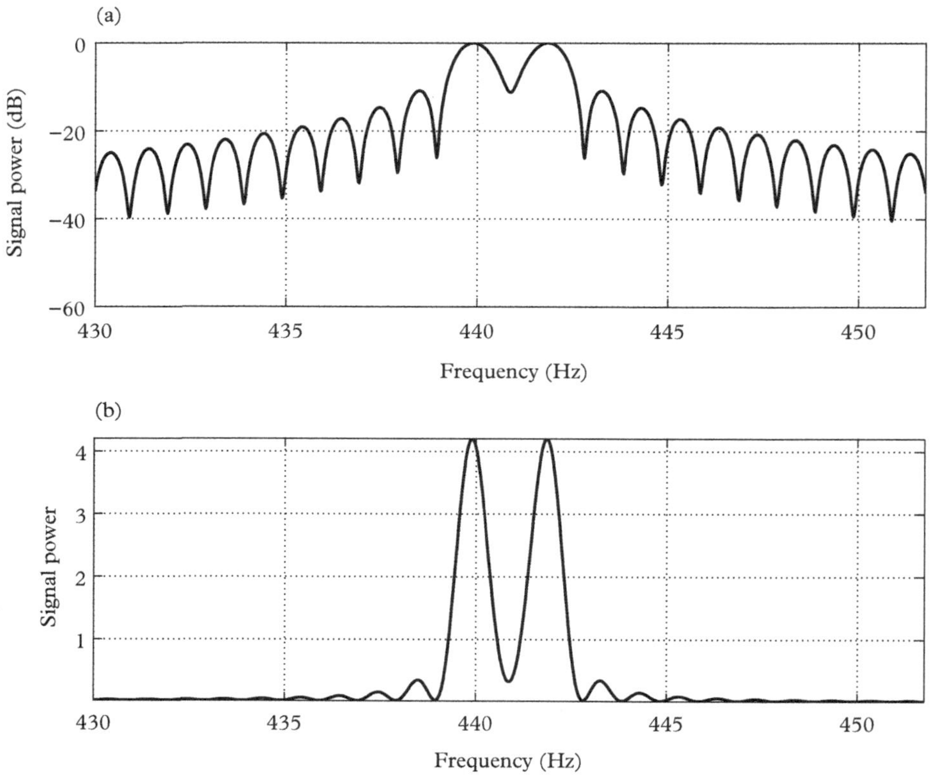

Figure A1.6 *Boxcar window applied to the test signal: (a) plotted in dB, and (b) plotted linearly*

The spectra of the boxcar-windowed signal are shown in Figure A1.6 in dB and linear units.

Note the side lobes in Figure A1.6(a), and that there is sufficient separation to identify two sinusoidal components.

The spectra of the Hamming-windowed signal and boxcar-windowed signal are shown in Figure A1.7.

Figure A1.7 shows that the separation using the Hamming window is not as great, but there are no side lobes in Figure A1.7(a). The boxcar or rectangular window is used exclusively in this book.

This last little exercise suggests that when using the fine grid DFT with the two windows the resolution is still limited by the length of the sample rather than the number of grid points in the frequency domain and that side lobes will appear. To support this contention the above experiment is repeated with a sampling period of 2.0 s instead of 1.0 s (so there are 2 × 44,100 samples).

Figure A1.8 shows that the boxcar window still has better separation but the Hamming window performs well and exhibits no lobes.

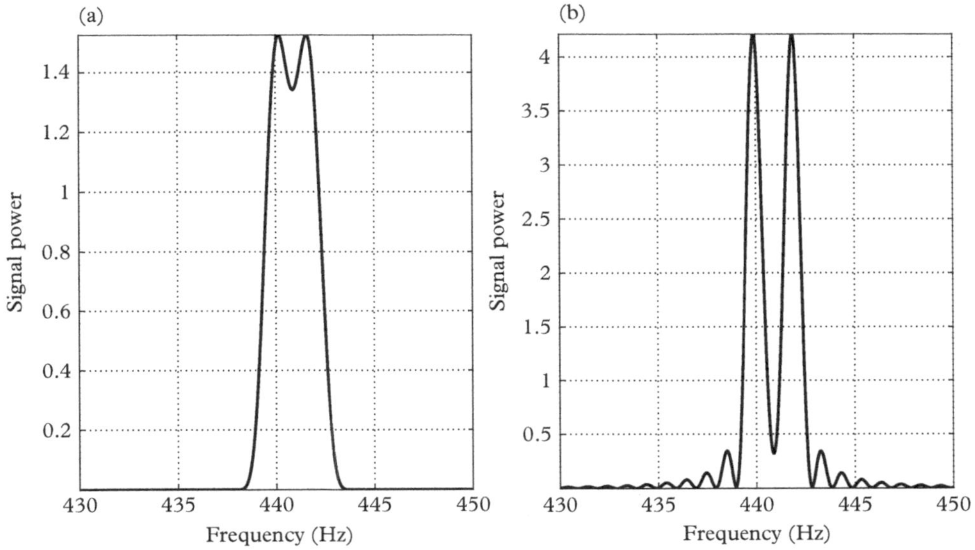

Figure A1.7 *Fine grid line spectrum of 1.0 s of test signal: (a) using the Hamming window, and (b) using the boxcar window*

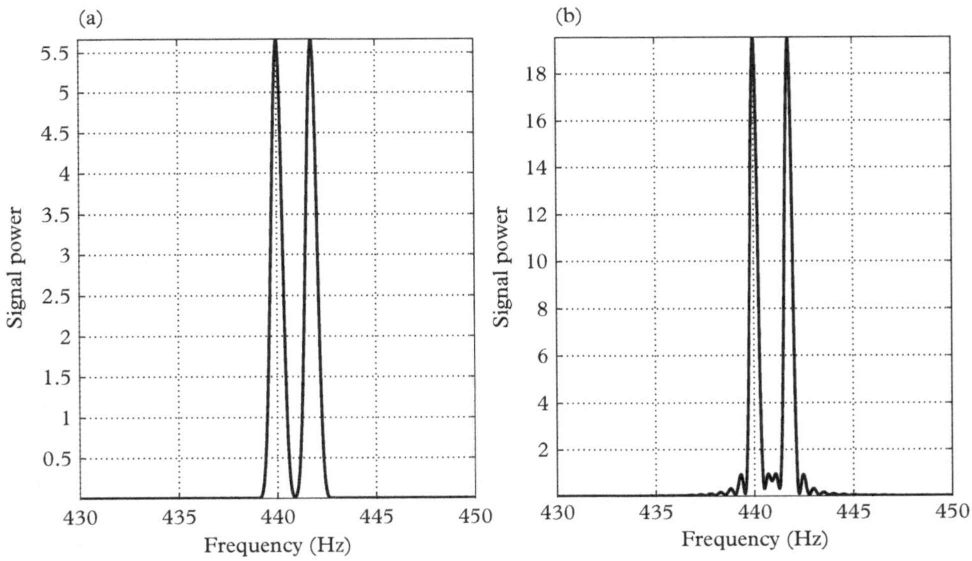

Figure A1.8 *Fine grid line spectrum of 2.0 s of test signal: (a) using the Hamming window, and (b) using the boxcar window*

The Fourier Series

In Section 1.3 we attempted to approximate a square wave with a sum of sines. Consider the more general case of approximating a wave $y(t),\ 0 \leq t \leq L$, by a sum of N sinusoids, but rather than writing the sum in terms of sines and cosines, lets use complex exponential sinusoids. Therefore, the approximation would look like

$$y(t) \cong \sum_{k=1}^{N} C_k\, e^{j2\pi f_k t}, \tag{A1.20}$$

where

$$e^{j2\pi f_k t} = \cos(2\pi f_k t) + j \sin(2\pi f_k t).$$

If the frequencies are represented by

$$f_k = \frac{k}{Nh}$$

and the time by

$$t \Rightarrow t_m = m\,h \qquad m = 0,\ 1,\ 2,\ \ldots,\ M-1,$$

then $y(t)$ becomes

$$y(t_m) = y_m \qquad m = 0,\ 1,\ 2,\ \ldots,\ M-1$$

and Eq. A1.20 becomes

$$y(t_m) = \sum_{k=1}^{N} C_k\, e^{j2\pi \frac{k}{Nh} mh} = \sum_{k=1}^{N} C_k e^{j2\pi \frac{k}{N} m}, \tag{A1.21}$$

where I have replaced the approximate equals sign with a regular equals sign.

To find the coefficients C_k, which in general will be complex, we multiply both sides of Eq. A1.21 by $e^{j2\pi \frac{p}{N} m}$ and sum over all of the sample points to get

$$\sum_{m=1}^{M} e^{j2\pi \frac{p}{N} m} y(t_m) = \sum_{m=1}^{M} \sum_{k=1}^{N} C_k\, e^{j2\pi \frac{p}{N} m}\, e^{j2\pi \frac{k}{N} m}. \tag{A1.22}$$

The orthogonality of the complex sinusoids means that

$$\sum_{k=1}^{N} C_k\, e^{j2\pi \frac{p}{N} m}\, e^{j2\pi \frac{k}{N} m} = N\delta_{pk},$$

which in turn means that Eq. A1.22 becomes

$$\frac{1}{N} \sum_{m=1}^{M} e^{j2\pi \frac{p}{N} m} y(t_m) = C_p. \tag{A1.23}$$

Therefore, the coefficients of the Fourier series approximation can be determined from a simple sum. Equation A1.23 could also have been derived using a least squares approach, meaning that the Fourier series is a least squares approximation.

Because of orthogonality the coefficients are independent of each other. Consequently, the following two approximations will have the same A_3, A_5, A_7 coefficients:

$$y(t) \cong A_1 \cos(2\pi f_1 t) + A_3 \cos(2\pi f_2 t) + A_5 \cos(2\pi f_5 t) + A_7 \cos(2\pi f_7 t)$$

$$y(t) \cong A_3 \cos(2\pi f_3 t) + A_5 \cos(2\pi f_5 t) + A_7 \cos(2\pi f_7 t).$$

This example is mentioned because of the discussion of the missing fundamental in Chapter 11. In the second approximation, the fundamental, or lowest frequency component, $A_1 \cos(2\pi f_1 t)$, has been removed from the first approximation. The remaining part of the approximation will still approximate $y(t)$ in a least squares sense.

Compare this situation with the following two least squares approximations:

$$y(t) \cong a_1 t + a_2 t^2 + a_3 t^3$$

$$y(t) \cong b_2 t^2 + b_3 t^3.$$

In this case, $b_2 \neq a_2$ and $b_3 \neq a_3$. Therefore, removing the lowest-power element in the first approximation without re-computing a_2 and a_3, leaving

$$y(t) \cong a_2 t^2 + a_3 t^3,$$

would not provide a least squares approximation.

Compare Eq. A1.23 with Eqs. A1.12 and A1.13. This suggests that finding the Fourier transform of a wave is analogous to approximating the wave with a series of sinusoids and that the absolute value of the coefficients, $|C_k|$, are related to the elements of the line spectrum that we use throughout the book.

If the wave has sharp corners in the time domain, as does a square wave, then N must be large, meaning that complex exponential sinusoids with high frequencies might be used. If those frequencies were greater than the Nyquist frequency $1/(2h) = F_s/2$, then their contribution would show up as aliases at frequencies inside the Nyquist interval. The line spectra in Figures 1.3.2 and 2.2.3 illustrate this point.

Snapshots of the Line Spectrum

The nature of a sound wave can vary with time. The spectrogram is one method of analyzing this phenomenon and we use it at several places in the book. It is essentially a calculation of the line spectrum from overlapping subsets of the sound wave. For example, Figure A1.9 shows a sequence of snapshots of the line spectrum of the Knabe C4 note. The first seven harmonics are shown as a function of time. Note how the power of each harmonic varies differently with time.

To get an idea of how this graph was constructed the reader should consult the Matlab script in the file SpectrogramHomeMade.m, which is located on the website mentioned in the preface.

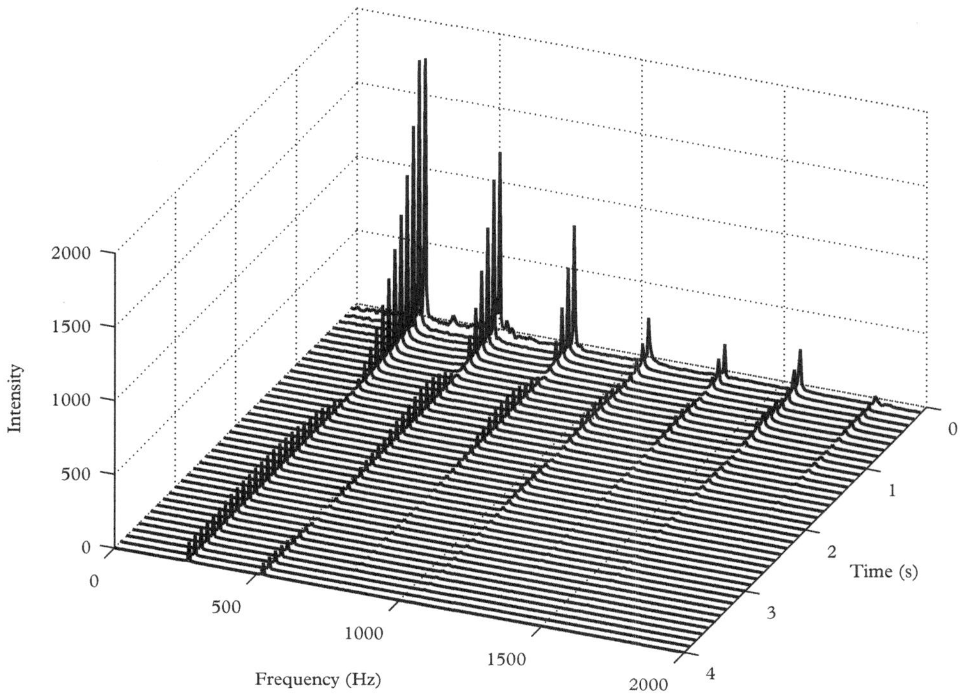

Figure A1.9 *Snapshots of the line spectrum for the Knabe C4 note*

Some of the spectrogram plots used in the book come from a built-in Matlab function and produce a slightly different-looking plot. The reader is directed to the Matlab Help feature to explore the intricacies of the spectrogram.m function.

Calculation of the Spectral Centroid

A less technical discussion of the spectral centroid was given in Section 2.11. Here we will go into the development of the formulas. But first, lets look at a simple example of computing an average.

Consider the case of N men having the following ages: $a_1, a_2, \ldots, a_N$. The un-weighted average age $\bar{a}$ is the sum of all the ages divided by the number of men, as in,

$$\bar{a} = \frac{\sum_{k=1}^{N} a_k}{N} = \frac{a_1 + a_2 + \cdots + a_N}{N},$$

where $\sum_{k=1}^{N}$ means "sum over whatever follows that has an index k from 1 to N."

If for some reason we wanted to weight each mans age by his height, where $h_1, h_2, \ldots, h_N$ are the heights of the men, the average age *weighted* by height would be

$$\hat{a} = \frac{\sum_{k=1}^{N} h_k a_k}{\sum_{k=1}^{N} h_k}.$$

This weighted average would be sensitive to extremely tall men and if there were a few tall and old men, $\hat{a}$ might be significantly larger than $\bar{a}$.

On the other hand, if all the men had the same height, say, H, then the weighted average would be

$$\hat{a} = \frac{\sum_{k=1}^{N} H a_k}{\sum_{k=1}^{N} H} = \frac{H \sum_{k=1}^{N} a_k}{NH} = \frac{\sum_{k=1}^{N} a_k}{N} = \bar{a}.$$

So, in the case of equal weights, the weighted average $\hat{a}$ is equal to the un-weighted average $\bar{a}$. In this case, we might say that the average is a weighted average where the weights are all unity.

At the other extreme, suppose that we were in the island country of Lilliput, that there were only ten men, that Lemuel Gulliver (the third man) was six feet tall and 50 years old and the other nine men were five inches tall and 20 years old, then the un-weighted average age of the men in Lilliput would be

$$\bar{a} = \frac{20 + 20 + 50 + 20 + 20 + 20 + 20 + 20 + 20 + 20}{10} = 23 \text{ years old.}$$

The average age weighted by height would be

$$\hat{a} = \frac{20 \times 5 + 20 \times 5 + 50 \times 72 + 20 \times 5 + 20 \times 5 + 20 \times 5 + 20 \times 5 + 20 \times 5 + 20 \times 5 + 20 \times 5}{5 + 5 + 72 + 5 + 5 + 5 + 5 + 5 + 5 + 5}$$

$$= 38.5 \text{ years old.}$$

The weighted average age is dominated by Gullivers age because of his height while the un-weighted age is nearly equal to that of the Lilliputians. Which is a better estimate of the age of the men on the island?

If we ask for the average frequency of a sound wave, we would probably want to use the strength at each frequency as a weight. For example, if we were interested in the average frequency of a wave that contained but one sinusoid having a frequency of 440 Hz, we would not want to take an un-weighted average over the whole Nyquist interval—that average would be equal to the sampling frequency divided by four and it would be independent of the nature of the sound wave. In this case, if the sampling frequency were 44,100 Hz, then the Nyquist frequency would be 22,050 Hz and the average frequency of the sound wave would be 11,025 Hz. On the other hand, if we weighted the frequencies by the strength at each frequency, the average frequency weighted by strength would be 440 Hz.

Therefore, if the strengths at each frequency in the Nyquist interval coming from an FFT are $(A_k)^2$, $k = 1, \ldots, \frac{N}{2}$, then the average frequency weighted by strength would be

$$\hat{f} = \frac{\sum_{k=1}^{N/2} (A_k)^2 f_k}{\sum_{k=1}^{N/2} (A_k)^2}. \tag{A1.24}$$

This is the spectral centroid. In Part 3 we see that the average frequency weighted by strength (spectral centroid) of a note like C1 with a fundamental frequency of 32.7 Hz is on the order of 400 Hz because so much power resides in the higher harmonics. On the other hand, the average frequency weighted by strength of the A5# note is in the neighborhood of 932 Hz, which is the fundamental frequency for that note because so little power resides in the higher harmonics.

The Autocorrelation

The autocorrelation was discussed in detail in Section 2.11 via an example. From that presentation, one perhaps can deduce that the autocorrelation is calculated from

$$Ac(m) = \frac{1}{F} \sum_{k=1}^{N-m} y(k)y(k+m),$$
$$F = \sum_{k=1}^{N} y(k)y(k). \tag{A1.25}$$

When the time delay index m is zero, $Ac(0) = 1$, which is another way of saying that a signal correlates with itself exactly.

Several authors, including Eric J. Heller (*Why You Hear What You Hear*), claim that the first and tallest isolated peak in the autocorrelation of a sound wave points to a time delay that is equal to the period of that waves pitch. If the sampling interval is h or $1/F_s$ and if the first and tallest peak is located at a time delay of $n\,h$ then it follows that the pitch estimate is $1/(n\,h)$.

The error in estimating the pitch frequency Δf can be found from

$$f + \Delta f = \frac{1}{h(n + \Delta n)}. \tag{A1.26}$$

That is, an error of Δn in estimating the time lag associated with an autocorrelation peak will cause an error of Δf in the associated frequency. It is perhaps safe to assume that one can estimate the time delay to within one sampling unit, so $\Delta n = 1$. Equation A1.26 can be manipulated to solve for Δf and the result is plotted in Figure A1.10.

The error plot shows that, at low frequencies when there are relatively many samples per period, the error is small. On the other hand, at high frequencies where there are relatively few samples per period the error is large—as much as 100 cents or more.

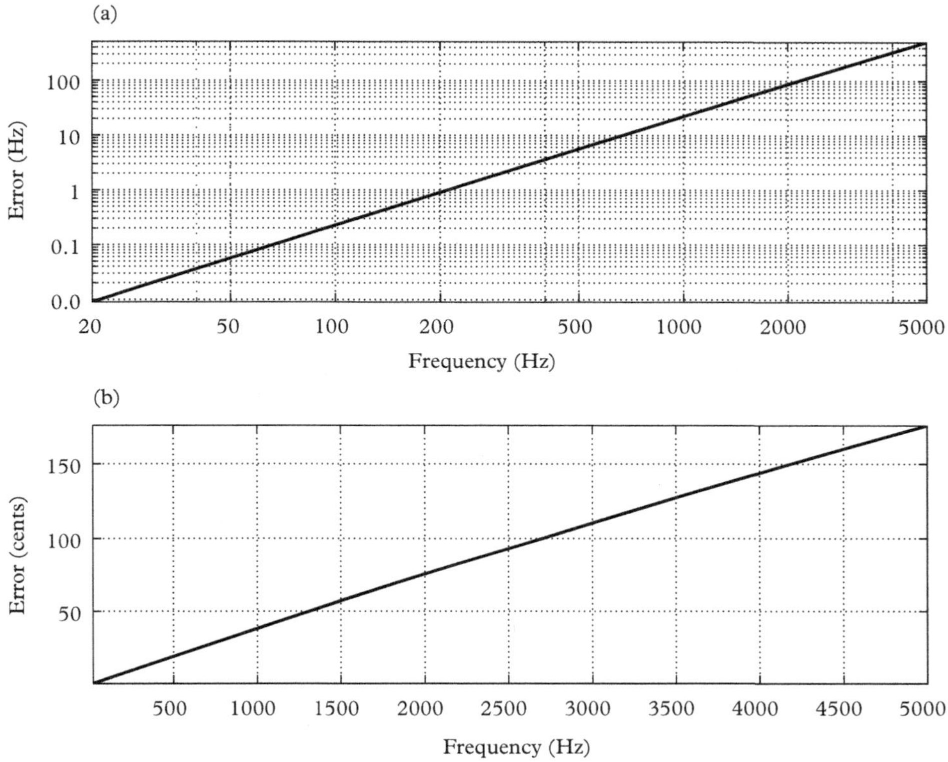

Figure A1.10 *Error in estimating frequencies when the sampling frequency is 44,100 Hz, plotted in (a) Hertz and (b) cents*

A Fine Grid Estimator of the Autocorrelation

The problem of errors in estimating the time lag associated with peaks in the autocorrelation can be addressed with a fine grid approach similar to that mentioned earlier in this appendix. However, to do so requires an alternative definition of the autocorrelation based on the famous WienerKhinchin theorem that allows the autocorrelation to be computed as an inverse cosine Fourier transform.

Although the mathematics involved is way beyond the scope of this appendix, suffice it say that Norbert Wiener proved in 1930 that the Fourier transform of the autocorrelation is equal to the power spectral density. We will not get into the formally correct definition of the power spectral density except to say that in the discrete time and frequency domains it is proportional to the line spectrum. Hence, we will posit that the Fourier transform of the autocorrelation can yield the line spectrum. The inverse is also true, i.e., the inverse Fourier transform of the line spectrum can yield the autocorrelation.

Since we are dealing with real quantities (the autocorrelation and the line spectrum do not have imaginary components) we can use the Fourier *cosine* transform pair instead of Eq. A1.1, which is

$$Y(f) = \int_{-\infty}^{\infty} y(t) \cos(2\pi f t)\, dt$$

$$y(t) = \int_{-\infty}^{\infty} Y(f) \cos(2\pi f t)\, df.$$

For our situation where we are interested in the line spectrum *LS* and the autocorrelation *AC*, this pair becomes

$$LS(f) = \int_{-\infty}^{\infty} AC(t) \cos(2\pi f t)\, dt$$

$$AC(t) = \int_{-\infty}^{\infty} LS(f) \cos(2\pi f t)\, df.$$

Earlier we have seen that a fine grid approach can be applied to estimating the line spectrum. The same thing can be done here, where a fine grid is set up in the frequency and time delay plane and the Matlab *ngrid* function is used to develop much finer estimates of the peak location in the autocorrelation than provided in Eq. A1.26. In this case, the line spectrum is computed using the conventional FFT algorithm. This provides a frequency grid interval of $1/(Nh)$ (as shown above). The integration in

$$AC(t) = \int_{-\infty}^{\infty} LS(f) \cos(2\pi f t)\, df$$

is carried out numerically with a fine grid in the time lag domain (t) centered about the time lag of interest. This means that the time lag maximum can be determined relatively accurately compared to the conventional approach. The following snippet of a Matlab script shows how it can be done:

```
% ===calc period for this freq (sec)
perExp=1/fExp; %(fExp is the frequency of the expected pitch)
%=== set the sampling frequency (Hz)
h=1/Fs;
% ===set up frequency grid
fL=0;
fR=Fs/2; %( the frequency interval is the entire Nyquist interval)

% ===calc the line spectrum (and generate the frequency grid)
%     (do it over the entire Nyquist interval)
[Y,f]=SimpleLineSpectrum2(y,h,fL, fR); %(use a homemade function to
       calc the line spectrum)
delFreq=f(3)-f(2); % (spacing in the frequency grid)
```

```
%=== pick time lag limits for the search
fMinSearch=fExp-freqInt; %(freqInt is the interval about the
           frequency of the expected pitch)
if fMinSearch<0
    fMinSearch=0;
    tLagMax=1/(fExp*.5);
else
    tLagMax=1/(fMinSearch);
end
fMaxSearch=fExp+freqInt;
tLagMin=1/(fMaxSearch);

% ===calc delta lag time based on nLags
delT=(tLagMax-tLagMin)/nLags; %(this is the spacing between time lags
      which we want to be small)

% ===calc the time lag grid
tLag=tLagMin:delT:tLagMax;
%=== calc the 2D grid
[T,F]=ndgrid(tLag,f); %( this is the crucial step which makes use of
      the matlab ndgrid function)

% ===calc the cosine factor on the 2D grid
G=cos(2*pi*F.*T)*delFreq; % (this is a matrix)

% ===calc the matrix product
[nrG,rcolG]=size(G);
[nrY,ncolY]=size(Y);
if ncolY>nrY
    Y=Y';
end
Q=G*Y; %(Q is the estimate of the AC which can be searched for
  a maximum)
```

The interested reader can pursue this by looking at the Matlab function in the file AutoCorrGrid3.m, which is available on the website mentioned in the preface.

As an example, consider a sound wave with a frequency of 4186 Hz. The line spectrum, the conventional autocorrelation, and the fine grid autocorrelation are shown in Figure A1.11. In Figure A1.11(b) the small circles show the spacing of the autocorrelation values available for choosing a maximum. In Figure A1.11(c) the circles are so densely packed that they appear as a solid curve.

Experience with this algorithm has shown that for low frequency waves where there are many samples per period, the conventional approach to calculating the autocorrelation is superior when there is noise riding on the signal. On the other hand, for high frequency waves like C8 the fine grid approach is better even in the face of noise.

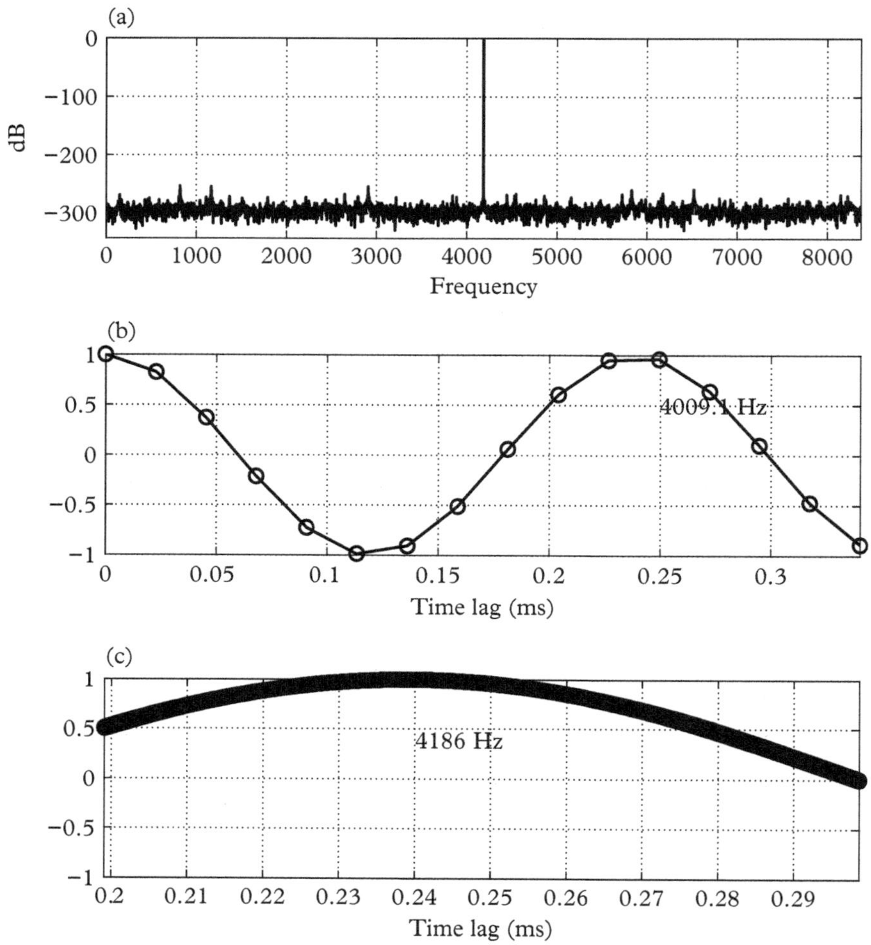

Figure A1.11 *A sine wave with a frequency of 4186 Hz: (a) line spectrum, (b) conventional autocorrelation, and (c) fine grid autocorrelation*

MP3 and WAV Files

The sound waves studied in this book are recorded in either wav or mp3 formats. The former uses a standard format that provides high fidelity. The latter has a compact format that requires much less storage space. The following graphs show line spectra of a Steinway C1 note in both formats. Figure A1.12 shows that for the low frequencies (01000 Hz) there is no significant difference, while Figure A1.13 shows that at about 16 kHz there is an apparent loss of information. Given that the signal level is approximately 100 dB down and that humans have difficulty discerning frequencies higher than 16 kHz there appears to be little advantage to using wav files over mp3 files.

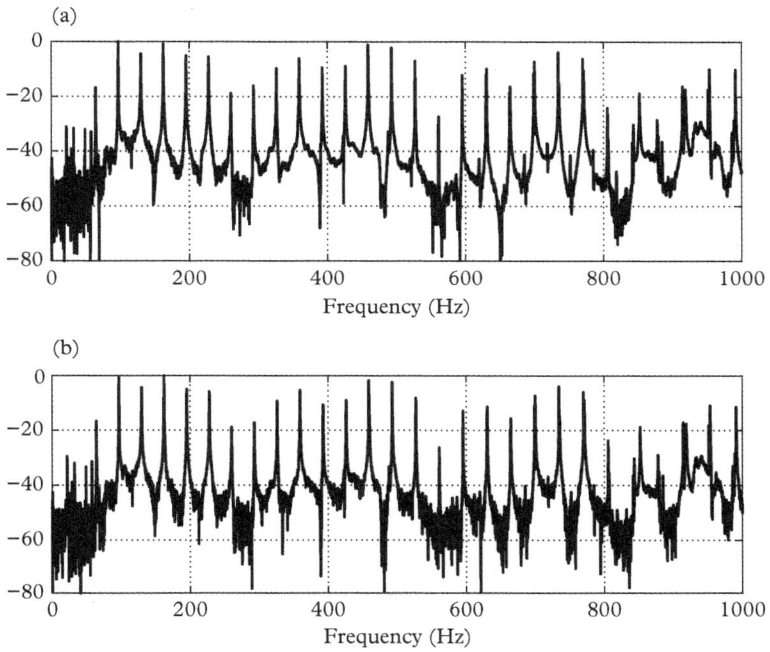

Figure A1.12 *Line spectra (0–1000 Hz) of the Steinway C1 note when collected as (a) wav, and (b) mp3*

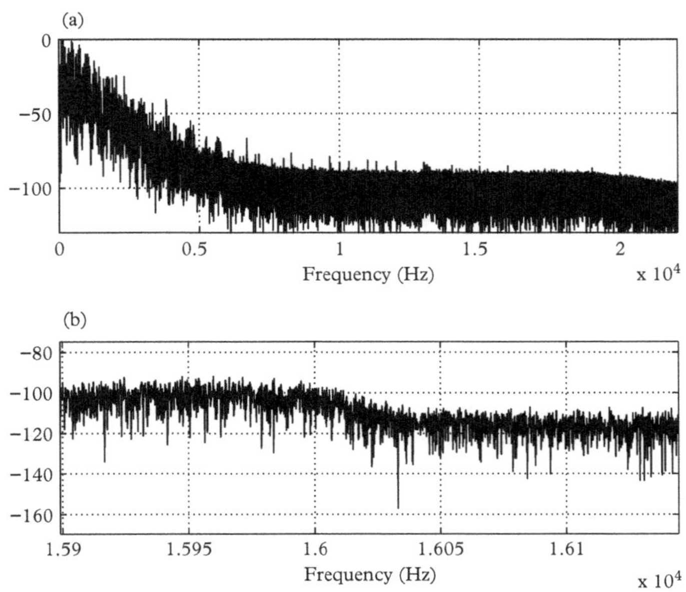

Figure A1.13 *Line spectra (over the entire Nyquist interval) of the Steinway C1 note when collected as (a) wav, and (b) mp3*

Appendix 2
Mathematical Basis for Part Two

The ideal vibrating string, bar, membrane, and plate are described by partial differential equations which are presented here. Some of the analytic and numerical methods used to solve them are reviewed. The concept of elasticity, Young's modulus, and Hooke's Law are briefly reviewed.

The Vibrating String

The wave equation

$$\frac{\partial^2 y}{\partial t^2} = c^2 \frac{\partial^2 y}{\partial x^2} \qquad c = \sqrt{\frac{\tilde{T}}{\tilde{\rho}}} \tag{A2.1}$$

subject to the initial conditions

$$y(x, 0) = f(x) \qquad \text{and} \qquad \left.\frac{\partial f}{\partial t}\right|_{x,0} = 0 \tag{A2.2}$$

and the boundary conditions

$$y(0, t) = 0 \qquad y(L, t) = 0 \tag{A2.3}$$

describes the behavior of the ideal flexible, frictionless, vibrating string, where $y(x)$ is the transverse displacement at a position x on the string, $\tilde{T}$ is the tension and $\tilde{\rho}$ is the linear density (g cm^{-1}). The transverse direction y is at a right angle to the longitudinal direction denoted by x. The string's length is L and c is the speed of sound along the string. The initial profile $y(x, 0) = f(x)$ can represent a pluck. The second part of condition A2.2 stipulates that initially the string is not moving. The two boundary conditions in A2.3 stipulate that the string is rigidly fixed at $x = 0$ and $x = L$.

An Analytical Approach—Eigenfunctions

The above problem can be solved by the separation of variables method that is covered in every textbook having a title like *Advanced Engineering Mathematics* or *Advanced Engineering Calculus*. For example, Chapter 11, Section 3 in Erwin Kreyszig's *Advanced Engineering Mathematics* is devoted to the solution of this problem and I will only mention the highlights.

The fundamental premise of the method is that one can factor the solution as follows:

$$y(x, t) = X(x)T(t). \tag{A2.4}$$

Feeding Eq. A2.4 into A2.1 yields two ordinary differential equations, one defined in the spatial domain and one in the time domain:

$$\frac{d^2X}{dx^2} = \alpha X = 0, \tag{A2.5a}$$

$$\frac{d^2T}{dt^2} - \alpha c^2 T = 0, \tag{A2.5b}$$

where the parameter α is an arbitrary constant that will be determined when the boundary conditions are applied. Both equations have the form

$$\frac{d^2}{dz^2} w = k\, w. \tag{A2.6}$$

That is, when $\frac{d^2}{dz^2}$ operates on w it generates a function that is proportional to w, namely, kw. The function w is called an "eigenfunction" and the proportionality constant k is called an "eigenvalue" of the operator $\frac{d^2}{dz^2}$. Thus, in Eqs. A2.5a and A2.5b there will be eigenfunction/eigenvalue pairs for the spatial and temporal domains. Those in the latter domain will lead to harmonics. In general, the eigenfunction of Eq. A2.6 can have the form of a complex exponential, as in

$$e^{jkz} = \cos(kz) + j\sin(kz).$$

After applying the separation of variables method and the two boundary conditions, the solution to Eq. A2.1 is

$$\begin{aligned} y(x,t) &= \sum_{n=1}^{\infty} B_n \cos(\lambda_n t)\, \sin\left(\frac{n\pi\, x}{L}\right) \\ \lambda_n &= \frac{n\pi\, c}{L} \\ n &= 1,\ 2,\ \ldots \end{aligned} \tag{A2.7}$$

The time domain eigenfunctions appearing in Eq. A2.7, namely,

$$\cos(\lambda_n t) = \cos\left(\frac{n\pi\, c\, t}{L}\right) = \cos(2\pi\, f\, t),$$

are the harmonics having frequencies of

$$f_n = \frac{nc}{2L} \qquad n = 1,\ 2,\ \ldots, \tag{A2.8}$$

and the λ_n are the eigenvalues. For future reference, note that the cosine function is the real part of the complex exponential. Let $n = 1$ for the fundamental frequency. Then Eq. A2.8, when combined with

$$c = \sqrt{\frac{\tilde{T}}{\tilde{\rho}}} \quad \text{and} \quad \tilde{\rho} = \frac{\rho}{S},$$

gives

$$f_1 = \frac{c}{2L} = \frac{\sqrt{\frac{T}{\tilde{\rho}}}}{2L} = \sqrt{\frac{T}{4L^2\tilde{\rho}}} = \sqrt{\frac{T}{4L^2\rho S}}, \tag{A2.9}$$

which can be used to estimate the tension needed in a piano string for a given diameter, frequency, density, and length.

At $t = 0$, Eq. A2.7 becomes

$$y(x,0) = f(x) = \sum_{n=1}^{\infty} B_n \sin\left(\frac{n\pi x}{L}\right), \tag{A2.10}$$

which is a Fourier series approximation to $f(x)$ which in turn is the initial shape of the string. This approximation uses the spatial domain eigenfunctions, $\sin(\frac{n\pi x}{L})$.

Taking advantage of the orthogonality of the eigenfunctions, the coefficients B_n can be determined from

$$B_n = \frac{2}{L}\int_0^L f(x) \sin\left(\frac{n\pi x}{L}\right) dx. \tag{A2.11}$$

For the plucked string, the initial profile is

$$f(x)\beta x \qquad 0 \le x \le p \tag{A2.12}$$

$$f(x) = \frac{L\beta p}{L-p} + \frac{\beta p}{p-L}x \qquad p \le x \le L, \tag{A2.13}$$

where the pluck point is at $x = p$ and β is an arbitrarily chosen slope. When Eqs. A2.12 and A2.13 are inserted into Eq. A2.11, the coefficients B_n can be determined and Eq. A2.7 can be used to calculate and plot the string profile as a function of time. This approach was used for several of the graphs in Section 2.1.

The initial string profile described by Eqs. A2.12 and A2.13 has a discontinuity in the slope at $x = p$ and Eq. A2.10 attempts to approximate that shape with an infinite number of terms. Since Eq. A2.10 cannot have an infinite number of terms, we must pick a finite number, say, N, for the actual calculations. In most of the calculations I picked $N = 20$. This means that the shape of the initial profile will be slightly smoother with a small wiggle, as will the subsequent wave shapes as time progresses.

An alternative approach, still separating the independent variables, attempts a solution of the form

$$y(x,t) = Ae^{jat}e^{jbx}. \tag{A2.14}$$

After substituting Eq. A2.14 into Eq. A2.1 one finds that $a = \pm c\,b$, so the form of the solution is

$$y(x,t) = A_1 e^{jb(ct-x)} + A_2 e^{jb(ct+x)}, \tag{A2.15}$$

which can be shown to represent waves moving to the right and left with the propagation speed c.

A Numerical Solution—Explicit Version

The terms in Eq. A2.1 can be replaced by finite difference approximations, as in

$$\frac{\partial^2 y}{\partial x^2} \to \frac{y(k+1,n) - 2y(k,n) + y(k-1,n)}{\Delta x^2},$$

$$\frac{\partial^2 y}{\partial t^2} \to \frac{y(k,n+1) - 2y(k,n) + y(k,n-1)}{\Delta t^2},$$

resulting in

$$c^2 \frac{y(k+1,n) - 2y(k,n) + y(k-1,n)}{\Delta x^2} = \frac{y(k,n+1) - 2y(k,n) + y(k,n-1)}{\Delta t^2}. \quad \text{(A2.16)}$$

Equation A2.16 can be solved for $y(k, n+1)$, resulting in

$$y(k,n+1) = 2\left(1-r^2\right) y(k,n) - y(k,n-1) + r^2 \left[y(k+1,n) + y(k-1,n)\right], \quad \text{(A2.17)}$$

where $r = \frac{c\Delta t}{\Delta x}$.

Equation A2.17 is started with $n = 2$. Then for $k = 1, \ldots, K$, the values of $y(k, 3)$ are calculated using the initial conditions at $y(k, 1)$ and $y(k, 2)$, which come from the initial pluck profile.

If a $k \times n$ grid is constructed with k being the abscissa and n being the ordinate, then the values of the initial pluck profile are placed in the first two rows and the left- and right-hand side boundary conditions are placed at $k = 1$ and $k = K$. The third row contains the string response $y(k, 3)$ at $t = \Delta t$ computed from Eq. A2.17.

The time index n is incremented for the fourth row ($t = 2\Delta t$) and the calculations are repeated for $y(k, 4)$ using $y(k, 2)$ and $y(k, 3)$. This cycle of calculations is continued until one reaches the desired final time.

Equation A2.17 shows that the x-axis and time axis have been discretized such that

$$x_k = (k-1)\,\Delta x, \quad k = 1, 2, \ldots, K,$$

where $\Delta x = \frac{L}{K-1}$ and

$$t_n = (n-2)\,\Delta t, \quad n = 1, 2, \ldots,$$

where Δt is chosen to keep r within a "satisfactory range." Regarding the satisfactory range for r, Eq. A2.17 suggests that if $r = 1$ then the coefficient of $y(k, n)$ will be zero. It turns out that if $r > 1$ the calculations will become unstable. I used $r < 1$. Further discussion of this problem can be found in Chapter 6 of Giordano and Nakanishi's *Computational Physics*.

A Numerical Solution—Implicit Version

If the left-hand side of Eq. A2.1 is rewritten as an average of two partial derivatives,

$$\left.\frac{\partial^2 y}{\partial t^2}\right|_{k,n} \to \frac{c^2}{2}\left(\left.\frac{\partial^2 y}{\partial x^2}\right|_{k,n+1} + \left.\frac{\partial^2 y}{\partial x^2}\right|_{k,n-1}\right), \quad \text{(A2.18)}$$

then when the partial derivatives are replaced with finite differences we get the so-called Crank–Nicholson form:

$$c^2\frac{y(k+1,n+1)-2y(k,n+1)+y(k-1,n+1)}{2\Delta x^2}+c^2\frac{y(k+1,n-1)-2y(k,n-1)+y(k-1,n-1)}{2\Delta x^2}$$
$$=\frac{y(k,n+1)-2y(k,n)+y(k,n-1)}{\Delta t^2}. \quad \text{(A2.19)}$$

If just the terms containing the time argument $n+1$ are gathered on the left-hand side we get

$$c^2\frac{y(k+1,n+1)-2y(k,n+1)+y(k-1,n+1)}{2\Delta x^2}+\frac{y(k,n+1)}{\Delta t^2}=$$
$$-c^2\frac{y(k+1,n-1)-2y(k,n-1)+y(k-1,n-1)}{2\Delta x^2}+\frac{-2y(k,n)+y(k,n-1)}{\Delta t^2}. \quad \text{(A2.20)}$$

After collecting terms, Eq. A2.20 becomes a linear equation in the three unknown values at time $t(n+1)$ in terms of values of y at times $t(n-1)$ and $t(n)$:

$$a_1y(k+1,n+1)+a_2y(k,n+1)+a_3y(k-1,n+1)=$$
$$b_1y(k+1,n-1)+b_2y(k,n-1)+b_3y(k-1,n-1)+b_4y(k,n). \quad \text{(A2.21)}$$

When Eq. A2.21 is evaluated for $k=2,\ldots,K-1$ a linear system in the unknowns $y(k,n+1)$, $k=2,\ldots,K-1$ is developed that can be solved by a variety of methods. As with the explicit formulation, one starts with $n=2$ and solves for $y(k,3), k=2,\ldots,K-1$. The index n is incremented and the linear system is solved again, and so on.

Applying the implicit method is more difficult but it is insensitive to the choices of Δx and Δt. I used it for several of the problems dealing with a right-hand side boundary condition associated with an approximation to a soundboard.

An Alternative Initial Condition

For the numerical solutions, I also used a Gaussian initial shape defined by

$$f(x)=\frac{1}{s}e^{-\frac{(x-x_c)^2}{s}}, \quad \text{(A2.22)}$$

as displayed in Figure 5.5.1.

An Approximation to a Soundboard

In Section 5.6 a mass/spring/dashpot system was appended to the wave equation to simulate the action of the soundboard. The differential equation describing the mass/spring/dashpot system is

$$m\frac{d^2y(L,t)}{dt^2}=-B\frac{dy(L,t)}{dt}-ky(L,t)+gF(t), \quad \text{(A2.23)}$$

where the dependent variable is the transverse displacement of the string at the right-hand end of the string, $y(L, t)$. The force $F(t)$ is the slope of the transverse displacement with respect to the position along the string at the end point:

$$F(t) = \left.\frac{dy}{dx}\right|_{x=L}. \tag{A2.24}$$

Upon replacing the derivatives in Eqs. A2.23 and A2.24 with finite differences, one obtains an equation that can be combined with those represented by Eq. A2.21 (or Eq. A2.17) to generate a set solvable for $y(k, n+1), \quad k = 2, \ldots, K$.

Forcing the String at One End

In Section 5.7 the string is forced at one end by replacing the right-hand side boundary condition in Eq. A2.3 with $y(L, t) = G(t)$ where $G(t)$ is chosen appropriately.

Striking the String with a Hammer

Two approaches were taken for this simulation. In the first, I simply evaluated the analytical solution presented on p. 102 of Morse's 1948 book *Vibration and Sound* which was obtained by separation of variables. Morse's starting point was Eq. A2.1, modified with a force term:

$$c^2 \frac{\partial^2 y}{\partial x^2} = \frac{\partial^2 y}{\partial t^2} + F(x, t). \tag{A2.25}$$

The solution method and the final result are rather involved and will not be repeated here. However, note that Morse obtained his result by solving a *linear* partial differential equation.

The second approach is taken from Section 11.2 of Giordano and Nakanishi's *Computational Physics*. Equation A2.25 is discretized to give an equation similar to Eq. A2.17:

$$y(k, n+1) = 2\left(1 - r^2\right) y(k, n) - y(k, n-1) + r^2 \left[y(k+1, n) + y(k-1, n)\right] + \frac{\Delta t^2}{\mu \Delta x} F_{k,n}, \tag{A2.26}$$

where μ is the string mass per unit length. A separate equation is presented for the mass of the hammer, derived from $F_{\mathrm{h}} = m_{\mathrm{h}} a_{\mathrm{h}}$. Consequently, there are three equations describing the motion of the hammer:

$$\begin{aligned} \frac{dz_{\mathrm{h}}}{dt} &= v_{\mathrm{h}}, \\ \frac{dv_{\mathrm{h}}}{dt} &= a_{\mathrm{h}}, \\ a_{\mathrm{h}} &= \frac{F_{\mathrm{h}}}{m_{\mathrm{h}}}, \end{aligned} \tag{A2.27}$$

where z_{h} is the vertical displacement of the hammer, a_{h} is the acceleration of the hammer, and m_{h} is the mass of the hammer. The force of the hammer on the string is proportional to the compression of the hammer's felt. Let $y(\tilde{k}, n)$ be the string's transverse displacement at position $\tilde{k}$ along the string at the n th instant of time, where $\tilde{k}$ denotes the spot along the string where the hammer strikes. Let $z_h(n)$ be the vertical displacement of the hammer at the n th instant of time. The force of the hammer on the string is given by

$$F_h(n) = K\left(y(\tilde{k}, n) - z_h(n)\right)^p, \tag{A2.28}$$

where $p = 3$ has been found to be suitable. These equations represent an idealized *nonlinear* model of the string–hammer dynamics and require a numerical method.

The same approach was used in the coupling of the vibrating string with the vibrating plate in Section 8.5.

The Stiff Vibrating String

A stiff string or wire is described by the following *fourth-order* partial differential equation:

$$\begin{aligned} -\varepsilon\frac{\partial^4 y}{\partial x^4} + \frac{\partial^2 y}{\partial x^2} &= \frac{1}{c^2}\frac{\partial^2 y}{\partial t^2}, \\ \varepsilon = \frac{YI}{T}, \qquad c &= \sqrt{\frac{T}{\rho S}}, \end{aligned} \tag{A2.29}$$

where Y is Young's modulus and I is the moment of inertia with respect to the longitudinal axis of the wire denoted by x:

$$I = \frac{\pi r^4}{4}.$$

The tension in the wire is T and the mass per unit length is ρS.

The initial conditions can be similar to those for Eq. A2.1. The boundary conditions must be expanded because of the fourth-order derivative with respect to distance along the wire. Usually, one specifies that, in addition to the string being fixed at each end (Eq. A2.3), the string is clamped such that

$$\frac{\partial^2 y}{\partial x^2} = 0 \qquad \text{at } x = 0 \text{ and } x = L.$$

Equation A2.29 is the same as Eq. A2.1 with the addition of the term $\varepsilon\frac{\partial^4 y}{\partial x^4}$ representing the impact of stiffness. Equation A2.29, like Eq. A2.1, has eigenfunctions and eigenvalues except that there are four rather than two. The eigenfunctions can still be complex exponentials but the eigenvalues are not integer multiples of a constant. Therefore, the overtones of the fundamental frequency are not harmonics as in the sense of the flexible string. In fact, it is shown in an article by Testa, Evangelista, and Cavaliere that the allowed frequencies f_n are given by

$$\frac{f_n}{f_1^0} \cong n\left(\sqrt{1 + n^2\pi^2\alpha^2}\right)\left(1 + 2\alpha + 4\alpha^2\right). \quad f_1^0 = \frac{c}{2L}, \quad \alpha = \frac{\sqrt{\varepsilon}}{L}. \tag{A2.30}$$

When $\alpha = 0$, i.e., when the stiffness goes away, Eq. A2.30 becomes Eq. A2.8 where the allowed frequencies f_n are integer multiples of the flexible string fundamental f_1^0. However, with stiffness, the allowed frequencies are increasingly larger than multiples of f_1^0.

Equation A2.29 can be solved numerically by replacing the partial derivatives with finite differences, as in

$$\frac{\partial^2 y}{\partial x^2} \cong \frac{y(k+1,n) - 2y(k,n) + y(k-1,n)}{\Delta x^2},$$

$$\frac{\partial^2 y}{\partial t^2} \cong \frac{y(k,n+1) - 2y(k,n) + y(k,n-1)}{\Delta t^2},$$

which were used in Eq. A2.8. The fourth-order partial derivative is replaced by a slightly more complicated expression:

$$\frac{\partial^4 y}{\partial x^4} \cong \frac{y(k+2,n) - 4y(k+1,n) + 6y(k,n) - 4y(k-1,n) + y(k-2,n)}{\Delta x^4}. \tag{A2.31}$$

After these expressions are substituted into Eq. A2.29, the user can develop an explicit expression for $y(k, n+1)$, or implicit expressions, as in

$$y(k+2,n+1), y(k+1,n+1), y(k,n+1), y(k-1,n+1), y(k-2,n+1). \tag{A2.32}$$

I tried both approaches and do not recommend the implicit method since it is complicated.

An interesting summary of the above mathematics is available at <http://www.physics.udel.edu/~jim/PHYS460_660_13S/Motion of a string/Motion of a realistic string.htm>, which is taken from the course notes for *Computational Physics*, by Prof. James McDonald of the University of Delaware.

The Vibrating Bar

There are *two* partial differential equations for the vibrating bar. The first deals with the *longitudinal* displacements along the direction of the bar:

$$\frac{\partial^2 \xi}{\partial t^2} = c^2 \frac{\partial^2 \xi}{\partial x^2}, \qquad c = \sqrt{\frac{Y}{\rho}}, \tag{A2.33}$$

where ξ is the longitudinal displacement (in the direction of x), Y is Young's modulus and ρ is the density (g cm^{-3}). Equation A2.33 has the same form as Eq. 2.1 except that the speed of sound c depends on Young's modulus rather than tension. This partial differential equation is not solved in this book. It is only mentioned here because passing reference is made to it in Chapter 5 where longitudinal waves in a vibrating string are dealt with in Section 5.11. However, since it has the same form as Eq. 2.1 and since the speed c is but a parameter, the solution could be obtained by the same methods as shown above.

The second equation deals with *transverse* displacements. For our purposes, the long thin bar is the same as a stiff, inflexible string without the tension. The describing equation is:

$$\begin{aligned} -\kappa^2 c^2 \frac{\partial^2 y(x,t)}{\partial x^4} &= \frac{\partial^2 y(x,t)}{\partial x^2}, \\ c &= \text{speed of sound} = \sqrt{\tfrac{Y}{\rho}}, \\ \kappa &= \text{radius of gyration}, \end{aligned} \tag{A2.34}$$

where, unlike the flexible string, the ends of the bar are not constrained. Instead, one specifies

$$\frac{\partial^2 y}{\partial x^2} = 0 \quad \text{and} \quad \frac{\partial^3 y}{\partial x^3} = 0 \quad \text{at} \quad x = 0 \,\text{and}\, x = L. \tag{A2.35}$$

The bending moment is represented by $\frac{\partial^2 y}{\partial x^2}$ and the shearing force is represented by $\frac{\partial^3 y}{\partial x^3}$ To simulate the behavior of a vibraphone or marimba, we specify that the transverse displacement $y\,(x, t)$ is zero at the two points

$$\begin{aligned} x &= \frac{L}{2} \pm d = x_{\mathrm{L}}, x_{\mathrm{R}}, \\ d &< \frac{L}{2}, \end{aligned} \tag{A2.36}$$

where the bar is supported. This is not treated as a boundary condition but is applied after the solution has been developed by subtracting $y\,(x_{\mathrm{L}}, t)$ from $y\,(x, t)$. There is no need to do anything at $x = x_{\mathrm{R}}$ because $y\,(x, t)$ is symmetrical about $= L/2$.

When the separation of variables method is applied to Eq. A2.34 the solution contains the term (an eigenfunction)

$$e^{j\omega(t - x/v)}.$$

The quantity $v = \sqrt{2\pi f c\kappa}$ can be interpreted as the propagation speed of a transverse displacement wave (see Kinsler and Frey for more details). When compared to the solution form of Eq. A2.15 one sees that the speed of propagation now is not just c, the speed of sound, but rather something that depends on the frequency of the wave, f, as well.

Equation A2.34 can be solved numerically by again replacing the partial derivatives with finite differences. Explicit or implicit methods can be derived. I used the latter but I do not necessarily recommend it over the explicit method.

The Vibrating Membrane

In an attempt to simulate a drum surface we extend the flexible string equation (wave equation) to two spatial dimensions and specify radial coordinates because the drum rim is circular. Therefore, the describing equation is:

$$\frac{\partial^2 y}{\partial t^2} = c^2 \left(\frac{\partial^2 y}{\partial r^2} + \frac{1}{r} \frac{\delta y}{\partial r} + \frac{1}{r^2} \frac{\partial^2 y}{\partial \theta^2} \right), \tag{A2.37}$$

$$\begin{aligned} y(R, \theta, t) &= 0 \quad 0 \le \theta \le 2\pi \quad t \ge 0, \\ y(r, \theta, 0) &= f(r, \theta) \quad 0 \le \theta \le 2\pi. \end{aligned} \tag{A2.38}$$

The first part of Eq. A2.38 specifies that the edge of the drum surface is constrained to the drum rim. The second part allows one to specify an initial displacement. In addition to Eq. A2.38, there are other boundary conditions that are necessary, especially at $r = 0$ where it is necessary to require that the solution be bounded or to require that there be symmetry, as in

$$\frac{\partial y}{\partial r} \big|_{r=0} = 0.$$

The eigenfunctions for Eq. A2.37 are no longer complex exponentials but rather Bessel functions, and the eigenvalues come from roots of the Bessel functions so, although there is still a fundamental frequency, there are no harmonics having frequencies that are integer multiples of the fundamental. A dynamic presentation of some of the vibrating modes is shown in <http://en.wikipedia.org/wiki/Vibrations_of_a_circular_drum>.

To solve Eq. A2.37 numerically, the circular surface is discretized into nodes defined by rings and sectors. A node is defined by the k th ring and the m th sector:

$$\begin{aligned} &y(r,\theta,t) \rightarrow y(k,m,n),\\ &r=(k-1)\Delta r \quad k=1,\ldots,K \quad \Delta r=\frac{R}{K-3},\\ &\theta=(m-1)\Delta\theta \quad m=1,\ldots,M \quad \Delta\theta=\frac{2\pi}{M-1},\\ &t=(n-1)\Delta t. \end{aligned} \tag{A2.39}$$

The partial derivatives are replaced by finite differences and the result is solved for $y(k,m,n+3)$. The initial conditions are applied where $n=1$ and $n=2$ and the boundary conditions are applied at the two outside rings where $k=K-1$ and $k=K-2$. The calculations would start with $n=2$ and $k=K-3$.

At $k=1$ (the center of the circle where $r=0$) there are problems stemming from the presence of the $1/r$ and $1/r^2$ terms in Eq. A2.37. This problem can be addressed by first realizing that

$$\frac{\partial^2 y}{\partial\theta^2}=0 \quad \text{at} \quad r=0.$$

That is, at the center of the circle there is no concept of dependence on the angle. This leaves only the $\frac{1}{r}\frac{\partial y}{\partial r}$ term in Eq. A2.37 to which the L'Hôpital's rule is applied:

$$\lim_{r\to 0}\frac{\dfrac{\partial y}{\partial r}}{r}=\lim_{r\to 0}\frac{\dfrac{\partial}{\partial r}\left(\dfrac{\partial y}{\partial r}\right)}{\dfrac{\partial}{\partial r}r}=\frac{\dfrac{\partial^2 y}{\partial r^2}}{1}=\frac{\partial^2 y}{\partial r^2}.$$

At $r=0$, Eq. A2.37 therefore becomes

$$\frac{\partial^2 y}{\partial t^2}=2c^2\frac{\partial^2 y}{\partial r^2}. \tag{A2.40}$$

When $k=1$, a discretized version of Eq. A2.40 is used.

For most of the simulations, the initial condition is

$$f(r,\theta)=\frac{1}{s}e^{-\frac{r^2}{s}}. \tag{A2.41}$$

The Vibrating Circular Plate

The describing equation for a circular plate is

$$\frac{K^2 Y}{\rho(1-\sigma^2)}\left(\frac{1}{r}\frac{\partial}{\partial r}\left(r\frac{\partial}{\partial r}\left(r\frac{\partial y}{\partial r}\right)\right)+\frac{2}{r^2}\frac{\partial^4 y}{\partial\theta^2\partial r^2}+\frac{1}{r^4}\frac{\partial^4 y}{\partial\theta^4}-\frac{2}{r^3}\frac{\partial^3 y}{\partial\theta^2\partial r}+\frac{4}{r^4}\frac{\partial^2 y}{\partial\theta^2}\right)=-\frac{\partial^2 y}{\partial t^2}, \tag{A2.42}$$

where $K = \frac{\delta}{\sqrt{12}}$ is the radius of gyration and δ is the thickness of the plate. The density is ρ, σ is the Poisson ratio, and Y is Young's modulus.

Using the same logic as with the circular membrane, the describing equation at the center is

$$\frac{K^2 Y}{\rho(1-\sigma^2)}\left(\frac{8}{3}\frac{\partial^4 y}{\partial r^4}\right) = -\frac{\partial^2 y}{\partial t^2}. \tag{A2.43}$$

After replacing the partial derivatives with finite differences, an expression for $y(k, m, n+3)$ can be derived in a manner similar to that for the circular membrane.

The initial displacement is similar to Eq. A2.41 for some of the simulations but a cone-like initial displacement was also used:

$$f(r,\theta) = s\left(1 - \frac{r}{R}\right).$$

The Vibrating Rectangular Plate

The rectangular plate is described by

$$\frac{K^2 Y}{\rho(1-\sigma^2)}\left(\frac{\partial^4 w}{\partial x^4} + \frac{\partial^4 w}{\partial y^4} + \frac{\partial^4 w}{\partial^2 x\, \partial^2 y}\right) = -\frac{\partial^2 w}{\partial t^2}, \tag{A2.44}$$

where the variable w represents the transverse displacement (rather than y since x and y denote the dimensions of the plate.

On some of the simulations the initial displacement is

$$f(x,y) = \frac{1}{s} e^{-\left((x-x_c)^2+(y-y_c)^2\right)/s}. \tag{A2.45}$$

Adding the String and the Hammer to the Plate

Equation A2.44 is augmented with Eq. A2.1 and with $f = ma$ for the hammer. The force is applied over a bridge that can be placed on a diagonal or parallel to one of the edges. The points $xp(i)$, $yp(i)$, $i = 1, p$ define a straight line lying in the x–y plane. These points are used to define a ridge-like surface that attempts to represent the bridge:

$$q(x,y) = \exp\left(\sum_{k=1}^{p}\left(\left(x(k) - xp(k)\right)^2 + \left(y(k) - yp(k)\right)^2\right)\right). \tag{A2.46}$$

The down-bearing force from the vibrating string is applied over this surface.

Elasticity and Stiffness

Hooke's Law

Elasticity is the ability of a solid object (such as a string) to be deformed through stress and return to its original size when the stress has been removed.

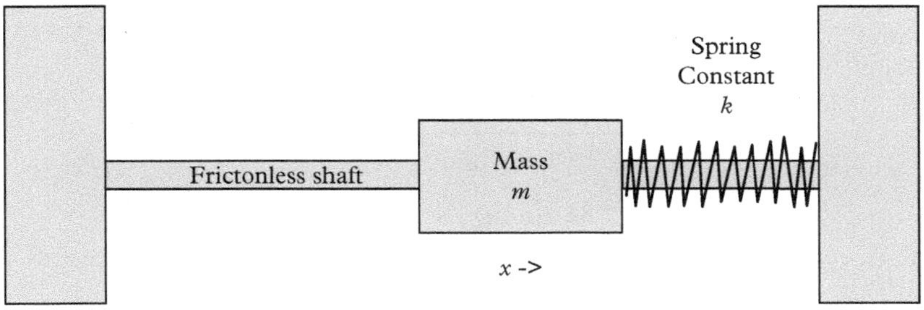

Figure A2.1 *Mass/spring system*

Measuring elasticity uses Hooke's Law, which states that deformation x (in m) is proportional to the stress S(in N) applied to the object:

$$S = -k\,x,$$

where k is Hooke's constant (N m^{-1}). The minus sign indicates that the displacement is in the opposite direction to the applied force.

For example, if the object is a spring with mass m on a shaft with no friction, as in Figure A2.1, then Newton's law would yield the following:

$$f = m\,a = m\frac{d^2x}{dt^2},$$
$$S = f = -k\,x,$$
$$m\frac{d^2x}{dt^2} = -kx,$$
$$\frac{d^2x}{dt^2} = \alpha\,x, \qquad \alpha = -\frac{k}{m}.$$

Using the terminology pointed out earlier in this appendix, the displacement x for this object is an eigenfunction and $-\frac{k}{m}$ is an eigenvalue of the $\frac{d^2}{dt^2}$ operator.

Young's Modulus

Young's modulus is defined as the ratio of stress to strain for a given material below the limit of proportionality (the elastic limit).

Stress is defined as the force per unit area of cross-section:

$$\text{force (in Newtons) /cross – sectional area (in square meters)} = \text{stress (Newtons per square meter)}.$$

Strain is defined as the fractional change in length produced when a body is subjected to stress. Its formula is:

$$\text{change in length produced (in meters) /original length (in meters)}.$$

There are no units as it is a ratio of two values that have the same units. Therefore, the units of Young's modulus are Nm^{-2}.

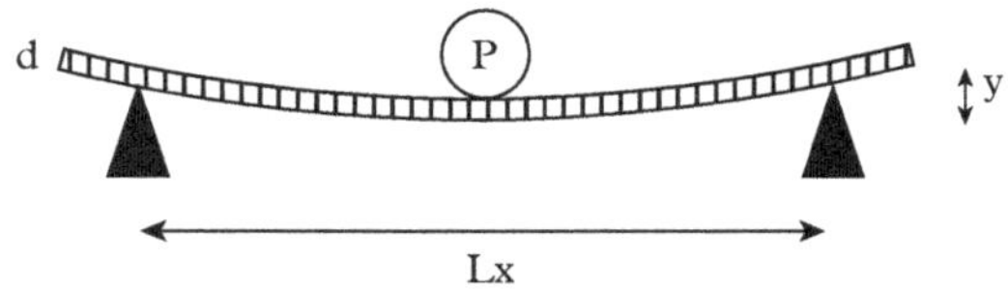

Figure A2.2 *Measuring the Young's modulus for a thin plate*

Young's Modulus of a Wood Plate

As an example, consider the estimation of Young's modulus for a thin plate of wood. Consider the diagram in Figure A2.2.

In Figure A2.2,

P = force applied to plate/beam in N
L_x = distance between fixed supports parallel to grain in m
y = deflection(s) of plate/beam in m
L_y = width of plate/beam perpendicular to L_x in m
d = thickness of plate/beam in m.

After the measurements y, L_x, L_y, d, P are made, the calculation of Young's modulus is

$$Y = \frac{K P L_x^3}{y L_y d^3},$$

where K is a constant for center-loaded beams.

Summary

As we move from the flexible vibrating string to the plate being activated by a bridge, which is in turn activated by a string being struck or plucked, one sees an increasing complexity in the mathematics involved. Also, the reader notices less detail being presented as the equations become more complicated. If the reader should be interested in some of the Matlab® scripts used to solve these equations and generate the graphs in the book, he/she can first look on the website and, if the script is not there, send me an email describing her interest and I will attempt to provide an applicable Matlab script.

Appendix 3
Mathematical Basis for Part Three

Part 3 used the conversion between Hertz and cents extensively so I will spend some time deriving the conversion relation. The cumulative sum of the difference of spectral centroids was used to determine if there was a significant difference between two pianos. Consequently, some time will be spent on trying to demonstrate the power of this tool.

Conversion of Hertz to Cents

The concept of cents was used in Chapter 1 and again in Chapter 3. The fundamental frequency of each note in the equal temperament system (in Hz) is defined by

$$\begin{aligned} f(n) &= 440r^{n-49}, \\ r &= 2^{\frac{1}{12}}, \end{aligned} \tag{A3.1}$$

where $n = 1, \ldots, 88$ with $n = 1$ denoting the lowest note on the piano, A0, and $n = 88$ denoting the highest note, C8.

By definition there are 100 cents per semitone or 200 cents per two semitones; therefore, the number of Hz per cent, *HpC*, can be written as

$$HpC\,(f(n)) = \frac{f(n+1) - f(n-1)}{200}. \tag{A3.2}$$

Combining Eqs. A3.1 and A3.2 gives

$$HpC = \frac{440r^{n-50}\,(r^2 - 1)}{200}. \tag{A3.3}$$

Using Eq. A3.1 in Eq. A3.3 gives

$$HpC = \frac{f(n)}{200}\,\frac{(r^2 - 1)}{r}.$$

Also, Eq. A3.1 can be solved for the note index when the frequency is given:

$$n = floor\left(49 + \frac{\log_{10}\left(\frac{f}{440}\right)}{\log_{10}(r)}\right). \tag{A3.4}$$

Cumulative Sum of a Difference

If x_k and y_k for $k = 1, \ldots, N$ are the samples of two signals, say, centroids of two pianos, their difference is

$$d_k = x_k - y_k \quad k = 1, \ldots, N.$$

The cumulative sum of the difference c_k is defined as

$$\begin{aligned} c_1 &= 0 \\ c_2 &= c_1 + d_1 \\ &\ldots \\ c_k &= c_{k-1} + d_{k-1} \\ &\ldots \\ c_N &= c_{N-1} + d_{N-1}, \end{aligned}$$

or

$$c_1 = 0 \qquad c_k = \sum_{m=2}^{k} d_m \quad k = 2, \ldots, N.$$

The continuous time analog is

$$c(t) = \int_0^t d(\tau)d\tau = \int_0^t (x(\tau) - y(\tau))\, d\tau.$$

(In Chapters 1 and 3 the cumulative line spectra used a similar construction; however, it was normalized while the cumulative sum is not.)

Consider the following hypothetical example where there is a signal that could be the difference of spectral centroids for two pianos. The difference is zero for the first 40 keys (A0 to C4). For keys 41 to 60 (C4# to G5#) there is a constant difference of 0.2 Hz. Finally, keys 61 to 88 (A5 to C8) again have zero difference. Figure A3.1 looks at the noiseless signal while Figure A3.2 looks at the same signal with normally distributed noise added. The added noise has a standard deviation of 0.4 Hz, which is double the actual difference in the centroids for keys 41 to 60.

When the difference between the spectral centroids is noiseless then it is clear from the plot of the difference itself when and what kind of changes have been made. When the difference is noisy (and the noise level is greater than the magnitude of the difference) it is sometimes not so clear and the cumulative sum plot can often be helpful in determining when changes take place.

In effect, the cumulative sum is a low pass filter that attenuates any signal that varies; the more it varies, the more it is attenuated. Since the sum operator and the integral operator are analogs, if the signal is constant (the difference is non-zero and constant), then the cumulative sum is a ramp. Furthermore, if the signal itself is a ramp, then the cumulative sum changes quadratically.

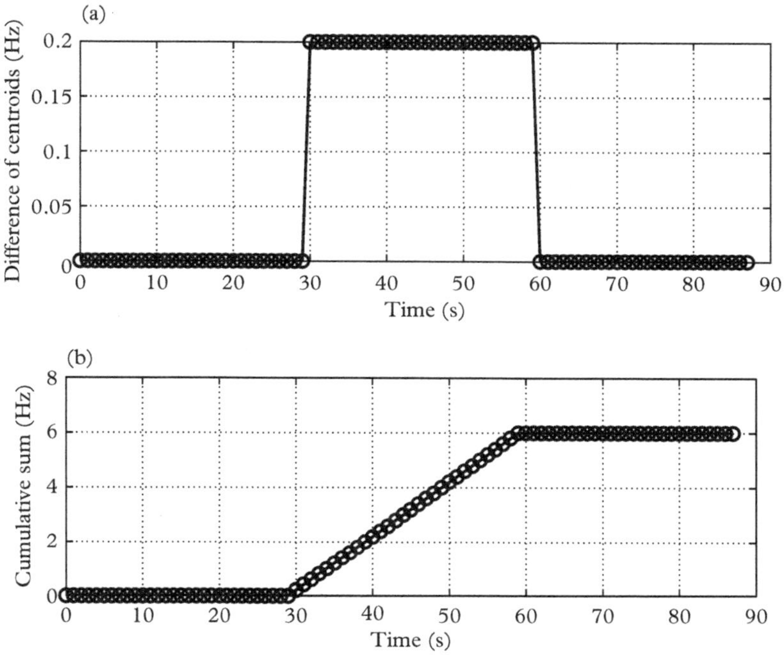

Figure A3.1 *(a) Noiseless difference of centroids, and (b) cumulative difference of the centroids*

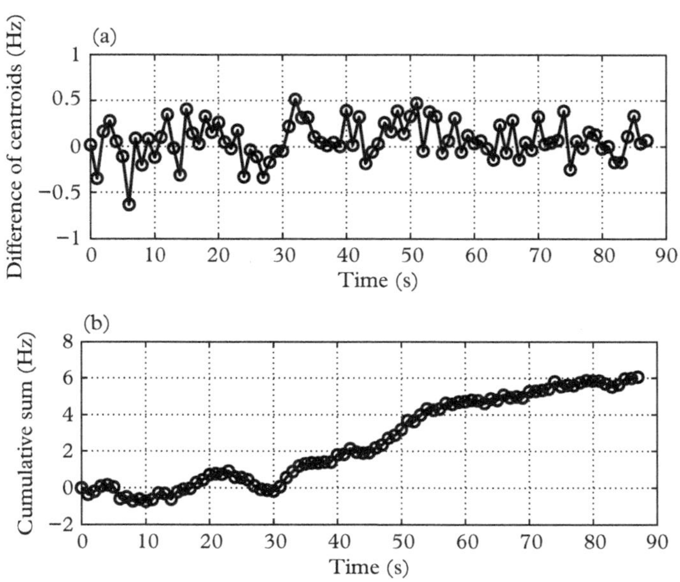

Figure A3.2 *(a) Noisy difference of centroids, and (b) cumulative difference of the centroids*

Azimuth and Elevation in Three-Dimensional Plots

Azimuth and elevation set the angle of the view from which an observer sees a three-dimensional plot. Azimuth is the horizontal rotation and elevation is the vertical elevation (both in degrees). Azimuth revolves about the z-axis, with positive values indicating counter-clockwise rotation of the viewpoint. Positive values of elevation correspond to moving above the object; negative values move below.

Appendix 4
Experimental Setup

Figure A4.1 shows the components of the data-gathering system for pianos:

1. A board with 88 holes aligned over the white and black keys and attached to the piano with two C-clamps. The holes have a $\frac{9}{16}''$ inner diameter and they are 67 mm long.
2. The Zoom H2 recorder placed on the music desk in between two speakers that are used to synchronize the data collection. An alternative approach used an Earthworks microphone mounted on a tripod with a Roland Duo-Capture Ex as an interface.

Figure A4.1 *Experimental setup for the Knabe grand: board with 88 holes attached to piano with C-clamps, Zoom H2 recorder attached to netbook, speakers beside the Zoom H2*

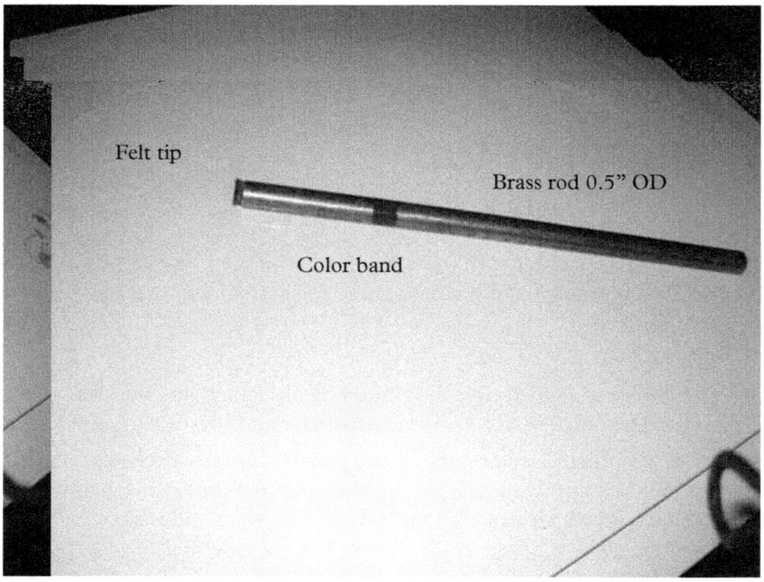

Figure A4.2 *Brass rod (0.5″ outer diameter) used for dropping through the 88-hole board to strike the keys, showing the felt tip and the colored band used to position the rod in the holes*

3. The netbook computer using Matlab® scripts designed to collect the data from the Zoom H2. A second approach uses the Matlab script in concert with Audacity, the Earthworks microphone, and the Roland Duo-Capture Ex. A third approach uses Octave scripts with Audacity, the Earthworks microphone, and the Roland Duo-Capture Ex.

Figure A4.2 shows the 0.5″ outer diameter brass rod, 258 mm in length, that is dropped through the holes in the above-mentioned board. The colored band is used to select the drop distance. For white keys, the rod is adjusted to a position such that the band is just *below* the level of the board before it is dropped. For black keys, the rod is raised to a position such that the band is just *above* the level of the board. This ensures that the brass rod drops 11 mm and hits the white and black keys with the same force and speed. For both the black and white keys the tip of the rod is all the way through the guiding hole, 11 mm above the surface of the key.

There are a variety of approaches, each slightly different. They are all presented here to perhaps suggest modifications that the reader might want to make should he/she want to collect data.

Approach 1

The data gathering uses the Matlab script in the file CollectAndSaveSyncTalk.m. The procedure is as follows. The Zoom H2 recorder is turned on. The laptop Enter button is pressed and, via the script, sends out a beep via the speakers. This beep is recorded by the Zoom H2 and is used as a reference point. For each key, the user presses the Enter button and, in response, the

Figure A4.3 *Experimental setup for the Knabe grand showing the Earthworks microphone*

computer speaks the name of the key to be struck. The brass rod is inserted into the correct hole and positioned such that the band is in the correct place. The rod is held in place manually.

The Enter button is pressed again. This generates a message and a colored circle on the screen prompting the user to get ready to release the brass rod. In 2 s the button changes color and the user releases the brass rod. The result of the strike is collected by the Zoom H2 for 10 s. The exact beginning and end of the wave can be determined relative to the time of the beep. The time domain plot, the line spectrum, and the cumulative line spectrum of the resulting wave are displayed on the laptop screen to help confirm that good data was collected. This sequence continues until two or three octaves of notes are collected (user's choice). Then the data collection is stopped, the Zoom H2 is connected to the USB port of the computer, and the file containing the key strike data is read into the computer.

The program Audacity is used to locate the beginning of the beep in the file that has just been read into the computer. This beep time will allow all the strikes to be synchronized. Once the time of the beep has been determined and entered, another Matlab® script from the file AnalyzeSyncdDataOffline2.m is used to extract the data for each key strike and save it in a matrix for subsequent analysis. During the extraction, the location of the peaks in the line spectra is used to verify that each key was struck and collected properly. It usually requires about 90 minutes to collect data for all 88 keys. Most of the time is consumed in the verification of the data.

Approach 2

Audacity is used to collect the data while the above Matlab script is running when the Earthworks microphone/Roland Duo-Capture Ex setup replaces the Zoom H2. Figure A4.3 shows the location of the microphone. After the data has been collected, Audacity can be used to determine the reference point. When using Audacity, the data collection script is in the file CollectAndSaveSyncTalkAud.m. During the collection, Audacity shows the wave being collected. This approach saves time over that using the Zoom H2 and may produce a higher quality recording.

Approach 3

When Octave, rather than Matlab, is used, then the appropriate script is in the file CollectPianoNotes.m. Either the Zoom H2 or the Earthworks/Roland Duo Ex can be used in this case but the latter combination is probably better. The Octave script executes much more slowly. There is no sound and no color prompt. The user has to watch the screen and strike the key when so prompted by a message. As in Approach 2, during the collection Audacity shows the wave being collected.

After the data has been collected, Audacity can be used to save the collected data as a wav file. It is also used to determine the reference point. The user then runs the script in the file AnalyzePianoNotes.m which stores all of the key's waves individually. This data collection and storing exercise can be carried out in several sections. For example, the user might want to collect for the first 44 keys and then for the remaining 44 keys. In this case, there would be two wav files and there would be 88 separate ASCII files for the keys.

Subsequent scripts can then be used to generate the graphical metrics discussed in Part 3, such as the spectral centroid map, the envelope map, and the cumulative line spectrum map. Consult the book's website for further help.

Approach 4

This approach is documented in detail on the book's website, <https://sites.google.com/site/analysisofsoundsandvibrations/>, where there is a YouTube video for guidance and where there are Octave scripts that can be downloaded to the computer to carry out the collection. Instead of a beep, this approach uses a staccato key strike to synchronize the Octave script with Audacity. There is no need for the Zoom H2 and the small speakers mounted on the music desk. Although the data can be effectively collected by this method (using a mobile computer, such as a netbook, that has Octave and Audacity), Octave is much too slow to analyze the data. For the analysis, the user can transfer the ASCII files, one for each key, to another computer that has Matlab. Then graphs, like the cumulative line spectrum map and the spectral centroid map, can be efficiently constructed.

It is not necessary to use the rod and the 88-hole board. Instead, one can simply place the striking finger on the key and press down to generate the sound. The key is not "struck" as much as it is "pressed." Although one can develop a consistent procedure for striking the key in this manner, it may not be as repeatable as the dropping-rod method. On the other hand, as shown in Chapter 16, the dropping rod generates a precursor wave that is not present when using the finger strike. A good bit of the data used in the book comes from the key-pressing method.

Appendix 5
A Brief Exposure to Matlab® and Octave

Matlab

In the late 1980s I got my first copy of Matlab. The slim owner's manual began with, "If you can't bother to read this manual, start here." In effect, Matlab was presented using "backward chaining." The manual quickly showed you what Matlab could do and motivated you to dig into the details to figure out how you could avail yourself of some awesome computing power. As with many circa 1980 engineers, encountering Matlab was a mind-blowing experience (remember, many of us were using BASIC, Quick Basic, and Fortran on 8 MHz IBM PC 386s). The whole Matlab system in 1989 came on one 5.25″ flexible disk. Today the Matlab system consists of a massive collection of toolboxes and can cost thousands of dollars. For the sake of full disclosure, I am a Matlab author, which means that I have complimentary access to the Matlab toolboxes needed to carry out the simulations and graph constructions in this book.

Matlab m-files consist of "scripts" which you write in a BASIC-like language and which allow you to call many built-in incredibly powerful routines or functions to carry out calculations. For example, the function *fft* calculates the fast Fourier transform of a vector of data. Fortunately, you do not have to understand the fast Fourier transform to use the function.

The following is an example script file, MatlabExample.m, with some comments (anything after a "%" is a comment). Try reading it line-by-line to see if it makes sense. The Matlab scripts for many of the figures in the book are available on the book's website.

```
% MatlabExample.m (anything after a % is a comment)
% This script calculates samples of a sound wave with a
% weak fundamental frequency of 32.7 Hz and a strong
% 2nd harmonic.
% It plots the samples in the time domain.
% It generates a sound via the computer speakers.
% It calculates the line spectrum using a function that I wrote.
% It plots the line spectrum.
Clear % clear the memory
close all % close all graphs

N=2*44100; % number of sample points (all statements end with
%            a semi-colon)
Fs=44100; % sampling frequency in Hz
h=1/Fs; % sampling interval in s
t=(0:N-1)*h; % the N time samples in s (a vector):
%              0,h,2h,3h,...,(N-1)h
```

```
% The above is an example of generating a vector of values with one
% simple statement.

f=32.7; % fundamental frequency of the wave in Hz
f=CallInput(f,'enter fundamental frequency of the wave'); % CallInput
%                                                            is one of my scripts

%=====generate the wave
y=0.1*sin(2*pi*t*f)+sin(2*pi*t*2*f); % construct the wave samples
%====add some background noise
y=y+0.001*randn(size(y)); % randn is a function that generates
%                           normally distributed
%                           random noise. The
%                           argument is the size
%                           of y
% (y is a vector, as is t. Both have N samples.)
% The above is another example of how a simple expression can
% generate a vector of values.

% ====== plot the wave in the time domain
fh1=figure; % define the first figure handle
set(fh1,'DefaultLineLineWidth',1.5) % specify the line width
plot(t,y),grid % plot the wave on a grid
title('A Wave with only Two Harmonics') % title for the graph
xlabel('time (sec)') % x-axis label
ylabel('amplitude')  % y-axis label
axis([0 .1 -inf inf]) % specify the extent of the axes as in
%                       axis([xmin xmax ymin ymax])
%                       plot t from 0 to .1 sec and put no
%                       limits on plotting y i.e., plot from -inf
%                       (the minimum value) to +inf (the
%                       maximum value
disp('press Enter to hear the sound of the wave ');pause

%%====listen to the sound of the wave (Octave does not have soundsc)
soundsc(y,Fs);

disp('press Enter to calculate and plot the line spectrum
of the wave ');pause

% ===set the parameters for the line spectrum
frL=0; % frequency in Hz of left-hand side of line spectrum
frR=3*f; % frequency in Hz of right-hand side of the line spectrum

% ===calculate the line spectrum
% Y is the magnitude of the FFT (a vector), F is the vector of
% frequencies
```

```
[Y, F]=SimpleLineSpectrum2(y,h,frL,frR); % SimpleLineSpectrum2 is one
%                                          of my scripts
Ymax=max(Y); % find the largest value of the line spectrum

% plot the line spectrum
fh2=figure; % define the second plot
set(fh2,'DefaultLineLineWidth',1.5) % specify the line width
plot(F,20*log10(Y/Ymax)),grid
title('Line Spectrum of a Wave')
xlabel('frequency (Hz)')
ylabel('dB')
axis([ frL frR -100 0])
disp('press Enter to calculate and plot the cumulative line spectrum
of the wave ');pause

%===calculate the cumulative line spectrum
fh3=figure;
[Cls,fCls]=CumLineSpectrum3(y,h); % CumLineSpectrum3 is one of
my scripts
plot(fCls,100*Cls),grid
title('Cumulative Line Spectrum of a Wave')
xlabel('frequency (Hz)')
ylabel('accumulated power in %')
axis([frL frR 0 100])
```

Use the Matlab editor to save this script and give it a name, such as MatlabExample.m. Then run it by entering the name at the Matlab prompt, as in

```
>> MatlabExample
```

To use the scripts, set up a folder, for example C:\APlaceForBookFiles, and download or copy the files on the website to it. You will need to copy SimpleLineSpectrum2.m and CumLineSpectrum3.m also. Modify your Matlab path to include this directory.

If you try to run the scripts that generate the figures you may run into trouble because many of them require large data files which are not on the website. In those cases, to get a feel for how the script works, you can substitute samples of a sine wave (or a sum of several sine waves) of your own making. If you do this, it is probably best to generate the wave in a manner similar to the following:

```
N=2*44100; % number of points for 2.0 s of data sampled at 44100 Hz
Fs=44100; % sampling frequency in Hz
h=1/Fs; % sampling interval in s
t=(0:N)*h; % the time samples
f=32.7; % frequency of the wave
y=sin(2*pi*t*f)+.5*sin(2*pi*t*2*f); % construct the wave
%                        (a fundamental and the 2nd harmonic)
```

You can save the wave as a mat-file (a Matlab data file) in C:\APlaceForBookFiles as follows:

```
Save testwave y Fs
```

Then in the script you can read or load the wave as follows:

```
Load testwave y Fs
```

You will have to comment out the parts of the script that read in the original long files.

Octave

There is a free software package called GNU Octave, or just Octave, that has some of the features of Matlab. The above example script will run under Octave, although about 3–5 times slower. The data collection software that I use to sample the pianos described in Appendix 4 runs under Octave in concert with Audacity, which is also free, but, again, much slower and with fewer features. Mat-files saved by Octave must be saved in ASCII format if they are to be read by both Matlab and Octave. At present, Octave version 3.6.4 has a problem constructing 3D graphs using the *plot3* command. This means that dealing with huge files resulting from sampling a sound wave at 44,100 Hz for several seconds is difficult. For example, constructing a CLSM for 88 keys where each key has been sampled for 10 s requires significant decimation. Alternatively, one can make CLSMs for a subset of the 88 keys, say, 30 keys at a time. As further versions are developed, this problem may be solved.

Having used Matlab for more than 20 years, I am spoiled by its features and its speed. I am just starting to use Octave. I will attempt to develop a suite of Octave scripts that will do the data collection and some of the 3D graphs and that will only require the user to understand how Windows Explorer works. The reader should look on the website for the latest versions. At the time of publication the suite consists of six Octave scripts:

CollectPianoNotes.m	Collects the waves from each piano note. See the YouTube video for guidance.
AnalyzePianoNotes.m	Parses the waves from the wav file collected by CollectPianoNotes and stores them as ASCII files, one per note.
CutOffAndPadOct.m	Reads the ASCII note files, removes the low amplitude trailing part, pads to 6 s, and saves the modified note files in another directory.
ConstructCLSMOct.m	Generates a CLSM from the note files. Has a limit of 30 keys at a time.
StudyNoteOct.m	Reads a selected ASCII note file and produces a time domain plot, a line spectrum and a cumulative line spectrum.
CalcSpecCenOct.m	Generates a spectral centroid map for a piano.

There is a YouTube video "Collecting Piano Data Using Octave and Audacity" that describes how to use the first two scripts. Note that *AnalyzePianoNotes* is called by *CollectPianoNotes* and is therefore transparent to the user. The third script, *CutOffAndPadOct*, makes use of whatever data was collected by the first two scripts. I recommend that you apply it immediately after you finish collecting your 88 notes. *StudyNoteOct* provides the user with an opportunity to look at the time domain plot, the line spectrum and the cumulative line spectrum of a specific note. *CalcSpecCenOct* provides a spectral centroid for the piano from which you collected the data. *ConstructCLSMOct* is a bit dicey and should be used with care. Depending on the computer, it will probably not produce a CLSM for more than 30 keys at a time. I would appreciate feedback on how well these scripts work (or not).

Please check the book's website at <https://sites.google.com/site/analysisofsoundsandvibrations/> for detailed instructions on how to download the necessary Octave scripts.

The scripts that I use to generate the snapshots of a vibrating membrane that are shown in the videos referenced on the website will not work under Octave (so far). Also, the simulation scripts that generate the response of, say, a vibrating string to a hammer strike are unacceptably slow under Octave. In general, Matlab has superior speed, debugging tools, editors, and graphics. However, since Octave is free, the reader should be patient and try to use it.

References

The references are mentioned in the order that I came across them.

1. John Powell, *How Music Works*, Little, Brown & Company, 2010.

John Powell's book got me interested in the subject. It is written on an elementary, introductory level and is enjoyable and insightful.

2. Arthur H. Benade, *Fundamentals of Musical Acoustics*, 2nd Revised Edition, Dover, 1990.

Arthur Benade's book came next. It avoids mathematics completely and is poorer for it. Benade attempts to explain things in a style that I find hard to follow. However, it does have a lot of information and, for me, is useful as a reference.

3. Philip M. Morse, *Vibration and Sound*, 2nd Edition, McGraw-Hill, 1948.

Philip Morse's book is elegant and insightful. Every topic is supported by mathematics as well as physics and therefore is quite useful to me. There is a later edition on this subject by the author.

4. Lawrence E. Kinsler and Austin R. Frey, *Fundamentals of Acoustics*, 2nd Edition, Wiley, 1962.

This book is an old edition that I got off the internet in used form. In several ways it is an elementary and introductory version of Morse's book. It is well written and supports most of the concepts with well-presented mathematics and physics. I have looked at later editions and I think the second edition is best.

5. Nicholas J. Giordano, *Physics of the Piano*, Oxford University Press, 2010.

The author, a physics professor from Purdue University, has written two books that I drawn on here for several topics. This book concentrates on the piano and contains little mathematics and is better for it.

6. Nicholas J. Giordano and Hisao Nakanishi, *Computational Physics*, 2nd Edition, Prentice Hall, 2006.

This book is written in an unimposing but thorough way. I have borrowed heavily from their method of modeling the string struck by a hammer.

7. John Askill, *Physics of Musical Sounds*, D. Van Nostrand, 1979.

An elementary and qualitative approach to musical instruments.

8. Neville H. Fletcher and Thomas D. Rossing, *The Physics of Musical Instruments*, 2nd Edition, Springer, 1998.

A comprehensive and mathematically detailed survey of the field. It appears to be the standard reference for this subject.

9. Harry F. Olson, *Music, Physics and Engineering*, 2nd Edition, Dover, 1967.

This is more of a handbook and is useful as a reference.

10. Erwin Kreyszig, *Advanced Engineering Mathematics*, 8th Edition, Wiley, 1999.

I mention Kreyszig's book because there is a section devoted to the solution of the wave equation.

11. Alfred Dolge, *Pianos and their Makers*, Dover, 1972 (originally published in 1911 by Covina Publishing Co.).

A non-technical but insightful and interesting book.

12. William C. Elmore and Mark A. Heald, *Physics of Waves*, Dover, 1985.

This is interesting because it takes a look at some of the more advanced problems in solving the wave equation.

13. G. A. Briggs, *Pianos, Pianists and Sonics*, Wharfedale Wireless Works, 1951.

A delightful, unpretentious technical book that actually covers a lot of the same material that I do in Part 3, except that Mr. Briggs did not have Matlab® or the Zoom H2 to aid him in his investigations. The more I read this little book, I am convinced that I am a descendant of Mr. Briggs. In his introduction he says,

> When I mentioned the idea for this book on pianos to various people I met with varying reactions – mostly unfavorable – of which the following remarks are typical:
>
> a. Few people will be interested.
>
> b. What will you find to say?
>
> c. What do you know about pianos?
>
> d. The musical instrument trade is conservative and will resent your intrusion.

Been there, done that.

14. Eric J. Heller, *Why You Hear What You Hear*, Princeton University Press, 2013.

This is the last textbook that I came across, mostly because it was published in 2013. It is a comprehensive and well-written book that contains extensive qualitative and insightful discussions of how and why sound waves behave. It is where I got the idea of trying to use the autocorrelation as a metric for pitch.

The following journal internet articles are somewhat specialized but quite useful.

1. I. Testa, G. Evangelista, and S. Cavaliere, "Physically Inspired Models for the Synthesis of Stiff Strings and Dispersive Waveguides," *Journal on Applied Signal Processing*, vol. 7, pp. 964–77, 2004.
2. J. McDonald, "Motion of a realistic string," available online at <http://www.physics.udel.edu/~jim/PHYS460_660_13S/Motion of a string/Motion of a realistic string.htm>.

An interesting paper by Prof. James McDonald of the University of Delaware.

3. Gabriel Weinreich, "The Coupled Motion of Piano Strings," which one can obtain by a Google search.
4. Lejaren Hiller and Pierre Ruiz, "Synthesizing Musical Sounds by Solving the Wave Equation for Vibrating Objects: Part I," *Journal of the Audio Engineering Society*, vol. 19, no. 6, June 1971.

Index